CONFRONTING TECHNOLOGY

SELECTED READINGS AND ESSAYS

Edited by

David Skrbina
The University of Michigan-Dearborn

Detroit
CREATIVE FIRE PRESS
2020

Library of Congress Cataloging-in-Publication Data

Confronting Technology: Selected Readings and Essays
edited by David Skrbina
          p.   cm.
Includes bibliographical references.
ISBN: 978-1732-3532-75 (pbk.: alk. paper)
Philosophy. I. Skrbina, David

Printing number: 9 8 7 6 5 4 3 2 1

Printed in the United States of America on acid-free paper.

CONTENTS

PART VI: TECHNOTOPIA?

APPENDICES

PREFACE

The phenomenon of technology is at once fascinating and troublesome. With us since the dawn of mankind, it has come to utterly pervade our lives. Conventional wisdom has it that technology exists only to serve our needs, and to better our lives. And yet there is a sense in which our lives are not really getting 'better' at all. In fact, many have the feeling that life is becoming more difficult, more demanding, more stressful, and less satisfying. Since technology is perpetually increasing in power, reach, and affordability, why are our lives not correspondingly better than ever? Why are environmental problems not quickly and permanently resolved? Why do hunger, disease, hardship, and suffering persist around the world? Why are we not on an unambiguously upward path, one in which life is always becoming richer, more fulfilling, healthier, and happier? There is this lingering sense in which technology gives on the one hand, and takes from the other. Thus it seems that technology also has its dark side, and this is an aspect that is too often ignored, misunderstood, and hence underestimated. In this text we will take a critical look at technology from the time of the Greeks to the present. Only by doing so can we fully appreciate the impact that modern technology has on our lives, our planet, and our future.

The entries in this book constitute the essential readings in historical and present-day critiques of technology. All prominent critics are represented here, with one notable exception: Jacques Ellul, and in particular his masterpiece *The Technological Society* (1954). The reason for this exclusion is straightfoward: this text is intended as a complementary work to his book. Ellul's ideas are too detailed and elaborate to be adequately captured in an anthology such as this.

—D.S.

CULTURE AND VALUE

LUDWIG WITTGENSTEIN (1947)

The truly apocalyptic view of the world is that things do *not* repeat themselves. It isn't absurd, e.g., to believe that the age of science and technology is the beginning of the end for humanity; that the idea of great progress is a delusion, along with the idea that the truth will ultimately be known; that there is nothing good or desirable about scientific knowledge and that mankind, in seeking it, is falling into a trap. It is by no means obvious that this is not how things are.

OZYMANDIUS

PERCY SHELLEY (1818)

I met a traveler from an antique land
Who said: Two vast and trunkless legs of stone
Stand in the desert. Near them, on the sand,
Half sunk, a shattered visage lies, whose frown
And wrinkled lip, and sneer of cold command,
Tell that its sculptor well those passions read
Which yet survive, stamped on these lifeless things,
The hand that mocked them and the heart that fed.

And on the pedestal these words appear:
"My name is Ozymandius, king of kings:
Look on my words, ye Mighty, and despair!"
Nothing beside remains. Round the decay
Of that colossal wreck, boundless and bare,
The lone and level sands stretch far away.

ODE TO WILLIAM H. CHANNING

RALPH WALDO EMERSON (1847)

What boots thy zeal,
O glowing friend,
That would indignant rend
The northland from the south?
Wherefore? To what good end?
Boston Bay and Bunker Hill
Would serve things still;
Things are of the snake.

The horseman serves the horse,
The neat-herd serves the neat,
The merchant serves the purse,
The eater serves his meat;
'Tis the day of the chattel,
Web to weave, and corn to grind;
Things are in the saddle,
And ride mankind.

There are two laws discrete,
Not reconciled—
Law for man, and law for thing;
The last builds town and fleet,
But it runs wild,
And doth the man unking.

PART I:
INTRODUCTION
ON THE PROBLEM OF TECHNOLOGY

CLOSE TO THE SINGULARITY

DANNY HILLIS

Danny Hillis (1956-) is a computer scientist and cofounder and chief scientist of Thinking Machines Corporation. He is the holder of more than 30 U.S. patents, and editor of several scientific journals, including *Artificial Life, Complexity, Complex Systems*, and *Future Generation Computer Systems*. He is the author of *The Connection Machine* (1985) and *The Pattern on the Stone* (1998). The following essay was written in 1995.

I like making things that have complicated behaviors. The ultimate thing that has a complicated behavior is, of course, a mind. The Holy Grail of engineering for the last few thousand years has been to construct a device that will talk to you and learn and reason and create. The first step in doing that requires a very different kind of computer from the simple sequential computers we deal with every day, because these aren't nearly powerful enough. The more they know, the slower they get—as opposed to the human mind, which has the opposite property. Most computers are designed to do things one at a time. For instance, when they look at a picture, they look at every dot in the picture one by one; when they look at a database, they search through the facts one by one. The human mind manages to look all at once at everything it knows and then somehow pull out the relevant piece of information. What I wanted to do was make a computer that was more like that.

It became clear that by using integrated-circuit technology you could build a computer that was structured much more like a human brain; it would do many simple things simultaneously, in parallel, instead of rapidly running through a sequence of things. That principle clearly works in the mind, because the mind manages to work with the hardware of the brain, and the hardware of the brain is actually very slow hardware compared with the hardware of the digital computer.

With modern integrated circuits, it's possible to replicate something over and over again very inexpensively, so I started building a computer by replicating simple processing circuits over and over again and then allowing them to connect with the other interrogatory patterns. Of course, the other thing about your mind is that if I slice up your brain, I see that it's almost all wires. It's all connections between the neurons. Putting into the computer the telephone system that will connect all those little processing elements is the hardest part. That's why my computer was called "the connection machine." I designed it at MIT, but I realized that it was much too big and complicated to be built at a university. It was going to require hundreds of people and tens of millions of dollars. So in 1983 I started the Thinking Machines Corporation, and we spent the next ten years becoming the company that made the world's biggest and fastest computers. The irony is that we were so distracted with all this scientific computing that I haven't made nearly as much progress on the thing I started out with, which is the thinking computer.

My view of what it's going to take to make a thinking machine has changed in recent years. When we started out, I naively believed that each of the pieces of intelligence could be engineered. I still believe that would be possible in principle, but it would take three hundred years to do it. There are so many different aspects to making an intelligent machine that if we used normal engineering methods the complexity would overwhelm us. That presents a great practical difficulty for me; I want to get this project done in my lifetime.

The other thing I've learned is how hard it is to get lots of people to work together on a project and manage the complexity. In some senses, a big connection machine is the most complicated machine humans have ever built. A connection machine has a few hundred billion active parts, all of which are working together, and the way they interact isn't really understood, even by its designers. The only way to design an object of this much complexity is to break it into parts. We decide it's going to have this box and that box and that box, and we send a group of people off to do each of those, and they have to agree on the interfaces before they go off and design their boxes.

Imagine engineering a thinking machine that way. Somebody like Marvin Minsky would say, "O.K., there's a vision box and a reason box and a grammar box," and so on. Then we might break the project up into parts and say, "O.K., Tommy"— Tomaso Poggio, at MIT—"you go off and do the vision box," and we'd get Steve Pinker to do the grammar box, and Roger Schank to do the story box. Then Poggio would take the vision box and say, "All right, we need a depth perception box, and we need a color-recognition box," and so on. Then the depth-perception team would say, "O.K., we need a box that perceives depth perception by focus clues and a box that perceives depth perception by binocular vision." Imagine a collection of tens of thousands of people doing these modules, which is how we'd have to engineer it. If you engineer something that way, it has to decompose, and it has to go through these fairly standardized interfaces. There's every reason to believe that the brain is not, in fact, that neatly partitioned. If you look at biological systems in general, while they're hierarchical at a gross level, there's a complex set of interactions between all the parts that doesn't follow the hierarchy. But I'm convinced that our standard methods of engineering wouldn't work very well for designing the brain, although not because of any physical principles we can't control. The brain is an information-processing device, and it does nothing that any universal information-processing device couldn't do.

There's another approach besides this strict engineering approach which can produce something of that complexity, and that's the evolutionary approach. We humans were produced by a process that wasn't engineering. We now have computers fast enough to simulate the process of evolution within the computer. So we may be able to set up situations in which we can cause intelligent programs to evolve within the computer.

I have programs that have evolved within the computer from nothing, and they do fairly complicated things. You begin by putting in sequences of random instructions, and these programs compete and interact with each other and have sex with each other and produce new generations of programs. If you put them in a world where they survive by solving a problem, then with each successive generation they get better and better at solving the problem, and after a few hundred thousand generations they solve the problem very well. That approach may actually be used to produce the thinking machine.

One of the most interesting things is that larger-order things emerge from the interaction of smaller things. Imagine what a multicellular organism looks like to a single-celled organism. The multicellular organism is dealing at a level that would be incomprehensible to a single-celled organism. I think it's possible that the part of our mind that does information processing is in large part a cultural artifact. A human who's not brought up around other humans isn't a very smart machine at all. Part of what makes us smart is our culture and our interactions with others. That's part of what would make a thinking machine smart, too. It would have to interact with humans and be part of that human culture.

On the biology side, how does this simple process of evolution organize itself into complicated biological organisms? On the engineering side, how do we take simple switching devices like transistors, whose properties we understand, and cause them to do something complex that we don't understand? On the physics side, we're studying the general phenomenon of emergence, of how simple things turn into complex things. All these disciplines are trying to get at essentially the same thing, but from different angles: how can the whole be more than the sum of the parts? How can simple, dumb things working together produce a complex thing that transcends them? That's essentially what Marvin Minsky's "society of mind" theory is about; that's what Chris Langton's "artificial life" is about; that's what Richard Dawkins' investigation of evolution is about; that's fundamentally what the physicists who are looking at emergent properties are studying; that's what Murray Gell-Mann's work on quarks is about; that is the thread that binds all these ideas together.

I am excited by the idea that we may find a way to exploit some general principles of organization to produce something that goes beyond ourselves. If you step back a zillion years, you can look at the history of life on Earth as fitting into this pattern. First, fundamental particles organized themselves into chemistry. Then chemistry organized itself into self-reproducing life. Then life organized itself into multicellular organisms and multicellular organisms organized themselves into societies bound together by language. Societies are now organizing themselves into larger units and producing something that connects them technologically, producing something that goes beyond them. These are all steps in a chain, and the next step is the building of thinking machines.

To me, the most interesting thing in the world is how a lot of simple, dumb things organize themselves into something much more complicated that has behavior on a higher level. Everything I'm interested in—whether it's the brain, or parallel computers, or phase transitions in physics, or evolution—fits into that pattern. Right now, I'm trying to reproduce within the computer the process of evolution, with the goal of getting intelligent behavior out of machines. What we do is put inside the machine an evolutionary process that takes place on a timescale of microseconds. For example, in the most extreme cases, we can actually evolve a program by starting out with random sequences of instructions—say, "Computer, would you please make a hundred million random sequences of instructions. Now, execute all those random sequences of instruction, all those programs, and pick out the ones that came closest to what I wanted." In other words, I defined what I wanted to accomplish, not how to accomplish it.

If I want a program that sorts things into alphabetical order, I'll use this simulated evolution to find the programs that are most efficient at alphabetizing. Of course, random sequences of instructions are unlikely to alphabetize, so none of them does it initially, but one of them may fortuitously put two words in the right order. Then I say to the computer, "Would you please take the 10 percent of those random programs that did the best job, save those, kill the rest, and have the ones that sorted the best reproduce by a process of recombination, analogous to sex. Take two programs and produce children by exchanging their subroutines." The "children" inherit the "traits," the subroutines, of the two programs. Now I have a new generation of programs, produced by combinations of the programs that did a superior job, and I say, "Please repeat that process, score them again, introduce some mutations, and repeat the process again and again, for many generations." Every one of those generations takes just a few milliseconds, so I can do the equivalent of millions of years of evolution within the computer in a few minutes—or, in complicated cases, in a few hours. Finally, I end up with a program that's absolutely perfect at alphabetizing, and it's much more efficient than any program I could ever have written by hand. But if I look at that program, *I'm unable to tell you how it works*. It's an obscure, weird program, but it does the job, because it comes from a line of hundreds of thousands of programs that did the job. In fact, those programs' lives depended on doing the job.

How do I really know the program will work? In the sorting case, I test it. What if it was something really important? What if this program was going to fly an airplane? Well, you might say, "Gee, it's really scary having a program flying an airplane when we don't have any idea how it works!" But that's exactly what you have with a human pilot; you have a program that was produced by a very similar method, and we have great confidence in it. I have much less confidence in the airplane itself, which was designed very precisely by a lot of very smart engineers. I remember riding in a 747 with Marvin Minsky once, and he pulls out this card from the seat pocket, which said, "This plane has hundreds of thousands of tiny parts, all working together to give you a safe flight." Marvin said, "Doesn't that make you feel confident?"

The engineering process doesn't work very well when it gets complicated. We're beginning to depend on computers that use a process very different from engineering—a process that allows us to produce things of much more complexity than we could with normal engineering. Yet we don't quite understand the possibilities of that process, so in a sense it's getting ahead of us. We're now using those programs to make much faster computers so that we will be able to run this process much faster. The process is feeding on itself. It's becoming faster. It's autocatalytic. We're analogous to the single-celled organisms when they were turning into multicellular organisms. We're the amoebas, and we can't quite figure out what the hell this thing is that we're creating. We're right at that point of transition, and there's something coming along after us.

It's haughty of us to think we're the end product of evolution. All of us are a part of producing whatever is coming next. We're at an exciting time. We're close to the singularity. Go back to that litany of chemistry leading to single-celled organisms, leading to intelligence. The first step took a billion years, the next step took a hundred million, and so on. We're at a stage where things change on the order of decades, and it seems to be speeding up. Technology has the autocatalytic effect of fast computers, which let us design better and faster computers faster. We're heading toward something which is going to happen very soon—in our lifetimes—and which is fundamentally different from anything that's happened in human history before.

People have stopped thinking about the future, because they realize that the future will be so different. The future their grandchildren are going to live in will be so different that the normal methods of planning for it just don't work anymore. When I was a kid, people used to talk about what would happen in the year 2000. Now, at the end of the century, people are still talking about what's going to happen in the year 2000. The future has been shrinking by one year per year, ever since I was born. If I try to extrapolate the trends, to look at where technology's going sometime early in the next century, there comes a point where something incomprehensible will happen. Maybe it's the creation of intelligent machines. Maybe it's telecommunications merging us into a global organism. If you try to talk about it, it sounds mystical, but I'm making a very practical statement here. I think something's happening now—and will continue to happen over the next few decades—which is incomprehensible to us, and I find that both frightening and exciting.

Skrbina to Hillis, 10/25/06:

I teach philosophy at the University of Michigan-Dearborn, and for a number of years now I have had my students read your essay "Close to the singularity." It always provokes a lively discussion, but it is rather dated at this point. We are wondering if you still subscribe to the 'singularity' notion, how we are progressing, and if/when you think it might arrive. Have the past 10 years accelerated as you anticipated? Any thoughts on the next 10 or 20 years?

Hillis to Skrbina, 10/26/06:

Yes, I do still believe the things that I said in that essay, although I have been surprised at how slow the progress has been in the last decade. So many experimenters have been distracted by the [business] application of computers that very few have been working on advancing the underlying science. (I am sorry to admit that I have been guilty of this too!) Still, I think the potential is there and we will make progress on it.

— Danny Hillis.

'BETTER HUMANS' AND THE NANOTECH FUTURE

DERRICK JENSEN AND GEORGE DRAFFAN

Jensen and Draffan are writers and environmental activists. They are co-authors of *Strangely Like War* (2003) and *Welcome to the Machine* (2004), from which the following is excerpted. Jensen has also written or co-written some 10 other books, including the anti-civilization works *Endgame* (2006) and *How Shall I Live my Life?* (2008).

We are all by now unfortunately familiar with biotechnology, the splicing of genes from one being into another. Nanotechnology goes far beyond biotech's manipulation of genes; it involves the manipulation of molecules and atoms. We can call it molecular engineering, we can call it the marriage of life and computers, we can call it the creation of green machines, or we can call it self-replicating biotech alchemy. It's been predicted to cure cancer, to end poverty by supplying the world, with an infinite supply of energy and self-assembling materials, and to keep your pants stain-free.

Nanotechnology works with small materials. Really small. A nanometer is a billionth of a meter. Ten hydrogen atoms side by side make a line a nanometer long. A DNA molecule is about two-and-a-half nanometers wide. A red blood cell is one-twentieth the width of a human hair, but it's 5,000 nanometers in diameter. It gives me a headache just to think about it: The individual components of a silicon transistor are about 130 nanometers across, but Intel can fit 42 million of them on one of its Pentium-4 computer chips.

So, that's all great! We can all have fast computers to play more complex games (while the real world burns) because the geniuses at Intel have figured out how to get all those transistors onto a single computer chip. But those nasty atomtech critics, including ETC Group's Jim Thomas, tell us that "nano-sized bits are so small that they can penetrate your skin, get into your lungs and travel through your body unmolested by the immune system."

The ultimate goal of nanotechnology, according to some of its proponents, is to meld humans and computers so that humans can at long last (insert mad scientist laugh here) live forever. As such, nanotechnology is motivated by a fear of death, and the consequent fear and hatred of the body, the natural. Here's how www.betterhumans.com puts it in an article entitled *Immortality*:

It is mortalist, deathist thinking that characterizes aging and dying as natural and good... Today, immortality—and the lesser goal of indefinite lifespan—seems entirely possible. The best way to think of it is that our 'soul' is to our bodies and brains what music is to a CD: an arrangement of information. The CD doesn't matter—it can be copied an indefinite number of times. So what could we do to achieve immortality? The same thing we would do to preserve the music on a rare, one-of-a-kind CD: make backup copies on a better medium.

While the technology to upload a mind doesn't exist yet, research in artificial intelligence, nanotechnology and cognitive science, as well as developments in computer hardware, are taking us in the right direction. Mind uploading is also a possible side-effect of improvements in human-computer interfaces, as direct links between the brain and computer hardware could lead to a gradual merging of biological and nonbiological components of the mind. At some point, enough information could exist in the nonbiological portion that destruction of the biological brain has no impact on personality or identity.

Destruction of the biological brain. Such a nice, clean way to say the killing of the animal.

Alexander Bolonkin, formerly of the US Air Force and the National Aeronautics and Space Administration, writes:

An immortal person made of chips and supersolid materials will have incredible advantages in comparison with common people. An 'E-man' will need no food, no dwelling, no air, no sleep, no rest, no ecologically pure environment [this latter is certain to come in handy, all things considered]. Such a being will be able to travel into space, or walk on the sea floor with no aqualungs. It will not be possible to destroy an artificial person with any kind of weapons, since it will be possible to copy the information of his mind and then keep it separately.

Bolonkin is, of course, describing the wet dream of those in charge of any police state. He believes that

such transition to immortality (E-creatures) will be possible in 10-20 years. The number of E-creatures will be growing and the number of people diminishing, until it gets to the minimum necessary for the zoos and small reservations. In all likelihood, the feelings that E-creatures may have toward humans as their ancestors will be fading away, until they become comparable to our own attitude toward apes or even bugs.

I cannot speak for his attitude toward apes, except to say that it seems to mirror the culture's attitude toward all of those it puts in zoos, in reservations, in the slash piles of clearcuts or back in the ocean dead as by-catch. Bolonkin continues,

> There is a hope that the ability to keep souls alive will be achieved by highly civilized countries first. In this case, they will prohibit torturing sinners, as they prohibit torturing criminals nowadays. Furthermore, criminal investigations will be simplified a lot, judicial mistakes will be excluded. It will be possible to access a soul consciousness and see every little detail of this or that action.

A simple scan of your computer disk, and those in power know everything about you.

The federal government pumps $700 million a year into the National Nanotechnology Initiative (one-third of which goes to the Pentagon)—making it one of the largest recipients of federal research money, along with the "war on cancer" and the militarization of space.

When something (inevitably) goes haywire, nanotechnology could, even according to some of its strongest boosters, lead to covering the entire planet with self-replicating "gray goo," what experts call "global ecophagy"—the eating of the Earth. As K. Eric Drexler, author of *Engines of Creation: The Coming Era of Nanotechnology* and one of the most vociferous supporters of the nanotech vision, puts it,

> Assembler-based replicators could beat the most advanced modern organisms. 'Plants' with 'leaves' no more efficient than today's solar cells could out-compete real plants, crowding the biosphere with an inedible foliage. Tough, omnivorous 'bacteria' could out-compete real bacteria: They could spread like blowing pollen, replicate swiftly and reduce the biosphere to dust in a matter of days. Dangerous replicators could easily be too tough, small and rapidly spreading to stop—at least if we made no preparation. We have trouble enough containing viruses and fruit flies.

We don't have to conjure gray goo to make nanotech deadly: Rats exposed to 20-nanometer particles of polytetrafluoroethylene all died within four hours, while those exposed to larger particles of the same chemical all survived. This shouldn't be surprising: Researchers have long known the dangers of nanoparticles—although they don't call them that, instead referring to them as "fines" and "ultrafines." As long ago as 1991, scientists at the Environmental Protection Agency estimated that fine particles kill 60,000 Americans per year. Ultrafines are estimated to be 10 to 50 times more dangerous, causing lung and cardiovascular disease, and probably promoting Alzheimer's and other forms of brain deterioration.

You and I both know that these dangers won't stop those in power from pursuing this course. The proximate danger of the deaths of tens of thousands of Americans per year (and uncounted non-Americans, and uncounted nonhumans, both of whom matter even less than American humans to those in power) won't stop them, nor will the ultimate danger of gray goo eating the Earth.

What will stop them is the end of civilization. Nanotechnology and the marriage of computers and life will not fulfill their potentials for increases in industrial production and therefore decreases in the capacity of the planet to support life. This marriage will fail to take place not because governments suddenly decide not to kill their own citizens, nor because activists suddenly become effective. Instead, the natural world will stop these possibilities. With any luck for the rest of the planet, civilization will crash long before humans are stored on hard drives or CDs.

THE WORST MISTAKE IN THE HISTORY OF THE HUMAN RACE

JARED DIAMOND

Diamond (1937-) is a scientist, professor (geography, physiology), and writer, currently teaching at UCLA. He is the author of *Guns, Germs, and Steel* (1997) and *Collapse* (2005). This article was published in *Discover* magazine in 1987.

To science we owe dramatic changes in our smug self-image. Astronomy taught us that our earth isn't the center of the universe but merely one of billions of heavenly bodies. From biology we learned that we weren't specially created by God but evolved along with millions of other species. Now archaeology is demolishing another sacred belief: that human history over the past million years has been a long tale of progress. In particular, recent discoveries suggest that the adoption of agriculture, supposedly our most decisive step toward a better life, was in many ways a catastrophe from which we have never recovered. With agriculture came the gross social and sexual inequality, the disease and despotism, that curse our existence.

At first, the evidence against this revisionist interpretation will strike twentieth century Americans as irrefutable. We're better off in almost every respect than people of the Middle Ages, who in turn had it easier than cavemen, who in turn were better off than apes. Just count our advantages. We enjoy the most abundant and varied foods, the best tools and material goods, some of the longest and healthiest lives, in history. Most of us are safe from starvation and predators. We get our energy from oil and machines, not from our sweat. What neo-Luddite among us would trade his life for that of a medieval peasant, a caveman, or an ape?

For most of our history we supported ourselves by hunting and gathering: we hunted wild animals and foraged for wild plants. It's a life that philosophers have traditionally regarded as nasty, brutish, and short. Since no food is grown and little is stored, there is (in this view) no respite from the struggle that starts anew each day to find wild foods and avoid starving. Our escape from this misery was facilitated only 10,000 years ago, when in different parts of the world people began to domesticate plants and animals. The agricultural revolution spread until today it's nearly universal and few tribes of hunter-gatherers survive.

From the progressivist perspective on which I was brought up, to ask "Why did almost all our hunter-gatherer ancestors adopt agriculture?" is silly. Of course they adopted it because agriculture is an efficient way to get more food for less work. Planted crops yield far more tons per acre than roots and berries. Just imagine a band of savages, exhausted from searching for nuts or chasing wild animals, suddenly grazing for the first time at a fruit-laden orchard or a pasture full of sheep. How many milliseconds do you think it would take them to appreciate the advantages of agriculture?

The progressivist party line sometimes even goes so far as to credit agriculture with the remarkable flowering of art that has taken place over the past few thousand years. Since crops can be stored, and since it takes less time to pick food from a garden than to find it in the wild, agriculture gave us free time that hunter-gatherers never had. Thus it was agriculture that enabled us to build the Parthenon and compose the B-minor Mass.

While the case for the progressivist view seems overwhelming, it's hard to prove. How do you show that the lives of people 10,000 years ago got better when they abandoned hunting and gathering for farming? Until recently, archaeologists had to resort to indirect tests, whose results (surprisingly) failed to support the progressivist view. Here's one example of an indirect test: Are twentieth century hunter-gatherers really worse off than farmers? Scattered throughout the world, several dozen groups of so-called primitive people, like the Kalahari bushmen, continue to support themselves that way. It turns out that these people have plenty of leisure time, sleep a good deal, and work less hard than their farming neighbors. For instance, the average time devoted each week to obtaining food is only 12 to 19 hours for one group of Bushmen, 14 hours or less for the Hadza nomads of Tanzania. One Bushman, when asked why he hadn't emulated neighboring tribes by adopting agriculture, replied, "Why should we, when there are so many mongongo nuts in the world?"

While farmers concentrate on high-carbohydrate crops like rice and potatoes, the mix of wild plants and animals in the diets of surviving hunter-gatherers provides more protein and a better balance of other nutrients. In one study, the Bushmen's average daily food intake (during a month when food was plentiful) was 2,140 calories and 93 grams of protein, considerably greater than the recommended daily allowance for people of their size. It's almost inconceivable that Bushmen, who eat 75 or so wild plants, could die of starvation the way hundreds of thousands of Irish farmers and their families did during the potato famine of the 1840s.

So the lives of at least the surviving hunter-gatherers aren't nasty and brutish, even though farmers have pushed them into some of the world's worst real estate. But modern hunter-gatherer societies that have rubbed shoulders with farming societies for thousands of years don't tell us about conditions before the agricultural revolution. The progressivist view is really making a claim about the distant past: that the lives of primitive people improved when they switched from gathering to farming. Archaeologists can date that switch by distinguishing remains of wild plants and animals from those of domesticated ones in prehistoric garbage dumps.

How can one deduce the health of the prehistoric garbage makers, and thereby directly test the progressivist view? That question has become answerable only in recent years, in part through the newly emerging techniques of paleopathology, the study of signs of disease in the remains of ancient peoples.

In some lucky situations, the paleopathologist has almost as much material to study as a pathologist today. For example, archaeologists in the Chilean deserts found well-preserved mummies whose medical conditions at time of death could be determined by autopsy. And feces of long-dead Indians who lived in dry caves in Nevada remain sufficiently well preserved to be examined for hookworm and other parasites.

Usually the only human remains available for study are skeletons, but they permit a surprising number of deductions. To begin with, a skeleton reveals its owner's sex, weight, and approximate age. In the few cases where there are many skeletons, one can construct mortality tables like the ones life insurance companies use to calculate expected life span and risk of death at any given age. Paleopathologists can also calculate growth rates by measuring bones of people of different ages, examine teeth for enamel defects (signs of childhood malnutrition), and recognize scars left on bones by anemia, tuberculosis, leprosy, and other diseases.

One straightforward example of what paleopathologists have learned from skeletons concerns historical changes in height. Skeletons from Greece and Turkey show that the average height of hunger-gatherers toward the end of the ice ages was a generous 5' 9" for men, 5' 5" for women. With the adoption of agriculture, height crashed, and by 3000 B. C. had reached a low of only 5' 3" for men, 5' for women. By classical times heights were very slowly on the rise again, but modern Greeks and Turks have still not regained the average height of their distant ancestors.

Another example of paleopathology at work is the study of Indian skeletons from burial mounds in the Illinois and Ohio river valleys. At Dickson Mounds, located near the confluence of the Spoon and Illinois rivers, archaeologists have excavated some 800 skeletons that paint a picture of the health changes that occurred when a hunter-gatherer culture gave way to intensive maize farming around A. D. 1150. Studies by George Armelagos and his colleagues then at the University of Massachusetts show these early farmers paid a price for their new-found livelihood. Compared to the hunter-gatherers who preceded them, the farmers had a nearly 50 per cent increase in enamel defects indicative of malnutrition, a fourfold increase in iron-deficiency anemia (evidenced by a bone condition called porotic hyperostosis), a threefold rise in bone lesions reflecting infectious disease in general, and an increase in degenerative conditions of the spine, probably reflecting a lot of hard physical labor. "Life expectancy at birth in the pre-agricultural community was about twenty-six years," says Armelagos, "but in the post-agricultural community it was nineteen years. So these episodes of nutritional stress and infectious disease were seriously affecting their ability to survive."

The evidence suggests that the Indians at Dickson Mounds, like many other primitive peoples, took up farming not by choice but from necessity in order to feed their constantly growing numbers. "I don't think most hunger-gatherers farmed until they had to, and when they switched to farming they traded quality for quantity," says Mark Cohen of the State University of New York at Plattsburgh, co-editor with Armelagos, of one of the seminal books in the field, *Paleopathology at the Origins of Agriculture*. "When I first started making that argument ten years ago, not many people agreed with me. Now it's become a respectable, albeit controversial, side of the debate."

There are at least three sets of reasons to explain the findings that agriculture was bad for health. First, hunter-gatherers enjoyed a varied diet, while early farmers obtained most of their food from one or a few starchy crops. The farmers gained cheap calories at the cost of poor nutrition. (Today just three high-carbohydrate plant—wheat, rice, and corn—provide the bulk of the calories consumed by the human species, yet each one is deficient in certain vitamins or amino acids essential to life.) Second, because of dependence on a limited number of crops, farmers ran the risk of starvation if one crop failed. Finally, the mere fact that agriculture encouraged people to clump together in crowded societies, many of which then carried on trade with other crowded societies, led to the spread of parasites and infectious disease. (Some archaeologists think it was the crowding, rather than agriculture, that promoted disease, but this is a chicken-and-egg argument, because crowding encourages agriculture and vice versa.) Epidemics couldn't take hold when populations were scattered in small bands that constantly shifted camp. Tuberculosis and diarrheal disease had to await the rise of farming, measles and bubonic plague the appearance of large cities.

Besides malnutrition, starvation, and epidemic diseases, farming helped bring another curse upon humanity: deep class divisions. Hunter-gatherers have little or no stored food, and no concentrated food sources, like an orchard or a herd of cows: they live off the wild plants and animals they obtain each day. Therefore, there can be no kings, no class of social parasites who grow fat on food seized from others. Only in a farming population could a healthy, non-producing elite set itself above the disease-ridden masses. Skeletons from Greek tombs at Mycenae c. 1500 B. C. suggest that royals enjoyed a better diet than commoners, since the royal skeletons were two or three inches taller and had better teeth (on the average, one instead of six cavities or missing teeth). Among Chilean mummies from c. A. D. 1000, the elite were distinguished not only by ornaments and gold hair clips but also by a fourfold lower rate of bone lesions caused by disease.

Similar contrasts in nutrition and health persist on a global scale today. To people in rich countries like the US, it sounds ridiculous to extol the virtues of hunting and gathering. But Americans are an elite, dependent on oil and minerals that must often be imported from countries with poorer health and nutrition. If one could choose between being a peasant farmer in Ethiopia or a bushman gatherer in the Kalahari, which do you think would be the better choice?

Farming may have encouraged inequality between the sexes, as well. Freed from the need to transport their babies during a nomadic existence, and under pressure to produce more hands to till the fields, farming women tended to have more frequent pregnancies than their hunter-gatherer counterparts — with consequent drains on their health. Among the Chilean mummies for example, more women than men had bone lesions from infectious disease.

Women in agricultural societies were sometimes made beasts of burden. In New Guinea farming communities today I often see women staggering under loads of vegetables and firewood while the men walk empty-handed. Once while on a field trip there studying birds, I offered to pay some villagers to carry supplies from an airstrip to my mountain camp. The heaviest item was a 110-pound bag of rice, which I lashed to a pole and assigned to a team of four men to shoulder together. When I eventually caught up with the villagers, the men were carrying light loads, while one small woman weighing less than the bag of rice was bent under it, supporting its weight by a cord across her temples.

As for the claim that agriculture encouraged the flowering of art by providing us with leisure time, modern hunter-gatherers have at least as much free time as do farmers. The whole emphasis on leisure time as a critical factor seems to me misguided. Gorillas have had ample free time to build their own Parthenon, had they wanted to. While post-agricultural technological advances did make new art forms possible and preservation of art easier, great paintings and sculptures were already being produced by hunter-gatherers 15,000 years ago, and were still being produced as recently as the last century by such hunter-gatherers as some Eskimos and the Indians of the Pacific Northwest.

Thus with the advent of agriculture an elite became better off, but most people became worse off. Instead of swallowing the progressivist party line that we chose agriculture because it was good for us, we must ask how we got trapped by it despite its pitfalls.

One answer boils down to the adage "Might makes right." Farming could support many more people than hunting, albeit with a poorer quality of life. (Population densities of hunter-gatherers are rarely over one person per ten square miles, while farmers average 100 times that. [And modern urban/suburban cities average *10,000 times* that – ed.]) Partly, this is because a field planted entirely in edible crops lets one feed far more mouths than a forest with scattered edible plants. Partly, too, it's because nomadic hunter-gatherers have to keep their children spaced at four-year intervals by infanticide and other means, since a mother must carry her toddler until it's old enough to keep up with the adults. Because farm women don't have that burden, they can and often do bear a child every two years.

As population densities of hunter-gatherers slowly rose at the end of the ice ages, bands had to choose between feeding more mouths by taking the first steps toward agriculture, or else finding ways to limit growth. Some bands chose the former solution, unable to anticipate the evils of farming, and seduced by the transient abundance they enjoyed until population growth caught up with increased food production. Such bands out-bred and then drove off or killed the bands that chose to remain hunter-gatherers, because a hundred malnourished farmers can still outfight one healthy hunter. It's not that hunter-gatherers abandoned their life style, but that those sensible enough not to abandon it were forced out of all areas except the ones farmers didn't want.

At this point it's instructive to recall the common complaint that archaeology is a luxury, concerned with the remote past, and offering no lessons for the present. Archaeologists studying the rise of farming have reconstructed a crucial stage at which we made the worst mistake in human history. Forced to choose between limiting population or trying to increase food production, we chose the latter and ended up with starvation, warfare, and tyranny.

Hunter-gatherers practiced the most successful and longest-lasting life style in human history. In contrast, we're still struggling with the mess into which agriculture has tumbled us, and it's unclear whether we can solve it. Suppose that an archaeologist who had visited from outer space were trying to explain human history to his fellow spacelings. He might illustrate the results of his digs by a 24-hour clock on which one hour represents 100,000 years of real past time. If the history of the human race began at midnight, then we would now be almost at the end of our first day. We lived as hunter-gatherers for nearly the whole of that day, from midnight through dawn, noon, and sunset. Finally, at 11:54 p. m. we adopted agriculture. As our second midnight approaches, will the plight of famine-stricken peasants gradually spread to engulf us all? Or will we somehow achieve those seductive blessings that we imagine behind agriculture's glittering façade, and that have so far eluded us?

MEGATECHNOLOGY

JERRY MANDER

From a career in advertising, Mander became a noted writer and environmental activist. At the moment he is a director at the Foundation for Deep Ecology. His most well-known books are *Four Arguments for the Elimination of Television* (1977) and *In the Absence of the Sacred* (1991). The following interview with Scott London occurred in 2001.

London: "Megatechnology" is a word that appears quite often in your writing. What does it mean?

Mander: I use the term "megatechnology" to describe a new reality where many technologies intertwine to create new technologies. For example, we talk about computers as if they were a single technology, but in fact they are intertwined with many, many other technologies. So it's no longer possible to relate to the computer on a one-to-one basis. We live in a new kind of global environment where everything is mediated by technology. We are interactive with it every minute of the day. We are in a car, or in an office with all kinds of machines, or relating to computers, or watching TV, or walking down the street (which is also a form of technology). So, we are constantly relating to it. We live in a technology environment, a technosphere.

What do you think that does to us?

Well, just as other creatures co-evolve with their environment, we are co-evolving with our technologies. In nature, creatures evolve by adjusting and reacting to other creatures. It used to be that way with human beings as well. But now we are co-evolving mainly with machines. Our compromise with them is that we start to become like them—we have to become a little like them in order to use them.

What do you mean?

I mean that if you're going to play a video game, for example, the point is to speed up your hand-eye coordination. The better you get at the video game, the faster your hand-eye connection. What you are doing with your hands and eyes is involving yourself in the computer program. So you are creating a cycle of actions and reactions with the computer technology. As your awareness and your nervous system become tuned to the computer, you are changed accordingly.

This is true of any technology. Look at television, for example. To watch television is to take in images that are artificially created for a specific purpose. By carrying these images, you begin to turn into them. That's basic to education and to all experience: as you ingest your environment you begin to evolve with it. In the case of television, you are evolving on the basis of carefully selected and programmed images, so you are getting acted on in a very aggressive manner. Television turns you into its own images. It rearranges your mind.

A point you make quite often is that technology is not neutral, as many people presume. It may seem neutral when we look at it in purely personal terms—the personal benefits of a computer, for example—but from a broader social and political perspective, technology actually changes our reality in dramatic and sometimes dangerous ways.

We need to understand how technology affects the whole system. Although we may find computers to be very helpful, or television to be entertaining, or cars able to move us rapidly where we need to go, these things also have serious effects on the environment, on the speed of life, on the way we think, on how we view ourselves, on how we react with nature, and on how power changes in the system. All of these are systemic changes; they are things that happen in the system as a whole.

It's not enough for us to say that cars are great because they will drive us someplace. Cars speed up life, they make oil wars happen, they create terrible waste problems, they require a lot of pavement, and they kill people. So you have to ask a series of systemic questions before you can make a judgment about whether cars are good. The same goes for computers, for TV, and for every technology.

How do you respond to the argument that technology is not inherently good or bad, it's how we use it that matters.

That's the major homily of our time. And it's a very serious mistake. The idea that technology is neutral—that it doesn't have social, political and environmental characteristics—is really dangerous.

Consider nuclear power and solar power. Both are energy forms, but they have entirely different effects on the system. Nuclear power is an inherently centralized technology. It requires centralized military-industrial institutions. Nobody knows what to do about 250,000 years of dangerous wastes. If we were to judge energy only in terms of who uses it, that would be like saying, "Well, if some good people got together and ran the nuclear power industry, the wastes wouldn't have to be safeguarded for 250,000 years." But these things are intrinsic to the technology. It's not a question of whether good people use them.

Solar technology is the exact opposite—it is inherently localizing. A couple of people can easily put it together, it's not expensive to use, the community can run it without having to hook up to the grid, and it has no lasting negative effects.

Some people feel that computers are inherently localizing, as you put it, since they decentralize institutions and empower and connect people in new ways.

Well, that is another example of failing to take a systemic viewpoint. People may edit their copy, communicate with their friends, connect with other like-minded people, and so on. But the computer doesn't change the fact that great centralized institutions—corporations, trade bureaucracies, militaries, governments and so on—are able to use those same computers with far greater connections and with far greater real power. So the Internet will not stop a forest from being cut down or global money speculation from affecting the fates of whole societies. These technologies have to be viewed in all their dimensions.

If computers enable you to do your work a little better, I don't argue with that. But it's an illusion for us to believe that our use of the computer will somehow change the centralized system of power. For those who would like to see equitable and sustainable systems develop, the use of the computer amounts to a net loss, not a net gain.

In your book, *In the Absence of the Sacred*, you make the case that we need to think more seriously and systematically about the effects of technologies before they are introduced.

Yes. If you give some time and thought to the potentialities of technologies and see all the directions that they can go in, it will reveal much about how they are going to be used. It's very necessary to understand those before a technology is used in a way that we don't approve of. Of course, people often say, "Well, we don't know how a technology is going to turn out." The point mentioned in the book is that we do know how technology is going to turn out. We can make lots of predictions about technology. Corporations live and die by how well they predict the uses of their products. A corporation is not going to put zillions of their dollars into creating a technology without studying every possible use (and every possible downside) of that technology. What could go wrong? What great catastrophe could it cause? What small catastrophes could it cause? A corporation wants to know everything. So they spend a lot of money on figuring out what these technologies are going to do. And what they tell us is only the good things that they are going to do. They don't tell us the downside of the story. We have to know the downside of the story now, before the technology is too far upon us.

Imagine how different America would look if we had predicted the effects of the automobile 70 or 80 years ago. Being stuck in traffic on a Los Angeles freeway always makes me wonder about the so-called benefits of technological progress.

Well, it's been documented that in the 1930s the automobile industry and the oil industry conspired to undermine and eventually destroy Los Angeles' public transportation system, which was very good at the time. San Francisco's public transportation system was also very good. There used to be trains going across the Bay Bridge and you could get to Oakland very fast and very comfortably. But the trains were wiped out on purpose. The automobile and oil industries conspired to destroy the commuter railroads and replace them with freeways. Now people are wondering about what to do with all the cars. They're studying how to put trains back on the bridge again.

One of the most striking passages in your book, *In the Absence of the Sacred*, is where you describe a trip to the MacKenzie Valley in Northwest Canada to speak with an Inuit women's association about the effects of television.

I was invited by an organization called the Native Women's Association of the Northwest Territories, an organization of Diné and Inuit women. The MacKenzie River Valley is where the Russian nuclear satellite came down some years ago. At the time, everybody was worried that it would fall on London or New York, but instead it fell on a so-called icy wasteland up in Canada. That's the place where I was invited to go. It was 40 degrees below zero the day I arrived.

The MacKenzie River Valley has 22 communities of native people. They are spread over an enormous area. They still have a very successful traditional economy based on hunting and fishing and live in a communal manner in log houses.

I was invited up there because television had begun to arrive in the area. The Women's Association was noticing startling changes in the communities where television had arrived. The men didn't go out on the ice to fish as often. The animals weren't being taken care of as well. The kids didn't want to go out and play traditional games. The kids were starting to want things—like cars (even though there are no roads there). The neighbors weren't hanging out together, working on the nets together, cooking together, eating together and so on. The community life was breaking down.

The most important thing, they told me, was the loss of story-telling. In the evenings, it used to be that the very old would gather with the very young in a corner of the house—several families together—and the old people would tell traditional stories and stories from their past. By hearing those stories, the young people could remember who they are, what's good about their people, and how to live in that very harsh environment. The stories were a window to their roots. Also, the process of young and old hanging out together in that way was very important. There was a lot of love flowing back and forth and the kids were proud to be connected to their grandparents.

Apparently, all of this has been wiped out by television. Story-telling has come to an end. Now families sit together silently—all these generations together—and watch *Dallas*, a bunch of white people standing around a swimming pool drinking martinis and plotting against each other.

What is the appeal of a program like *Dallas*?

Human beings are genetically programmed to pay attention to anything that is new. It goes back to our time when we lived in jungles and had to depend on the information coming in through our senses. It's part of our survival technique: we pay attention to anything new that takes place in our environment. But in this case it's not an animal hiding behind a bush, it's a whole technology speaking into our heads. It's very hard to change ourselves genetically to keep up with the technological changes.

As human beings, we are supposed to believe what we see. Our system is constructed for seeing-is-believing. If we see birds flying south, we depend for survival on the fact that the birds are in fact flying south. But we've moved out of the forest and into the city and now we depend strictly on what is delivered to us as information. When we see images on television, we don't know how not to believe them. Television is very powerful and compelling.

So, of course they watch Dallas. It's completely exotic. If they are watching television, of course they're going to want to see something different than a bunch of Indians sitting around. They are going to want to see something different, and television provides them with that. It's a dream-machine.

I find it interesting that it was the women, not the men, who invited you to come.

I don't know why it was the women and not the men. I presume it is because they are the keepers of the family and they are worried about the breakdown of family-life.

I've been told that there is now an Inuit version of *Sesame Street* that is quite excellent.

Oh well. I don't think that's necessarily a good development. I'm not a fan of *Sesame Street*. It's a very arch-Western-style of television. The way the imagery is presented on the show is tantamount to advertising. It was originally created by advertising people, that's why it's edited in that rapid-fire, repetitive fashion. It's an advertising technique that people really respond to.

But this is presented as an example of how native communities are adapting educational programming to their own unique culture and tradition.

It's not true. The fact that native people are producing *Sesame Street* doesn't mean that it has been integrated into their culture. It's a sign that the native culture is being destroyed and integrated into Western culture.

Native people are getting trained in all kinds of technologies now. The stuff is being brought in and promoted like crazy. The Canadian government very strongly wants them to watch television. They want them to turn from Indians into Canadians and therefore become workers and consumers. They want them to be inculcated into the culture. They want to get in there and show how it can be better. So they say things like, "Keep the native culture, but use computers." Or, "Keep the native culture, but use television." I think the government understands that by introducing these technologies, they are doing exactly the opposite—not preserving the native culture, but assimilating native people.

Tell me something about your own relationship to technology. Do you have a car?

I have lots of modern technologies. It's impossible to function and not have some relationship to technology.

Do you own a car? Yes, I have a car.

How about a computer?

No, I don't use a computer. My feeling is that computers really strongly change the way we think. I think computers are changing the world more rapidly and more negatively than any other around. So I would really like to maintain a disconnection from that.

On the other hand, that's a false distinction because I'm a writer. Eventually, everything I write is entered on a computer in order for somebody to publish it. So, there is no way around it. The car that I drive has computers in it. The microphones that we're talking through have computers in them. The office I work in is full of computers.

Do you have a television? There is a television in the house, but I don't watch it.

You actually have a television? The way you're asking me that, it's like you're trying to catch me in some hypocrisy.

Well, you wrote a book called *Four Arguments for the Elimination of Television* but you still have one in the house. I'm curious about that.

That's not as significant as it may sound. I'm sitting here in a room with two microphones talking on the radio. I sometimes fly to meetings on airplanes. I wear machine-made shoes. I have lights in my house. I use some home appliances. What is important is for every person to develop a relationship to technology that expresses their views of what is important and what is not important.

So, in the case of television, I don't have a relationship to it. For example, I don't appear on television. By not participating in that medium, I am making a statement about not supporting its output of information.

I'm glad you agreed to this radio interview.

I do go on radio, but there is not a whole lot of difference. There are some benefits: it's a much less expensive medium, it's much less centralized, and it keeps listeners' imagery operating. But as a mass medium, it still has many problems connected with it.

The thing that people need to do is decide which technologies they want to relate to, and why, and try to stick by that as much as they possibly can. In my case, I could go live on a farm and use oxen but I'm not sure that would be a net gain in the end. I have to decide what is necessary in order to stay in the process and share my ideas. We all live in a compromised situation. So we do the best that we can.

Wendell Berry is right: the idea is to minimize as much as you possibly can all the relationships that you have. If some lesser level of technology will get the job done, you're obliged to use that. That's the principle that I go by.

The Illusion of Neutral Technology
—from *Four Arguments for the Elimination of Television* (1978)

Most Americans, whether on the political left, center, or right, will argue that technology is neutral, that any technology is merely a benign instrument, a tool, and depending upon the hands into which it falls, it may be used one way or another. There is nothing that prevents a technology from being used well or badly; nothing intrinsic in the technology itself or the circumstances of its emergence which can predetermine its use, its control or its effects upon individual human lives or the social and political forms around us.

The argument goes that television is merely a window or a conduit through which any perception, any argument or reality may pass. It therefore has the potential to be enlightening to people who watch it and is potentially useful to democratic processes.

It will be the central point of this book that these assumptions about television, as about other technologies, are totally wrong.

If you once accept the principle of an army—a collection of military technologies and people to run them—all gathered together for the purpose of fighting, overpowering, killing and winning, then it is obvious that the supervisors of armies will be the sort of people who desire to fight, overpower, kill and win, and who are also good at these assignments: generals. The fact of generals, then, is predictable by the creation of armies. The kinds of generals are also predetermined. Humanistic, loving, pacifistic generals, though they may exist from time to time, are extremely rare in armies. It is useless to advocate that we have more of them.

If you accept the existence of automobiles, you also accept the existence of roads laid upon the landscape, oil to run the cars, and huge institutions to find the oil, pump it and distribute it. In addition you accept a sped-up style of life and the movement of humans through the terrain at speeds that make it impossible to pay attention to whatever is growing there. Humans who use cars sit in fixed positions for long hours following a narrow strip of gray pavement, with eyes fixed forward, engaged in the task of driving. As long as they are driving, they are living within what we might call "roadform." Slowly they evolve into car-people. McLuhan told us that cars "extended" the human feet, but he put it the wrong way. Cars *replaced* human feet.

If you accept nuclear power plants, you also accept a techno-scientific-industrial-military elite. Without these people in charge, you could not have nuclear power. You and I getting together with a few friends could not make use of nuclear power. We could not build such a plant, nor could we make personal use of its output, nor handle or store the radioactive waste products which remain dangerous to life for thousands of years. The wastes, in turn, determine that *future* societies will have to maintain a technological capacity to deal with the problem, and the military capability to protect the wastes. So the existence of the technology determines many aspects of the society.

If you accept mass production, you accept that a small number of people will supervise the daily existence of a much larger number of people. You accept that human beings will spend long hours, every day, engaged in repetitive work, while suppressing any desires for experience or activity beyond this work. The workers' behavior becomes subject to the machine. With mass production, you also accept that huge numbers of identical items will need to be efficiently distributed to huge numbers of people and that institutions such as advertising will arise to do this. One technological process cannot exist without the other, creating symbiotic relationships among technologies themselves.

If you accept the existence of advertising, you accept a system designed to persuade and to dominate minds by interfering in people's thinking patterns. You also accept that the system will be used by the sorts of people who like to influence people and are good at it. No person who did not wish to dominate others would choose to use advertising, or choosing it, succeed in it. So the basic nature of advertising and all technologies created to serve it will be consistent with this purpose, will encourage this behavior in society, and will tend to push social evolution in this direction.

In all of these instances, the basic form of the institution and the technology determines its interaction with the world, the way it will be used, the kind of people who use it, and to what ends.

And so it is with television.

Far from being "neutral," television itself predetermines who shall use it, how they will use it, what effects it will have on individual lives, and, if it continues to be widely used, what sorts of political forms will inevitably emerge.

PART II: HISTORICAL ORIGINS

Milestones in Technology

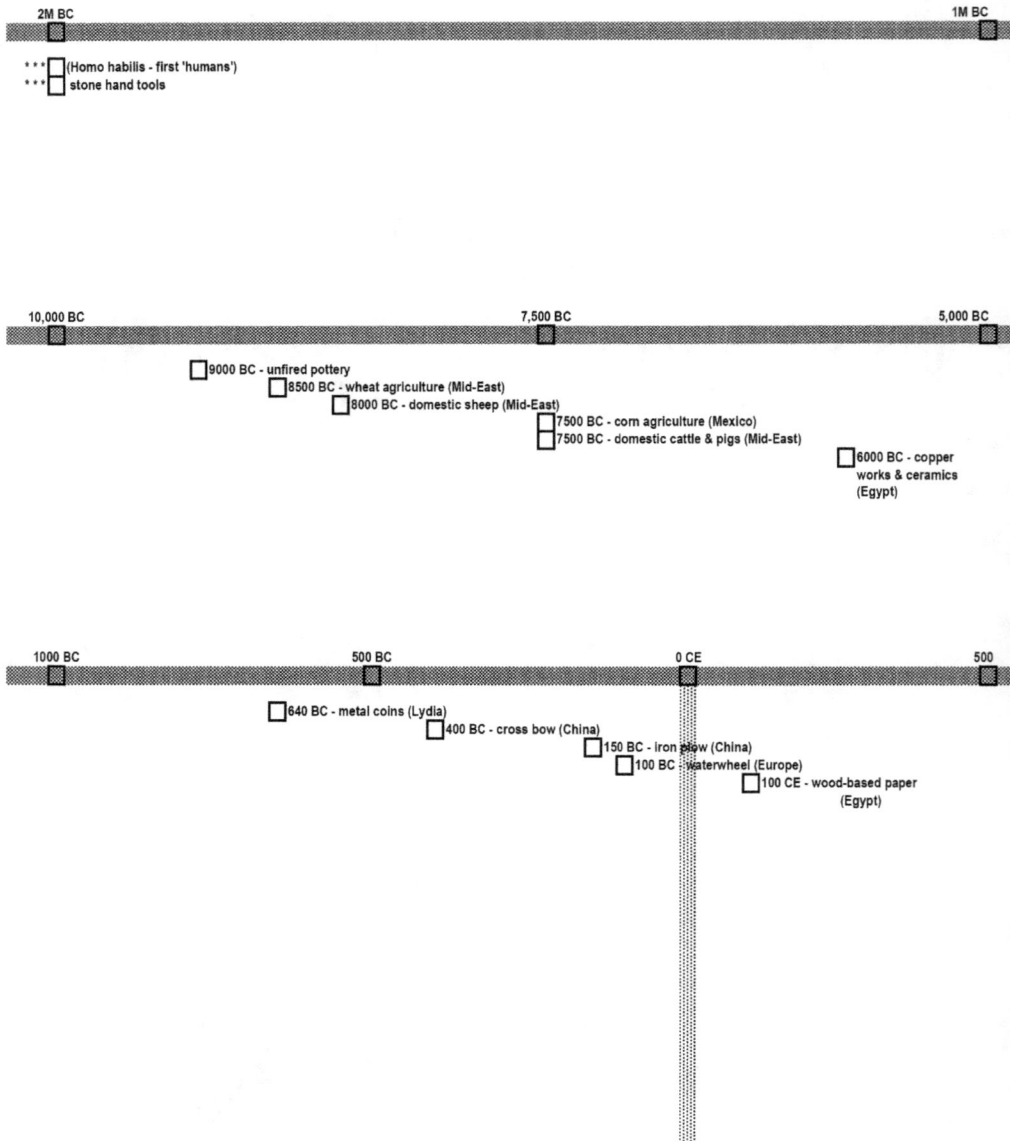

2M BC   1M BC

* * * (Homo habilis - first 'humans')
* * * stone hand tools

10,000 BC   7,500 BC   5,000 BC

9000 BC - unfired pottery
8500 BC - wheat agriculture (Mid-East)
8000 BC - domestic sheep (Mid-East)
7500 BC - corn agriculture (Mexico)
7500 BC - domestic cattle & pigs (Mid-East)
6000 BC - copper
works & ceramics
(Egypt)

1000 BC   500 BC   0 CE   500

640 BC - metal coins (Lydia)
400 BC - cross bow (China)
150 BC - iron plow (China)
100 BC - waterwheel (Europe)
100 CE - wood-based paper
(Egypt)

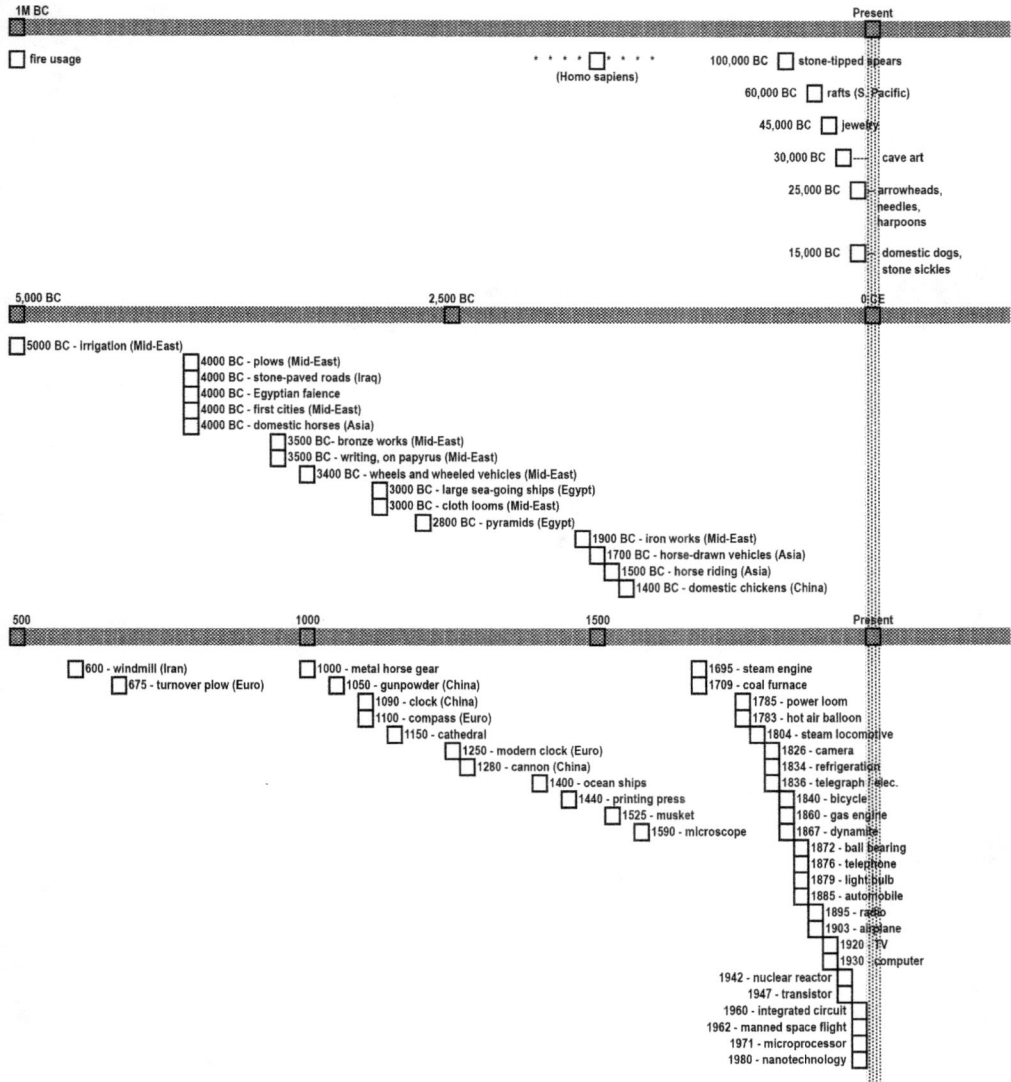

1M BC ... **Present**

fire usage

• • • • ☐ • • • •
(Homo sapiens)

100,000 BC ☐ stone-tipped spears

60,000 BC ☐ rafts (S. Pacific)

45,000 BC ☐ jewelry

30,000 BC ☐ ---- cave art

25,000 BC ☐ -- arrowheads, needles, harpoons

15,000 BC ☐ -- domestic dogs, stone sickles

5,000 BC ... **2,500 BC** ... **0 CE**

☐ 5000 BC - irrigation (Mid-East)

☐ 4000 BC - plows (Mid-East)
☐ 4000 BC - stone-paved roads (Iraq)
☐ 4000 BC - Egyptian faience
☐ 4000 BC - first cities (Mid-East)
☐ 4000 BC - domestic horses (Asia)
☐ 3500 BC- bronze works (Mid-East)
☐ 3500 BC - writing, on papyrus (Mid-East)
☐ 3400 BC - wheels and wheeled vehicles (Mid-East)
☐ 3000 BC - large sea-going ships (Egypt)
☐ 3000 BC - cloth looms (Mid-East)
☐ 2800 BC - pyramids (Egypt)
☐ 1900 BC - iron works (Mid-East)
☐ 1700 BC - horse-drawn vehicles (Asia)
☐ 1500 BC - horse riding (Asia)
☐ 1400 BC - domestic chickens (China)

500 ... **1000** ... **1500** ... **Present**

☐ 600 - windmill (Iran)
☐ 675 - turnover plow (Euro)
☐ 1000 - metal horse gear
☐ 1050 - gunpowder (China)
☐ 1090 - clock (China)
☐ 1100 - compass (Euro)
☐ 1150 - cathedral
☐ 1250 - modern clock (Euro)
☐ 1280 - cannon (China)
☐ 1400 - ocean ships
☐ 1440 - printing press
☐ 1525 - musket
☐ 1590 - microscope
☐ 1695 - steam engine
☐ 1709 - coal furnace
☐ 1785 - power loom
☐ 1783 - hot air balloon
☐ 1804 - steam locomotive
☐ 1826 - camera
☐ 1834 - refrigeration
☐ 1836 - telegraph / elec.
☐ 1840 - bicycle
☐ 1860 - gas engine
☐ 1867 - dynamite
☐ 1872 - ball bearing
☐ 1876 - telephone
☐ 1879 - light bulb
☐ 1885 - automobile
☐ 1895 - radio
☐ 1903 - airplane
☐ 1920 - TV
☐ 1930 - computer
1942 - nuclear reactor ☐
1947 - transistor ☐
1960 - integrated circuit ☐
1962 - manned space flight ☐
1971 - microprocessor ☐
1980 - nanotechnology ☐

TECHNICS AND THE NATURE OF MAN

LEWIS MUMFORD

Lewis Mumford (1895-1990) was an historian of technology and science. His analyses of cities and modern architecture were extremely influential. He was the author of several books, including *Technics and Civilization* (1934), *The City in History* (1961), and *The Myth of the Machine* (1967-1970). The present essay dates from 1965.

The last century, we all realize, has witnessed a radical transformation in the entire human environment, largely as a result of the impact of the mathematical and physical sciences on technology. This shift from an empirical, tradition-bound technics to an experimental scientific mode has opened up such now realms as those of nuclear energy, supersonic transportation, cybernetic intelligence, and instantaneous planetary communication.

In terms of the currently accepted picture of the relation of man to technics, our age is passing from the primeval state of man, marked by his invention of tools and weapons for the purpose of achieving mastery over the forces of Nature, to a radically different condition, in which he will not only have conquered Nature but detached himself completely from the organic habitat. With this new megatechnology, man will create a uniform, all-enveloping structure, designed for automatic operation. Instead of functioning actively as a tool-using animal, man will become a passive, machine-serving animal whose proper functions will either be fed into a machine, or strictly limited and controlled for the benefit of depersonalized collective organizations. The ultimate tendency of this development was correctly anticipated by Samuel Butler more than a century ago; but it is only now that his playful fantasy shows many signs of becoming a far-from-playful reality.

My purpose in this paper is to question both the assumptions and the predictions on which our commitment to the present form of technical and scientific progress, as an end itself, has been based. In particular, I find it necessary to cast doubts on the generally accepted theories of man's basic nature which have been implicit during the last century in our constant over-rating of the role of tools and machines in the human economy. I shall suggest that not only was Karl Marx in error in giving the instruments of production a central place and a directive function in human development, but even the seemingly benign interpretation by Teilhard de Chardin reads back into the whole story of man the narrow technological rationalism of our own age, and projects into the future a final state in which all the further possibilities of human development would come to an end, because nothing would be left of man's original nature that had not been absorbed into, if not suppressed by, the technical organizations of intelligence into a universal and omnipotent layer of mind.

Since the conclusions I have reached require, for their background, a large body of evidence I have been marshaling in a still unpublished book, I am aware that the following summary must, by its brevity, seem superficial and unconvincing. At best, I can only hope to show that there are serious reasons for reconsidering the whole picture of both human and technical development on which the present organization of Western society is based.

Now, we cannot understand the role that technics has played in human development without a deeper insight into the nature of man: yet that insight has itself been blurred, during the past century, because it has been conditioned by a social environment in which a mass of now mechanical inventions has suddenly proliferated, sweeping away many ancient processes and institutions, and altering our very conception of both human limitations and technical possibilities.

For more than a century man has been habitually defined as a *tool-using animal*. This definition would have seemed strange to Plato, who attributed man's rise from a primitive state as much to Marsyas and Orpheus as to Prometheus and Hephaestus, the blacksmith-god. Yet the description of man as essentially a tool-using and tool-making animal has become so firmly accepted that the mere finding of the fragments of skulls, in association with roughly shaped pebbles, as with Dr. L. S. B. Leakey's australopithecines, is deemed sufficient to identify the creature as a proto-human, despite marked physical divergences from both earlier apes and later men.

By fastening attention on the surviving stone artifacts, many anthropologists and ethnologists have gratuitously attributed to the shaping and using of tools the enlargement of man's higher intelligence, though motor-sensory coordination involved in this elementary manufacture do not call for or evoke any considerable mental acuteness. Since the sub-hominids of South Africa had a brain capacity about a third that of *Homo sapiens*, no greater indeed than that of many apes, the capacity to make tools neither called for nor generated early man's rich cerebral equipment, as Dr. Ernst Mayr of Harvard has recently pointed out.

An error in interpreting man's nature arises from the present-day tendency to read back into prehistoric times modern man's overwhelming interest in tools and machines, to the exclusion of equally important items of technical equipment. Tools and weapons are specialized extrapolations of man's own organs for pushing, pounding, crunching, cutting, stabbing—all basic motor activities. No one can doubt that these dynamic processes, which man shares with many other species, formed an essential part of his earliest technical complex.

But just because man's need for tools is so obvious, we must guard against overemphasizing the role of tools hundreds of thousands of years before they became functionally efficient. In treating tool-making as central to the Paleolithic economy, ethnologists have underplayed, or neglected, a mass of activities in which many other species were for long far more knowledgeable than man. Despite the contrary evidence put forward, there is still a tendency to identify tools and machines with technology: to substitute a part for the whole.

In any comprehensive definition of technics, it should be plain that many insects, birds and mammals had made far more radical innovations in the fabrication of containers, with their intricate nests and bowers, their geometric beehives, their urbanoid anthills and termitaries, than man's ancestors had achieved in the making of tools until the emergence of *Homo sapiens*. In short, if technical proficiency were alone sufficient to identify potential intelligence, man would, for long, have been rated a hopeless duffer alongside many other species. The consequences of this perception should be plain; namely, that there was nothing uniquely human in early technology until it was modified by linguistic symbols, social organization, and aesthetic designs. At that point symbol-making leaped far ahead of tool-making, and in turn fostered neater technical facility.

At the beginning, then, I suggest that the human race had achieved no special position by reason of its tool-using or tool-making propensities alone. Or rather, man at the beginning possessed one primary all-purpose tool that was more important than any later assemblage: namely, his own mind-activated body, every part of it, not just those motor activities that produced hand axes and wooden spears. To compensate for his extremely primitive working gear, early man had a much more important asset that widened his whole technical horizon: a body not specialized for any single activity, but, precisely because of its extraordinary plasticity, more effective in using a larger portion of both his external environment and his internal psychical resources.

Through man's over-developed and incessantly active brain, he had more mental energy to tap than he needed for survival at a purely animal level; and he was, accordingly, under the necessity of canalizing that energy, not just into food-getting and reproduction, but into modes of living that would convert this energy more directly and constructively into appropriate cultural—that is, *symbolic*—forms. Cultural 'work,' by necessity, took precedence over manual work; this involved far more than the discipline of hand, muscle and eye in making and using tools. It likewise demanded a control of all man's biological functions, including his bodily organs, his emotions, his sexual activities, his dreams. Even the hand was no mere work-tool: it stroked a lover's body, held a baby close to the breast, made significant gestures, or expressed in ordered dance and shared ritual some otherwise inexpressible sentiment about life or death, a remembered past or an anxious future. Tool-technics is but a fragment of *bio-technics*: man's total equipment for life.

On this interpretation, one may well hold it an open question whether the standardized patterns and the repetitive order which came to play such an effective part in the development of tools from an early period, as Braidwood has pointed out, derive solely from tool-making. Do they not rather derive even more, perhaps, from the forms of ritual, song and dance—forms which exist in a state of perfection among primitive peoples, often in a far more exquisitely finished state than their tools? There is, in fact, widespread evidence, first noted by Hocart, that ritual exactitude in ceremony preceded mechanical exactitude in work: that the first rigorous division of labour came through specialization in ceremonial offices.

These facts help to explain why simple peoples who easily get bored by purely mechanical tasks which might improve their physical well-being, will, nevertheless, repeat a meaningful ritual often to the point of physical exhaustion. The debt of technics to play and to play toys, to myth and fantasy, to magic rite and religious rote, which, called attention to in my book *Technics and Civilization*, has still to be sufficiently recognized, though J. Huizinga, in *Homo Ludens*, went so far as to treat play itself as the basic formative element in all culture.

Tool-making, in the narrow technical sense, may, indeed, go back to our hominid African ancestors. But the technical equipment of Chellean and Acheulian times remained extremely limited until a more richly endowed creature, with a nervous system nearer to that of *Homo sapiens* than to any primeval hominid predecessors, had come into existence, and brought into operation not alone his hands and legs but his entire body and mind, projecting them, not just in tools and utensils but in more purely symbolic non-utilitarian forms.

———————————

In this revision of the accepted technological stereotypes, I would go even further: for I submit that at every stage, man's technological expansions and transformations were less for the purpose of increasing the food supply or controlling Nature, than for utilizing his immense actual resources, and expressing his super-organic potentialities. When not threatened by a hostile environment, the elaboration of symbolic culture was a more imperious need than control over the external environment—and, as one must infer, largely predated it and out-paced it.

On this reading, the invention of *language*—a culmination of man's more elementary forms of expressing and transmitting meaning—was incomparably more important to further human development than the chipping of a mountain of hand axes. Beside the relatively simple co-ordinations required for tool-using, the delicate interplay of the many organs needed for the creation of articulate speech was a far more striking advance, and must have occupied a great part of early man's time, energy and mental concentration, since its collective product, language, was infinitely more complex and sophisticated at the dawn of civilization than the Egyptian or Mesopotamian kit of tools. For only when knowledge and practice could be stored in symbolic forms, and passed on by word of mouth from generation to generation, was it possible to keep each fresh cultural acquisition from dissolving with the passing moment or the dying generation. Then, and then only, did the domestication of plants and animals become possible. Need I remind you that this decisive technical transformation was achieved with no better tools than the digging stick, the axe, the mattock? The plow, like the cart-wheel, came later as a specialized adaptation to the large-scale field cultivation of grain.

To consider man as primarily a tool-using animal, then, is to overlook the main chapters of human prehistory. Opposed to this stereotype is the present view that man is preeminently a *mind-using, self-mastering* animal; and the primary locus of all his activities is his own organism. Until he had made something of himself, he could make little of the world around him.

In this process of self-discovery and self-transformation, technics, in the narrow sense, served well as a subsidiary instrument, but not as the main operative agent in man's development; for technics was never until our own age dissociated from the larger cultural whole. Early man's original development was based on what Andre Varagnac happily called "the technology of the body": the utilization of man's highly plastic bodily capacities for the expression of his still unformed and uninformed mind, before that mind had yet achieved, through the development of symbols and images, its own more etherealized technical instruments. From the beginning, the creation of significant modes of symbolic expression, rather than more effective tools, was the basis of the further development of *Homo sapiens*.

Unfortunately, so firmly were the prevailing nineteenth-century conceptions committed to the notion of man as primarily *Homo faber*, the tool-maker, rather than *Homo sapiens*, the mind-maker, that, as is known, the first discovery of the art of the Altamira eaves was dismissed as a hoax, because the leading palaeo-ethnologists would not admit that the Ice Age hunters, whose weapons and tools they had recently discovered, could have had either the leisure or the mental inclination to produce art—not crude forms, but images that showed powers of observation and abstraction of a high order.

But when we compare the carvings and paintings of the Aurignacian or Magdalenian finds, with their surviving technical equipment, who shall say whether it is art or technics that shows the highest development? Even the finely finished Solutrean laurel leaf points were plainly a gift of aesthetically sensitive artisans. The classic Greek usage for 'technics' makes no distinction between industrial production and symbolic art; and for the greater part of human history these aspects were inseparable, one side respecting objective conditions and functions, the other responding to subjective needs.

Our age has not yet overcome the peculiar utilitarian bias that regards technical invention as primary, and aesthetic expression as secondary or superfluous; and this means that we have still to acknowledge that technics derives from the whole man, in his intercourse with every part of the environment, utilizing every aptitude in himself to make the most of his own biological and ecological potentials.

Even at the earliest stage, trapping and foraging called less for tools than for sharp observation of animal habits and habitats, backed by a wide experimental sampling of plants and shrewd interpretation of the effects of various foods, medicines and poisons on the human organism. And in those horticultural discoveries which, if Oakes Ames was right, must have preceded by many thousands of years the active domestication of plants, taste and formal beauty played a part no less than their food value; so that the earliest domesticates, other than the grains, were often valued for the color and form of their flowers, for their perfume, their texture, their spiciness, rather than merely for nourishment. Edgar Anderson has suggested that the Neolithic garden, like the gardens in many simple cultures today, was probably a mixture of food plants, dye plants and ornamentals—all treated as equally essential for life.

Similarly, some of early man's most daring technical experiments had nothing whatever to do with the mastery of the external environment: they were concerned with the anatomical modification or the superficial decoration of the human body, for sexual emphasis, self-expression, or group identification. The Abbe Breuil found evidence of such practices as early as the Mousterian culture, which served equally in the development of ornament and surgery.

Plainly, tools and weapons, so far from dominating man's technical equipment, as the stone artifacts too glibly suggested, constituted a small part of the biotechnical assemblage; and the struggle for existence, though sometimes severe, did not engross the energy and vitality of early man, or divert him from his more central need to bring order and meaning into every part of his life. In that larger effort, ritual, dance, song, painting, carving and, above all, discursive language must for long have played a decisive role.

At its points of origin, then, technics was related to the whole nature of man. Primitive technics was life-centered, not work-centered or power-centered. As in all ecological complexes, a variety of human interests and purposes, and other organic needs, restrained the overgrowth of any single component. As for the greatest technical feat before our own age — the domestication of plants and animals — this advance owed almost nothing to new tools, though it encouraged the development of clay containers. But it owed much, we now begin to realize, since Edouard Hahn, to an intense subjective concentration on sexuality in all its manifestations, abundantly visible in cult objects and symbolic art. Plant selection, hybridization, fertilization, manuring, seeding and castration were the products of an imaginative cultivation of sexuality, the first evidence of which one finds tens of thousands of years earlier in the emphatically sexual carvings of Paleolithic woman: the so-called Venuses.

But at the point where history, in the form of the written record, becomes visible, that life-centered economy, a true polytechnics, was challenged and in part displaced in a series of radical technical and social innovations. About five thousand years ago, a *monotechnics*, devoted to the increase of power and wealth by the systematic organization of workaday activities in a rigidly mechanical pattern, came into existence. At this moment, a new conception of the nature of man arose, and with it a new stress on the exploitation of physical energies, cosmic and human, apart from the processes of growth and reproduction, came to the fore. In Egypt, Osiris symbolizes the older, *life*-oriented techniques; Atum-Re, the Sun God, who characteristically created the world out of his own semen without female cooperation, stands for the *machine*-centered one. The expansion of power, through ruthless human coercion and mechanical organization, took precedence over the enhancement of life.

The chief mark of this change was the construction of the first complex, high-powered machines; and therewith the beginning of a new regimen, accepted by all later civilized societies — though reluctantly by more archaic cultures — in which work at a single specialized task, segregated from other biological and social activities, not only occupied the entire day, but increasingly engrossed the entire lifetime. That was the fundamental departure which, during the past few centuries, has led to the increasing mechanization and automation of all production. With the assemblage of the first collective machines, work, by its systematic dissociation from the rest of life, became a curse, a burden, a sacrifice, a form of punishment: and by reaction this new regimen soon awakened compensatory dreams of effortless affluence, emancipated not only from slavery but from work itself. These ancient dreams now dominate our age.

The machine I refer to was never discovered in any archaeological diggings, for a simple reason: it was composed almost *entirely of human parts*. These parts were brought together in a hierarchical organization under the rule of an absolute monarch, the commands of whom, supported by a coalition of the priesthood, the armed nobility, and the bureaucracy, secured a corpse-like obedience from all the components of the machine.

Let us call this archetypal collective machine—the human model for all later specialized machines—the '*Megamachine*.' This new kind of machine was far more complex than the contemporary potter's wheel or bow-drill, and it remained the most advanced type of machine until the invention of the mechanical clock in the fourteenth century.

Only through the deliberate invention of such a high-powered machine could the colossal works of engineering that marked the Pyramid Age in both Egypt and Mesopotamia have been brought into existence, often in a single generation. This new technics came to an early climax in the Great Pyramid at Giza. That structure, as J. H. Breasted pointed out, exhibited a watchmaker's standard of exact measurement. By operating as a mechanical unit, the 100,000 men who worked on that pyramid generated ten thousand horse-power. This human mechanism alone made it possible to raise that colossal structure with the use of only the simplest stone and copper tools— without the aid of such otherwise indispensable machines as the wheel, the wagon, the pulley, the derrick, or the winch.

Two things must be noted about this new mechanism, because they identify it through its historic course down to the present. The first is that the organizers of this machine derived their power and authority from a cosmic source. The exactitude in measurement, the abstract mechanical order, the compulsive regularity of this Megamachine sprang directly from astronomical observations and abstract scientific calculations: this inflexible, predictable order, incorporated in the calendar, was then transferred to the regimentation of the human components. By a combination of divine command and ruthless military coercion, a large population was made to endure grinding poverty and forced labor at dull repetitive tasks, in order to ensure "life, prosperity and health" for the divine or semi-divine ruler and his entourage.

The second point is that the grave social defects of the human machine were partly offset by its superb achievements in flood control and grain production, which plainly benefited the whole community. This laid the ground for an enlargement in every area of human culture: in monumental art, in codified law, and in systematically pursued and permanently recorded thought. Such order, such collective security and abundance as was achieved in Mesopotamia and Egypt, later in India, China, in the Andean and Mayan cultures, were never surpassed until the Megamachine was re-established in a new form in our own time. But conceptually the machine was already detached from other human functions and purposes than the increase of mechanical power and order. With mordant symbolism, the Megamachine's ultimate products in Egypt were tombs and mummies, while later in Assyria the chief testimonial to its dehumanized efficiency was, again typically, a waste of destroyed cities and poisoned soils.

In a word, what modern economists lately termed the Machine Age had its origin, not in the eighteenth century, but at the very outset of civilization. All its salient characteristics were present from the beginning in both the means and the ends of the collective Megamachine. So Keynes's acute prescription of pyramid building as an essential means of coping with the insensate productivity of a highly mechanized technology, applies both to the earliest manifestations and the present ones; for what is a space rocket but the precise dynamic equivalent, in terms of our present-day theology and cosmology, of the static Egyptian pyramid? Both are devices for securing, at an extravagant cost, a passage to heaven for the favored few.

Unfortunately, though the labor machine lent itself to vast constructive enterprises, which no small-scale community could even contemplate, much less execute, the most conspicuous result has been achieved through *military machines*, in colossal acts of destruction and human extermination; acts which monotonously soil the pages of history, from the rape of Sumer to the blasting of Warsaw and Hiroshima. Sooner or later, I suggest, we must have the courage to ask ourselves: Is this association of inordinate power and productivity with equally inordinate violence and destruction a purely accidental one?

Now the misuse of Megamachines would have proved intolerable had they not also brought genuine benefits to the whole community by raising the ceiling of collective human effort and aspiration. The most dubious of these advantages was the *gain in efficiency* derived from concentration on rigorously repetitive motions in work, already, indeed, introduced in the grinding and polishing processes of Neolithic tool-making. This inured civilized man to long spans of regular work, with higher productive efficiency per unit. But the social by-product of this new discipline was, perhaps, even more significant; for some of the psychological benefits, hitherto confined to religious ritual, were transferred to *work*. The sterile repetitive tasks imposed by the Megamachine, which in a pathological form we associate with a compulsion neurosis, nevertheless served, like all ritual and restrictive orders, to lessen anxiety and to defend the worker himself from the often demonic promptings of the unconscious, no longer held in check by the traditions and customs of the Neolithic village.

In short, mechanization and regimentation, through labor armies, military armies, and ultimately through the derivative modes of industrial and bureaucratic organization, supplemented and increasingly replaced religious ritual as a means of coping with anxiety and promoting psychical stability in mass populations. Orderly, repetitive work provided a daily means of self-control: more pervasive, more effective, more universal than either ritual or law. This hitherto unnoticed psychological contribution was possibly more important than those gains in productive efficiency, which were too often offset by absolute losses in war and conquest. Unfortunately, the ruling classes, which claimed immunity from manual labor, were not subject to this discipline: hence, as the historic record testifies, their disordered fantasies too often found an outlet in reality through destruction and extermination.

Having indicated the beginnings of this process, I must regrettably pass over the actual institutional forces which have been at work during the past five thousand years, and leap, all too suddenly, into the present age, in which the ancient forms of biotechnics are being either suppressed or supplanted, and in which the continued enlargement of the Megamachine itself has become, with increasing compulsiveness, the condition of scientific and technical advance. This unconditional commitment to the Megamachine is now regarded by many as the main purpose of human existence.

But if the clues I have been attempting to expose prove helpful, many aspects of the scientific and technical transformation of the last three centuries will call for reinterpretation and judicious reconsideration. For, at the very least, we are now bound to explain why the whole process of technical development has become increasingly coercive, totalitarian, and—subjectively speaking—compulsive and grimly irrational.

Before accepting the ultimate translation of all organic processes, biological functions, and human aptitudes into an externally controllable mechanical system, increasingly automatic and self-expanding, it might be well to reexamine the ideological foundations of this whole system, with its overconcentration on centralized power and external control. We must, in fact, ask ourselves if the probable destination of this system is compatible with the further development of specifically human potentialities.

Consider the alternatives now before us. If man were actually, as current theory still supposes, a creature whose use of tools alone played the largest formative part in his development, on what valid grounds do we now propose to strip mankind of the wide variety of autonomous activities historically associated with agriculture and manufacture, leaving the residual mass of workers with the trivial tasks of watching buttons and dials, and responding to one-way communication and remote control? If man actually owes his intelligence mainly to his tool-using propensities, by what logic do we now take his tools away, so that he will become a functionless, workless being, conditioned to accept only what the Megamachine offers him: an automaton within a larger system of automation, condemned to compulsory consumption, as he was once condemned to compulsory production? What in fact will be left of human life, if one function after another is either taken over by the machine, or else surgically removed—perhaps genetically altered—to fit the Megamachine?

But if the analysis of human development in relation to technics proves sound, there is an even more fundamental criticism to be made. For we must then go on to question the basic soundness of the current scientific and educational ideology, which is now pressing to shift the locus of human activity from the organic environments and the human group to the Megamachine, and eventually reduce all forms of life and culture to those that can be translated into the current system of scientific abstractions, and transferred on a mass basis to machines and electronic apparatus. We are now in a position to question the dubious assumptions that have too long been treated as axioms, for the system of thought on which they are still based, antedated by three centuries anything like the present comprehension—scientific, humanistic and historic—of man's nature and special gifts.

From our present vantage point, we can see that the inventors and controllers of the Megamachine have been haunted by delusions of omniscience and omnipotence; and these delusions are not less irrational now that they have at their disposal all the formidable resources of exact science and a high-energy technology. The Nuclear Age conception of 'absolute power' and infallible intelligence, exercised by a military-scientific elite, corresponds to the Bronze Age conception of divine kingship. Such power, to succeed on its own terms, must destroy the symbiotic cooperations between all species and communities essential to man's survival and development. Both ideologies belong to the same infantile magico-religious scheme as ritual human sacrifice.

Living organisms can use only limited amounts of energy, as living personalities can utilize only limited quantities of knowledge and experience. 'Too much' or 'too little' is equally fatal to organic existence. Even too much abstract knowledge, insulated from feeling, from moral evaluation, from historic experience, from responsible, purposeful action, can produce a serious unbalance in both the personality and the community. Organisms, societies and human persons are nothing less than delicate devices for regulating energy and putting it at the service of life.

To the extent that our Megatechnics ignores these fundamental insights into the nature of living organisms, it is actually pre-scientific, even when not actively irrational: a dynamic agent of arrest and regression. When the implications of this weakness are taken in, a deliberate large-scale dismantling of the Megamachine, in all its institutional forms, must surely take place, with a re-distribution of power and authority to smaller units, open to direct human control.

If technics is to be brought back again into the service of human culture, the path of advance will lead, not to the further expansion of the Megamachine, but to the development of all those parts of the organic environment and the human personality that have been suppressed in order to magnify the offices of the Megamachine.

The deliberate expression and fulfillment of human potentialities requires a quite different approach from that bent solely on the control of natural forces, and the modification of human nature in order to facilitate and expand the system of control. We know now that play and sport and ritual and dream-fantasy, no less than organized work, have exercised a formative influence on human culture and even on technics. But make-believe cannot for long be a sufficient substitute for productive work: only when play and work form part of a larger cultural whole, as in Tolstoy's pictures of the mowers in *Anna Karenina*, can the many-sided requirements for full human growth be satisfied. Without serious responsible work, man progressively loses his grip on reality.

Instead of liberation *from* work being the chief contribution of mechanization and automation, I would suggest that liberation *for* work, for educative, mind-forming work, self-rewarding work, may become the most salutary contribution of a life-centered technology. This may prove an indispensable counter-balance to universal automation: partly by protecting the displaced worker from boredom and suicidal desperation, only temporarily relievable by anesthetics, sedatives, and narcotics, partly by giving play to constructive impulses, autonomous functions, and meaningful activities.

Relieved from abject dependence on the Megamachine, the whole world of biotechnics will at once become open to man; and those parts of his personality that have been crippled or paralyzed by insufficient use should again come into play. Automation is indeed the proper end of a purely mechanical system; and in its place, subordinate to other human purposes, automation will serve the human community no less effectively than the reflexes, the hormones, and the autonomic nervous system — Nature's earliest experiment in automation — serve the human body. But *autonomy* is the proper end of organisms; and further technical development must aim at re-establishing autonomy at every stage of human growth by giving play to every part of the human personality, not merely to those functions which serve the Megamachine.

I realize that in opening up these difficult questions I am not in a position to provide ready-made answers, nor do I suggest that such answers will be easy to fabricate. But it is time that our present wholesale commitment to the machine, which arises largely out of our one-sided interpretation of man's early technical development, should be replaced by a fuller picture of both human nature and the technical milieu, as both have evolved together.

GORGIAS

PLATO

Plato (428-348 BC) is often regarded as one of the greatest philosophers in the Western tradition. Through his interactions with his
teacher, Socrates, and his primary student, Aristotle, he laid the
groundwork for all philosophy to come. In the following excerpt from
the beginning of the dialogue *Gorgias* (circa 380 BC), Plato makes
the very first connection between the concepts of *techne* and *logos*.

[447a]   **Callicles.** The wise man, as the proverb says, is late for a fray, but not for a feast.

Socrates. And are we late for a feast?

Cal. Yes, and a delightful one; for Gorgias has just been presenting to us many fine
things.

Soc. It's not my fault, Callicles; our friend Chaerephon is to blame; for he would keep
us loitering in the Agora.

Chaerephon. Never mind, Socrates; the misfortune of which I have been the cause I
will also repair; for Gorgias is a friend of mine, and I will make him give the presentation again either now, or, if you prefer, at some other time.

Cal. What is this, Chaerephon? — Does Socrates want to hear Gorgias?

Chaer. Yes, that was our intention in coming.

Cal. Come to my house, then; for Gorgias is staying with me, and he will give you a
presentation there.

Soc. Very good, Callicles; but will he answer our questions? For I want to hear from
him what is the power of his *technê*, and what it is which he professes and teaches; he
may, as you [Chaerephon] suggest, defer the presentation to some other time.

Cal. There is nothing like asking him, Socrates; and indeed to answer questions is a part of his presentation, for he was saying only just now, that anyone might put any question to him, and that he would answer.

Soc. How fortunate! Will you ask him, Chaerephon?

Chaer. What shall I ask him?

Soc. Ask him what he is.

Chaer. What do you mean?

Soc. I mean such a question as would elicit from him, if he had been a maker of shoes, the answer that he is a cobbler. Do you understand?

Chaer. I understand, and will ask him: Tell me, Gorgias, is our friend Callicles right in saying that you undertake to answer any questions that you are asked?

Gorgias. Quite right, Chaerephon: I was saying as much only just now; and I may add, that many years have elapsed since anyone has asked me a new one.

Chaer. Then you must be very ready, Gorgias.

Gor. Here's your chance to try me, Chaerephon.

Polus. Yes, indeed, and if you like, Chaerephon, you may make trial of me too! For I think that Gorgias, who has been talking a long time, is tired.

Chaer. And do you, Polus, think that you can answer better than Gorgias?

Pol. What does that matter if I answer well enough for you?

Chaer. Not at all: — and you shall answer if you like.

Pol. Ask—

Chaer. My question is this: If Gorgias had the *technê* of his brother Herodicus, what ought we to call him? Ought he not to have the name that is given to his brother?

Pol. Certainly.

Chaer. Then we should be right in calling him a physician?

Pol. Yes.

Chaer. And if he had the *technê* of Aristophon the son of Aglaophon, or of his brother Polygnotus, what ought we to call him?

[448c]   **Pol.** Clearly, a painter.

Chaer. But now what shall we call him—what is his *technê*?

Pol. O Chaerephon, there are many *technai* among mankind which have their origin in experience, for experience makes the days of men to proceed according to *technê*, and inexperience according to chance, and different persons in different ways are proficient in different *technai*, and the best persons in the best *technai*. And our friend Gorgias is one of the best, and the *technê* in which he is proficient is the noblest.

Soc. Polus has been taught how to wonderfully express his *logos*, Gorgias; but he is not fulfilling the promise that he made to Chaerephon.

Gor. What do you mean, Socrates?

Soc. I mean that he has not exactly answered the question that he was asked.

Gor. Then why not ask him yourself?

Soc. But I would much rather ask *you*, if you are disposed to answer: For I see, from the few words which Polus has uttered, that he has attended more to the *technê* which is called rhetoric [oratory] than to dialectic [discussion].

Pol. What makes you say so, Socrates?

Soc. Because, Polus, when Chaerephon asked you what was the *technê* that Gorgias knows, you praised it as if you were answering someone who found fault with it, but you never said what the *technê* was.

Pol. Why, did I not say that it was the noblest of them all?

Soc. Yes, indeed, but that was no answer to the question: Nobody asked what was the quality, but what was the nature, of the *technê*, and by what name we were to describe Gorgias. And I would still beg you briefly and clearly, as you answered Chaerephon when he asked you at first, to say what this *technê* is, and what we ought to call Gorgias: Or rather, Gorgias, let me turn to you, and ask the same question: What are we to call you, and what is the *technê* which you profess?

Gor. Rhetoric, Socrates, is my *technê*.

Soc. Then I am to call you a rhetorician?

Gor. Yes, Socrates, and a good one too, if you would call me that which, in Homeric language, "I boast myself to be."

Soc. Of course I do.

Gor. Then call me that.

[449b] **Soc.** And are we to say that you are able to make other men rhetoricians?

Gor. Yes, that is exactly what I profess to make them, not only at Athens, but in all places.

Soc. And will you continue to ask and answer questions, Gorgias, as we are at present doing, and reserve for another occasion the longer expression of *logos* which Polus was attempting? Will you keep your promise, and answer briefly the questions that are asked of you?

Gor. Some answers, Socrates, are of necessity longer; but I will do my best to express my *logos* as briefly as possible; for a part of my profession is that I can be as short as anyone.

Soc. That is what is wanted, Gorgias! Present the shorter method now, and the longer one at some other time.

Gor. Very well, I will; and you will certainly say, that you never heard a man use fewer words.

Soc. Very good then; as you profess to be a rhetorician, and a maker of rhetoricians, let me ask you, with what is rhetoric concerned? For example, I might ask with what is weaving concerned, and you would reply, would you not, 'with the making of garments'? – Yes.

Soc. And music is concerned with the composition of melodies? – It is.

Soc. By Hera, Gorgias, I admire the outstanding brevity of your answers!

Gor. Yes, Socrates, I do think myself good at that.

Soc. I am glad to hear it; answer me in like manner about rhetoric: with what is rhetoric concerned?

[449e] **Gor.** With *logos*.

Soc. What sort, Gorgias? Such *logos* as would teach the sick how they might get well? – No.

Soc. Then rhetoric does not treat of all kinds of *logoi*? – Certainly not.

Soc. And yet rhetoric makes men able to speak? – Yes.

Soc. And to understand that about which they speak? – Of course.

Soc. But does not the art of medicine, which we were just now mentioning, *also* make men able to understand and speak about the sick? – Certainly.

Soc. Then medicine is also concerned with *logos*? – Yes.

Soc. Concerning diseases? – Just so.

Soc. And isn't physical training also concerned with *logos*, regarding good and bad condition of the body? – Very true.

[450b] **Soc.** And the same, Gorgias, is true of all the other *technai*: All of them are concerned with a particular *logos*, namely, the one that corresponds to each particular *technê*. – Clearly.

Soc. Then why don't you call all the other *technai* 'rhetoric'? They are concerned with *logos* too!

Gor. Because, Socrates, the knowledge of the other *technai* has only to do with some sort of external action, such as with the hands; but there is no such manual action in my rhetoric, which works and takes effect only through the *logos*. And therefore I am justified in saying that rhetoric is the particular *technê* that treats of *logos*.

RHETORIC

ARISTOTLE

The most famous student of Plato, Aristotle (384-322 BC) is among the greatest and most influential philosophers of all time. In *Rhetoric*, circa 340 BC, Aristotle takes up where Plato's *Gorgias* leaves off. Since rhetoric (oratory) is the *technê* that addresses the *logos*, it is a short step to the concept of 'rules of art/rhetoric' (*technologousin*, or *technologein*). Aristotle is thus the first person to coin the term that would become our 'technology.'

BOOK I

I. [1354a] Rhetoric is the counterpart of Dialectic; for both have to do with matters that are in a manner within the understanding of all men and not confined to any special science. Hence all men in a manner have a share of both; for all, up to a certain point, endeavor to criticize or uphold an argument, to defend themselves or to accuse.

Now, the majority of people do this either at random or with a familiarity arising from habit. But since both these ways are possible, it is clear that matters can be reduced to a system, for it is possible to examine the reason why some attain their end by familiarity and others by chance; and all would admit such an examination to be the function of a *technê* [art].

Now, previous compilers of the **technas tôn logôn** ['Arts' of Rhetoric] have provided us with only a small portion of this art, for proofs are the only things in it that come within the province of art; everything else is merely an accessory. And yet they say nothing about enthymemes [i.e. syllogisms] which are the body of proof, but chiefly devote their attention to matters outside the subject; for the arousing of prejudice, compassion, anger, and similar emotions has no connection with the matter in hand, but is directed only to the one who judges. The result would be that, if all trials were now carried on as they are in some States, (20) especially those that are well administered, there would be nothing left for the rhetorician to say. For all men either think that all the laws ought so to prescribe, or in fact carry out, the principle and forbid speaking outside the subject, as in the court of Areopagus, and in this they are right. For it is wrong to warp the judge's feelings, to arouse him to anger, jealousy or compassion, which would be like making the rule crooked which one intended to use.

Further, it is evident that the only business of the litigant is to prove that the fact in question is or is not so, that it has happened or not; whether it is important or unimportant, just or unjust; in all cases in which the legislator has not laid down a ruling, it is a matter for the judge himself to decide; it is not the business of the litigants to instruct him.

First of all, therefore, it is proper that laws, properly enacted, should themselves define the issue of all cases as far as possible, and leave as little as possible to the discretion of the judges; in the first place, because it is easier to find one or a few men of good sense, [1354b] capable of framing laws and pronouncing judgments, than a large number; secondly, legislation is the result of long consideration, whereas judgments are delivered on the spur of the moment, so that it is difficult for the judges properly to decide questions of justice or expediency. But what is most important of all is that the judgment of the legislator does not apply to a particular case, but is universal and applies to the future, whereas the member of the public assembly and the judge have to decide present and definite issues, and in their case love, hate, or personal interest is often involved, so that they are no longer capable of discerning the truth adequately, their judgment being obscured by their own pleasure or pain.

All other cases, as we have just said, should only be left to the authority of the judge as seldom as possible, except where it is a question of a thing having happened or not, of its going to happen or not, of being or not being so; this must be left to the discretion of the judges, for it is impossible for the legislator to foresee such questions. If this is so, it is obvious that all those who definitely lay down, for instance, what should be the contents of the preface, or the narrative, or of the other parts of the discourse, are bringing under the **technologousin** [rules of art] what is outside the subject; for the only thing to which their attention is devoted is how to put the judge into a certain frame of mind. They give no account of the artificial proofs, which make a man a master of rhetorical argument.

Hence, although the method of 'deliberative' and 'forensic' Rhetoric is the same, and although the pursuit of the former is nobler and more worthy of a statesman than that of the latter, which is limited to transactions between private citizens, they say nothing about the former, but without exception endeavor to bring forensic speaking under the **technologein** [rules of art]. The reason of this is that in public speaking it is less worthwhile to talk of what is outside the subject, and that deliberative oratory lends itself to trickery less than forensic, because it is of more general interest. For in the Assembly the judges decide upon their own affairs, so that the only thing necessary is to prove the truth of the statement of one who recommends a measure, but in the law courts this is not sufficient; there it is useful to win over the hearers, for the decision concerns other interests than those of the judges, who, having only themselves to consider and listening merely for their own pleasure, surrender to the pleaders but do not give a real decision. [1355a] That is why, as I have said before, in many places the law prohibits speaking outside the subject in the law courts, whereas in the Assembly the judges themselves take adequate precautions against this.

It is obvious, therefore, that a system arranged according to the **entechnos methodos** [rules of art] is only concerned with *proofs*; that proof is a sort of demonstration, since we are most strongly convinced when we suppose anything to have been demonstrated; that rhetorical demonstration is an enthymeme, which, generally speaking, is the strongest of rhetorical proofs and lastly, that the enthymeme is a kind of syllogism. Now, as it is the function of Dialectic as a whole, or of one of its parts, to consider every kind of syllogism in a similar manner, it is clear that he who is most capable of examining the matter and forms of a syllogism will be in the highest degree a master of rhetorical argument, if to this he adds a knowledge of the subjects with which enthymemes deal and the differences between them and logical syllogisms. For, in fact, "the true" and that which resembles it come under the purview of the same faculty, and at the same time men have a sufficient natural capacity for the truth and indeed in most cases attain to it; wherefore one who divines well in regard to the truth will also be able to divine well in regard to probabilities.

It is clear, then, that all other rhetoricians bring under the **technologousi** [rules of art] what is outside the subject, (20) and have rather inclined to the forensic branch of oratory.

Nevertheless, Rhetoric is useful, because the true and the just are naturally superior to their opposites, so that, if decisions are improperly made, they must owe their defeat to their own advocates; which is reprehensible. Further, in dealing with certain persons, even if we possessed the most accurate 'scientific knowledge' [*epistemen logos*], we should not find it easy to persuade them by the employment of such knowledge. For scientific discourse is concerned with instruction, but in the case of such persons instruction is impossible; our proofs and arguments must rest on generally accepted principles, as we said in the _Topics_, when speaking of conversation with the masses. Further, the orator should be able to prove opposites, as in logical arguments; not that we should do both (for one ought not to persuade people to do what is wrong), but that the real state of the case may not escape us, and that we ourselves may be able to counteract false arguments, if another makes an unfair use of them. Rhetoric and Dialectic alone of all the arts prove opposites; for both are equally concerned with them. However, it is not the same with the subject matter, but, generally speaking, that which is true and better is naturally always easier to prove and more likely to persuade. Besides, it would be absurd if it were considered disgraceful not to be able to defend oneself with the help of the body, [1355b] but not disgraceful as far as *logôi* [speech] is concerned, whose use is more characteristic of man than that of the body. If it is argued that one who makes an unfair use of such *dynamei tôn logôn* [faculty of speech] may do a great deal of harm, this objection applies equally to all good things except *arête* [virtue], and above all to those things which are most useful, such as strength, health, wealth, generalship; for as these, rightly used, may be of the greatest benefit, so, wrongly used, they may do an equal amount of harm.

It is thus evident that Rhetoric does not deal with any one definite class of subjects, but, like Dialectic, is of general application; also, that it is useful; and further, that its function is not so much to persuade, as to find out in each case the existing means of persuasion. The same holds good in respect to all the other *technais* [arts]. For instance, it is not the function of medicine to restore a patient to health, but only to promote this end as far as possible; for even those whose recovery is impossible may be properly treated. It is further evident that it belongs to Rhetoric to discover the real and apparent means of persuasion, just as it belongs to Dialectic to discover the real and apparent syllogism. For what makes the sophist is not the *dynamei* [faculty] but the moral purpose. But there is a difference: in Rhetoric, one who acts in accordance with sound argument, and one who acts in accordance with moral purpose, (20) are both called rhetoricians;

but in Dialectic it is the moral purpose that makes the sophist, the dialectician being one whose arguments rest, not on moral purpose, but on the *dynamin* [faculty].

Let us now endeavor to treat of the method itself, to see how and by what means we shall be able to attain our objects. And so let us as it were start again, and having defined Rhetoric anew, pass on to the remainder of the subject.

II. Rhetoric then may be defined as *the* dynamis *[faculty] of discovering the possible means of persuasion in reference to any subject whatever*. This is the function of no other of the *technês* [arts], each of which is able to instruct and persuade in its own special subject; thus, medicine deals with health and sickness, geometry with the properties of magnitudes, arithmetic with number, and similarly with all the other *technôn* [arts] and *epistêmôn* [sciences]. But Rhetoric, so to say, appears to be able to discover the means of persuasion in reference to any given subject. That is why we say that as an art its rules are not applied to any particular definite class of things.

As for proofs, some are *entechnoi* [artificial], others *atechnoi* [natural]. By the latter I understand all those which have not been furnished by ourselves but were already in existence, such as witnesses, tortures, contracts, and the like; by the former, all that can be constructed by system and by our own efforts. Thus we have only to make use of the latter, whereas we must invent the former.

[1356a] Now the proofs furnished by the *logou* [speech] are of three kinds. The first depends upon the moral character of the speaker, the second upon putting the hearer into a certain frame of mind, the third upon the *logoi* [speech] itself, in so far as it proves or seems to prove.

The orator persuades by moral character when his speech is delivered in such a manner as to render him worthy of confidence; for we feel confidence in a greater degree and more readily in persons of worth in regard to everything in general, but where there is no certainty and there is room for doubt, our confidence is absolute. But this confidence must be due to the speech itself, not to any preconceived idea of the speaker's character; for it is not the case, as some writers of rhetorical treatises lay down in their **technologounton** ['Art'], that the worth of the orator in no way contributes to his powers of persuasion; on the contrary, *moral character*, so to say, *constitutes the most effective means of proof*. The orator persuades by means of his hearers, when they are roused to emotion by his speech; for the judgments we deliver are not the same when we are influenced by joy or sorrow, love or hate; and it is to this alone that, as we have said, the present-day writers of treatises endeavor to devote their attention. (We will discuss these matters in detail when we come to speak of the emotions.) Lastly, persuasion is produced by the speech itself, when we establish the true (20) or apparently true from the means of persuasion applicable to each individual subject.

NICOMACHEAN ETHICS

ARISTOTLE

Written some 10 years after his *Rhetoric*, Aristotle's *Nicomachean Ethics* addresses many philosophical topics of relevance to ethics. Here, in Book 6, he offers the first formal definition of *technê*.

III. [1139b] Let us then discuss these virtues anew, going more deeply into the matter.

Let it be assumed that there are five qualities through which the mind [*psyche*] achieves truth in affirmation or denial, namely: *technê* [Art or technical skill], *epistêmê* [Scientific Knowledge], *phronêsis* [practical wisdom], *sophia* [philosophic wisdom], and *nous* [comprehension]. Judgment and Opinion are capable of error.

The nature of Scientific Knowledge [*epistêmê*] may be made clear as follows. We all conceive that a thing which we know scientifically cannot vary; when a thing that can vary is beyond the range of our observation, we do not know whether it exists or not. An object of Scientific Knowledge, therefore, exists of necessity. It is therefore eternal, for everything existing of absolute necessity is eternal; and what is eternal does not come into existence or perish. Again, it is held that all Scientific Knowledge can be communicated by teaching, and that what is scientifically known must be learnt. But all teaching starts from facts previously known, as we state in the book <u>Analytics</u>, since it proceeds either by way of induction, or else by way of deduction. Now induction supplies a first principle or universal, deduction works *from* universals; therefore there are first principles from which deduction starts, which cannot be proved by deduction; therefore they are reached by induction.

Scientific Knowledge, therefore, is the quality whereby we demonstrate, with the further qualifications included in our definition of it in the <u>Analytics</u>, namely, that a man knows a thing scientifically when he possesses a conviction arrived at in a certain way, and when the first principles on which that conviction rests are known to him with certainty—for unless he is more certain of his first principles than of the conclusion drawn from them he will only possess the knowledge in question accidentally. Let this stand as our definition of Scientific Knowledge.

IV. [1140a] The class of changeable things includes both things *made* and actions *done*. But making is different from doing (a distinction that even the common man accepts). Hence the 'rational quality concerned with doing' [*meta logou hexis praktikê*] is different from the 'rational quality concerned with making' [*meta logou hexis poiêtikês*]. Nor is one of them a part of the other, for doing is not a form of making, nor making a form of doing.

Now architecture [*oikodomikê*] is a *technê*, and it is also a 'rational quality concerned with making'. Now, there is no *technê* which is not a 'rational quality concerned with making', nor any such quality which is not a *technê*. It follows, therefore, that **technê is the same thing as a 'rational quality, concerned with making, that reasons truly'** [*hexis meta logou alêthous poiêtikê*].

All *technê* deals with bringing something into existence; and to pursue *technê* means to study how to bring into existence a thing which may either exist or not, and the *archê* [first cause] of which lies in the maker and not in the thing made; because *technê* does not deal with things that exist or come into existence 'of necessity', or 'according to nature', since these have their first cause in themselves. But as doing and making are distinct, it follows that *technê* must be a matter of making, not of doing.

And in a sense *technê* deals with the same objects as 'chance', as Agathon says:

"Chance is beloved by *technê*, and *technê* by Chance."

Technê, therefore, as has been said, is a 'rational quality, concerned with making, that reasons truly' [*hexis meta logou alêthous poiêtikê*]. Its opposite, Lack of *Technê* [*atechnia*, or 'natural'], is a 'rational quality, concerned with making, that reasons *falsely*'. Both deal with that which is changeable.

GENESIS

THE BIBLE

The Judeo-Christian worldview has an ambiguous status with respect to technology. On the one hand, it contains *warnings* and *cautions* about worldly power. Technology is associated with the original sin, and human pride:

- The "knowledge" represented by the forbidden apple probably included both moral and practical (technical) aspects.
- The very first act of Adam and Eve, after eating the apple, was a technical act: the *sewing of garments*. "Then the eyes of both of them were opened, and they realized they were naked; so they sewed fig leaves together and made coverings for themselves." (Gen 3:6)
- The sinful and murderous Cain's role in the world was to *build cities*: "Cain lay with his wife, and she became pregnant and gave birth to Enoch. Cain was then building a city, and he named it after his son Enoch." (Gen 4:17)
- The only approved technology of the Old Testament was the ark, which was designed by God, not man: "So make yourself an ark of cypress wood; make rooms in it and coat it with pitch inside and out. This is how you are to build it: The ark is to be 450 feet long, 75 feet wide and 45 feet high. Make a roof for it and finish the ark to within 18 inches of the top. Put a door in the side of the ark and make lower, middle and upper decks." (Gen 6:14-17)
- The one major technological achievement of the Old Testament, the Tower of Babel, was condemned by God as a sign of hubris:

> "Now the whole world had one language and a common speech. As men moved eastward, they found a plain in Shinar and settled there. They said to each other, 'Come, let's make bricks and bake them thoroughly.' They used brick instead of stone, and tar for mortar. Then they said, 'Come, let us build ourselves a city, with a tower that reaches to the heavens, so that we may make a name for ourselves and not be scattered over the face of the whole earth.' But the Lord came down to see the city and the tower that the men were building. The Lord said, 'If as one people speaking the same language they have begun to do this, then nothing they plan to do will be impossible for them. Come, let us go down and confuse their language so they will not understand each other.' So the Lord scattered them from there

over all the earth, and they stopped building the city. That is why it was called Babel—because there the Lord confused the language of the whole world. From there the Lord scattered them over the face of the whole earth." (Gen 11:1-9)

- Technology deals with 'spiritless' matter, and hence with that which is temporal, mundane, unclean. A preoccupation with material things can be a sign of a fallen soul.

On the other hand, we find evidence that God *accepts* and even *endorses* technology and technical acts:

- God gave man 'dominion' over nature, and ordered him to 'subdue' it— with the implicit use of tools and technology:

  Then God said, 'Let Us make man in Our image, according to Our likeness; let them have dominion over the fish of the sea, over the birds of the air, and over the cattle, over all the earth and over every creeping thing that creeps on the earth.' ... Then God blessed them, and God said to them, 'Be fruitful and multiply; fill the earth and subdue it; have dominion over the fish of the sea, over the birds of the air, and over every living thing that moves on the earth.' (Gen 1:26-28)
  And the fear of you and the dread of you shall be on every beast of the earth, on every bird of the air, on all that move on the earth, and on all the fish of the sea. They are given into your hand. (Gen 9:2)

- Adam was placed in the Garden of Eden in order to 'work' it: "Then the Lord God took the man and put him in the Garden of Eden to till and keep it." (Gen 2:15) [alternate translations for 'till': 'cultivate', 'work', 'tend']
- Christianity in general opposed the 'pagan animism' of earlier times, in which the things of nature possessed spirits or souls. Thus the good Christian was free "to exploit nature in a mood of indifference to the feelings of natural objects" (see Lynn White article—following essay). As with the 'dominion' command, exploiting nature requires the use of tools and technology.
- Man was made "in the image of God"—a God who was a *creator*, a *maker*. Thus, Christians have inferred that they have a divine right to be creators and makers in this world.
- Jesus himself was a *carpenter*—a tool-user, and a technician! ("Is this not the carpenter, the Son of Mary...?" [Mark 6:3])

Thus, Christianity has been both blamed for the excesses of modern technology, and been invoked as a restraint against it.

THE HISTORICAL ROOTS OF OUR ECOLOGIC CRISIS

LYNN WHITE, JR.

White (1907-1987) was a professor of medieval history at UCLA. He was an expert on the technology of the Middle Ages. This, his most influential article, was originally published in 1967.

A conversation with Aldous Huxley not infrequently put one at the receiving end of an unforgettable monologue. About a year before his lamented death he was discoursing on a favorite topic: Man's unnatural treatment of nature and its sad results. To illustrate his point he told how, during the previous summer, he had returned to a little valley in England where he had spent many happy months as a child. Once it had been composed of delightful grassy glades; now it was becoming overgrown with unsightly brush because the rabbits that formerly kept such growth under control had largely succumbed to a disease, myxomatosis, that was deliberately introduced by the local farmers to reduce the rabbits' destruction of crops. Being something of a Philistine, I could be silent no longer, even in the interests of great rhetoric. I interrupted to point out that the rabbit itself had been brought as a domestic animal to England in 1176, presumably to improve the protein diet of the peasantry.

All forms of life modify their contexts. The most spectacular and benign instance is doubtless the coral polyp. By serving its own ends, it has created a vast undersea world favorable to thousands of other kinds of animals and plants. Ever since man became a numerous species he has affected his environment notably. The hypothesis that his firedrive method of hunting created the world's great grasslands and helped to exterminate the monster mammals of the Pleistocene from much of the globe is plausible, if not proved. For six millennia at least, the banks of the lower Nile have been a human artifact rather than the swampy African jungle which nature, apart from man, would have made it. The Aswan Dam, flooding 5000 square miles, is only the latest stage in a long process. In many regions terracing or irrigation, overgrazing, the cutting of forests by Romans to build ships to fight Carthaginians or by Crusaders to solve the logistics problems of their expeditions, have profoundly changed some ecologies. Observation that the French landscape falls into two basic types, the open fields of the north and the *bocage* of the south and west, inspired Marc Bloch to undertake his classic study of medieval agricultural methods. Quite unintentionally, changes in human ways often affect nonhuman nature. It has been noted, for example, that the advent of the automobile eliminated huge flocks of sparrows that once fed on the horse manure littering every street.

The history of ecologic change is still so rudimentary that we know little about what really happened, or what the results were. The extinction of the European aurochs as late as 1627 would seem to have been a simple case of overenthusiastic hunting. On more intricate matters it often is impossible to find solid information. For a thousand years or more the Frisians and Hollanders have been pushing back the North Sea, and the process is culminating in our own time in the reclamation of the Zuider Zee. What, if any, species of animals, birds, fish, shore life, or plants have died out in the process? In their epic combat with Neptune have the Netherlanders overlooked ecological values in such a way that the quality of human life in the Netherlands has suffered? I cannot discover that the questions have ever been asked, much less answered.

People, then, have often been a dynamic element in their own environment, but in the present state of historical scholarship we usually do not know exactly when, where, or with what effects man-induced changes came. As we enter the last third of the 20th century, however, concern for the problem of ecologic backlash is mounting feverishly. Natural science, conceived as the effort to understand the nature of things, had flourished in several eras and among several peoples. Similarly there had been an age-old accumulation of technological skills, sometimes growing rapidly, sometimes slowly. But it was not until about four generations ago that Western Europe and North America arranged a marriage between science and technology, a union of the theoretical and the empirical approaches to our natural environment. The emergence in widespread practice of the Baconian creed that scientific knowledge means technological power over nature can scarcely be dated before about 1850, save in the chemical industries, where it is anticipated in the 18th century. Its acceptance as a normal pattern of action may mark the greatest event in human history since the invention of agriculture, and perhaps in nonhuman terrestrial history as well.

Almost at once the new situation forced the crystallization of the novel concept of ecology; indeed, the word *ecology* first appeared in the English language in 1873. Today, less than a century later, the impact of our race upon the environment has so increased in force that it has changed in essence. When the first cannons were fired, in the early 14th century, they affected ecology by sending workers scrambling to the forests and mountains for more potash, sulphur, iron ore, and charcoal, with some resulting erosion and deforestation. Hydrogen bombs are of a different order: a war fought with them might alter the genetics of all life on this planet. By 1285 London had a smog problem arising from the burning of soft coal, but our present combustion of fossil fuels threatens to change the chemistry of the globe's atmosphere as a whole, with consequences which we are only beginning to guess. With the population explosion, the carcinoma of planless urbanism, the now geological deposits of sewage and garbage, surely no creature other than man has ever managed to foul its nest in such short order.

There are many calls to action, but specific proposals, however worthy as individual items, seem too partial, palliative, negative: ban the bomb, tear down the billboards, give the Hindus contraceptives and tell them to eat their sacred cows. The simplest solution to any suspect change is, of course, to stop it, or better yet, to revert to a romanticized past: make those ugly gasoline stations look like Anne Hathaway's cottage or (in the Far West) like ghost-town saloons. The "wilderness area" mentality invariably advocates deep-freezing an ecology, whether San Gimignano or the High

Sierra, as it was before the first Kleenex was dropped. But neither atavism nor pretti-fication will cope with the ecologic crisis of our time.

What shall we do? No one yet knows. Unless we think about fundamentals, our specific measures may produce new backlashes more serious than those they are designed to remedy.

As a beginning we should try to clarify our thinking by looking, in some histor-ical depth, at the presuppositions that underlie modern technology and science. Science was traditionally aristocratic, speculative, intellectual in intent; technology was lower-class, empirical, action-oriented. The quite sudden fusion of these two, towards the middle of the 19th century, is surely related to the slightly prior and con-temporary democratic revolutions which, by reducing social barriers, tended to assert a functional unity of brain and hand. Our ecologic crisis is the product of an emerging, entirely novel, democratic culture. The issue is whether a democratized world can sur-vive its own implications. Presumably we cannot unless we rethink our axioms.

The Western Traditions of Technology and Science

One thing is so certain that it seems stupid to verbalize it: both modern technology and modern science are distinctively Occidental. Our technology has absorbed elements from all over the world, notably from China; yet everywhere today, whether in Japan or in Nigeria, successful technology is Western. Our science is the heir to all the sciences of the past, especially perhaps to the work of the great Islamic scientists of the Middle Ages, who so often outdid the ancient Greeks in skill and perspicacity: al-Razi in med-icine, for example; or ibn-al-Haytham in optics; or Omar Khayyam in mathematics. Indeed, not a few works of such geniuses seem to have vanished in the original Arabic and to survive only in medieval Latin translations that helped to lay the foundations for later Western developments. Today, around the globe, all significant science is Western in style and method, whatever the pigmentation or language of the scientists.

A second pair of facts is less well recognized because they result from quite recent historical scholarship. The leadership of the West, both in technology and in science, is far older than the so-called Scientific Revolution of the 17th century or the so-called Industrial Revolution of the 18th century. These terms are in fact outmoded and obscure the true nature of what they try to describe—significant stages in two long and separate developments. By A.D. 1000 at the latest—and perhaps, feebly, as much as 200 years earlier—the West began to apply water power to industrial processes other than milling grain. This was followed in the late 12th century by the harnessing of wind power. From simple beginnings, but with remarkable consistency of style, the West rapidly expanded its skills in the development of power machinery, labor-saving devices, and automation. Those who doubt should contemplate that most monumental achievement in the history of automation: the weight-driven mechanical clock, which appeared in two forms in the early 14th century. Not in craftsmanship but in basic technological capacity, the Latin West of the later Middle Ages far out-stripped its elaborate, sophisticated, and esthetically magnificent sister cultures, Byzantium and Islam. In 1444 a great Greek ecclesiastic, Bessarion, who had gone to

Italy, wrote a letter to a prince in Greece. He is amazed by the superiority of Western ships, arms, textiles, glass. But above all he is astonished by the spectacle of water-wheels sawing timbers and pumping the bellows of blast furnaces. Clearly, he had seen nothing of the sort in the Near East.

By the end of the 15th century the technological superiority of Europe was such that its small, mutually hostile nations could spill out over all the rest of the world, conquering, looting, and colonizing. The symbol of this technological superiority is the fact that Portugal, one of the weakest states of the Occident, was able to become, and to remain for a century, mistress of the East Indies. And we must remember that the technology of Vasco da Gama and Albuquerque was built by pure empiricism, drawing remarkably little support or inspiration from science.

In the present-day vernacular understanding, modern science is supposed to have begun in 1543, when both Copernicus and Vesalius published their great works. It is no derogation of their accomplishments, however, to point out that such structures as the *Fabrica* and the *De revolutionibus* do not appear overnight. The distinctive Western tradition of science, in fact, began in the late 11th century with a massive movement of translation of Arabic and Greek scientific works into Latin. A few notable books—Theophrastus, for example—escaped the West's avid new appetite for science, but within less than 200 years effectively the entire corpus of Greek and Muslim science was available in Latin, and was being eagerly read and criticized in the new European universities. Out of criticism arose new observation, speculation, and increasing distrust of ancient authorities. By the late 13th century Europe had seized global scientific leadership from the faltering hands of Islam. It would be as absurd to deny the profound originality of Newton, Galileo, or Copernicus as to deny that of the 14th century scholastic scientists like Buridan or Oresme on whose work they built. Before the 11th century, science scarcely existed in the Latin West, even in Roman times. From the 11th century onward, the scientific sector of Occidental culture has increased in a steady crescendo.

Since both our technological and our scientific movements got their start, acquired their character, and achieved world dominance in the Middle Ages, it would seem that we cannot understand their nature or their present impact upon ecology without examining fundamental medieval assumptions and developments.

Medieval View of Man and Nature

Until recently, agriculture has been the chief occupation even in "advanced" societies; hence, any change in methods of tillage has much importance. Early plows, drawn by two oxen, did not normally turn the sod but merely scratched it. Thus, cross-plowing was needed and fields tended to be squarish. In the fairly light soils and semiarid climates of the Near East and Mediterranean, this worked well. But such a plow was inappropriate to the wet climate and often sticky soils of northern Europe. By the latter part of the 7th century after Christ, however, following obscure beginnings, certain northern peasants were using an entirely new kind of plow, equipped with a vertical knife to cut the line of the furrow, a horizontal share to slice under the sod, and a moldboard to turn it over. The friction of this plow with the soil was so great that it normally required not two but eight oxen. It attacked the land with such violence that cross-plowing was not needed, and fields tended to be shaped in long strips.

In the days of the scratch-plow, fields were distributed generally in units capable of supporting a single family. Subsistence farming was the presupposition. But no peasant owned eight oxen: to use the new and more efficient plow, peasants pooled their oxen to form large plow-teams, originally receiving (it would appear) plowed strips in proportion to their contribution. Thus, distribution of land was based no longer on the needs of a family but, rather, on the capacity of a power machine to till the earth. Man's relation to the soil was profoundly changed. Formerly man had been part of nature; now he was the exploiter of nature. Nowhere else in the world did farmers develop any analogous agricultural implement. Is it coincidence that modern technology, with its ruthlessness toward nature, has so largely been produced by descendants of these peasants of northern Europe?

This same exploitive attitude appears slightly before A.D. 830 in Western illustrated calendars. In older calendars the months were shown as passive personifications. The new Frankish calendars, which set the style for the Middle Ages, are very different: they show men coercing the world around them—plowing, harvesting, chopping trees, butchering pigs. Man and nature are two things, and man is master.

These novelties seem to be in harmony with larger intellectual patterns. What people do about their ecology depends on what they think about themselves in relation to things around them. Human ecology is deeply conditioned by beliefs about our nature and destiny—that is, by *religion*. To Western eyes this is very evident in, say, India or Ceylon. It is equally true of ourselves and of our medieval ancestors.

The victory of Christianity over paganism was the greatest psychic revolution in the history of our culture. It has become fashionable today to say that, for better or worse, we live in the "post-Christian age." Certainly the forms of our thinking and language have largely ceased to be Christian, but to my eye the substance often remains amazingly akin to that of the past. Our daily habits of action, for example, are dominated by an implicit faith in perpetual progress which was unknown either to Greco-Roman antiquity or to the Orient. It is rooted in, and is indefensible apart from, Judeo- Christian theology. The fact that Communists share it merely helps to show what can be demonstrated on many other grounds: that Marxism, like Islam, is a Judeo-Christian heresy. We continue today to live, as we have lived for about 1700 years, very largely in a context of Christian axioms.

What did Christianity tell people about their relations with the environment?

While many of the world's mythologies provide stories of creation, Greco-Roman mythology was singularly incoherent in this respect. Like Aristotle, the intellectuals of the ancient West denied that the visible world had a beginning. Indeed, the idea of a beginning was impossible in the framework of their cyclical notion of time. In sharp contrast, Christianity inherited from Judaism not only a concept of time as nonrepetitive and linear but also a striking story of creation. By gradual stages a loving and all-powerful God had created light and darkness, the heavenly bodies, the earth and all its plants, animals, birds, and fishes. Finally, God had created Adam and, as an afterthought, Eve to keep man from being lonely. Man named all the animals, thus establishing his dominance over them. God planned all of this explicitly for man's benefit and rule: no item in the physical creation had any purpose save to serve man's purposes. And, although man's body is made of clay, he is not simply part of nature: he is made in God's image.

Especially in its Western form, Christianity is the most anthropocentric religion the world has seen. As early as the 2nd century both Tertullian and Saint Irenaeus of Lyons were insisting that when God shaped Adam he was foreshadowing the image of the incarnate Christ, the Second Adam. Man shares, in great measure, God's transcendence of nature. Christianity, in absolute contrast to ancient paganism and Asia's religions (except, perhaps, Zorastrianism), not only established a dualism of man and nature but also insisted that it is God's will that man exploit nature for his proper ends.

At the level of the common people this worked out in an interesting way. In Antiquity every tree, every spring, every stream, every hill had its own *genius loci*, its guardian spirit. These spirits were accessible to men, but were very unlike men; centaurs, fauns, and mermaids show their ambivalence. Before one cut a tree, mined a mountain, or dammed a brook, it was important to placate the spirit in charge of that particular situation, and to keep it placated. By destroying pagan animism, Christianity made it possible to exploit nature in a mood of indifference to the feelings of natural objects.

It is often said that for animism the Church substituted the cult of saints. True; but the cult of saints is functionally quite different from animism. The saint is not *in* natural objects; he may have special shrines, but his citizenship is in heaven. Moreover, a saint is entirely a man; he can be approached in human terms. In addition to saints, Christianity of course also had angels and demons inherited from Judaism and perhaps, at one remove, from Zorastrianism. But these were all as mobile as the saints themselves. The spirits *in* natural objects, which formerly had protected nature from man, evaporated. Man's effective monopoly on spirit in this world was confirmed, and the old inhibitions to the exploitation of nature crumbled.

When one speaks in such sweeping terms, a note of caution is in order. Christianity is a complex faith, and its consequences differ in differing contexts. What I have said may well apply to the medieval West, where in fact technology made spectacular advances. But the Greek East, a highly civilized realm of equal Christian devotion, seems to have produced no marked technological innovation after the late 7th century, when Greek fire was invented. The key to the contrast may perhaps be found in a difference in the tonality of piety and thought which students of comparative theology find between the Greek and the Latin Churches. The Greeks believed that sin was intellectual blindness, and that salvation was found in illumination, orthodoxy — that is, clear thinking. The Latins, on the other hand, felt that sin was moral evil, and that salvation was to be found in right conduct. Eastern theology has been intellectualist. Western theology has been voluntarist. The Greek saint contemplates; the Western saint acts. The implications of Christianity for the conquest of nature would emerge more easily in the Western atmosphere.

The Christian dogma of creation, which is found in the first clause of all the Creeds, has another meaning for our comprehension of today's ecologic crisis. By revelation, God had given man the Bible, the Book of Scripture. But since God had made nature, nature also must reveal the divine mentality. The religious study of nature for the better understanding of God was known as natural theology. In the early Church, and always in the Greek East, nature was conceived primarily as a symbolic system through which God speaks to men: the ant is a sermon to sluggards; rising flames are the symbol of the soul's aspiration. The view of nature was essentially artistic rather than scientific. While Byzantium preserved and copied great numbers of ancient Greek scientific texts, science as we conceive it could scarcely flourish in such an ambience.

However, in the Latin West by the early 13th century natural theology was following a very different bent. It was ceasing to be the decoding of the physical symbols of God's communication with man and was becoming the effort to understand God's mind by discovering how his creation operates. The rainbow was no longer simply a symbol of hope first sent to Noah after the Deluge: Robert Grosseteste, Friar Roger Bacon, and Theodoric of Freiberg produced startlingly sophisticated work on the optics of the rainbow, but they did it as a venture in religious understanding. From the 13th century onward, up to and including Leibniz and Newton, every major scientist, in effect, explained his motivations in religious terms. Indeed, if Galileo had not been so expert an amateur theologian he would have got into far less trouble: the professionals resented his intrusion. And Newton seems to have regarded himself more as a theologian than as a scientist. It was not until the late 18th century that the hypothesis of God became unnecessary to many scientists.

It is often hard for the historian to judge, when men explain why they are doing what they want to do, whether they are offering real reasons or merely culturally acceptable reasons. The consistency with which scientists during the long formative centuries of Western science said that the task and the reward of the scientist was "to think God's thoughts after him" leads one to believe that this was their real motivation. If so, then modern Western science was cast in a matrix of Christian theology. The dynamism of religious devotion shaped by the Judeo-Christian dogma of creation, gave it impetus.

An Alternative Christian View

We would seem to be headed toward conclusions unpalatable to many Christians. Since both *science* and *technology* are blessed words in our contemporary vocabulary, some may be happy at the notions, first, that viewed historically, modern science is an extrapolation of natural theology and, second, that modern technology is at least partly to be explained as an Occidental, voluntarist realization of the Christian dogma of man's transcendence of, and rightful master over, nature. But, as we now recognize, somewhat over a century ago science and technology—hitherto quite separate activities—joined to give mankind powers which, to judge by many of the ecologic effects, are out of control. If so, Christianity bears a huge burden of guilt.

I personally doubt that disastrous ecologic backlash can be avoided simply by applying to our problems more science and more technology. Our science and technology have grown out of Christian attitudes toward man's relation to nature which are almost universally held not only by Christians and neo-Christians but also by those who fondly regard themselves as post-Christians. Despite Copernicus, all the cosmos rotates around our little globe. Despite Darwin, we are *not*, in our hearts, part of the natural process. We are superior to nature, contemptuous of it, willing to use it for our slightest whim. The newly elected Governor of California [Ronald Regan], like myself a churchman but less troubled than I, spoke for the Christian tradition when he said (as is alleged), "when you've seen one redwood tree, you've seen them all." To a Christian a tree can be no more than a physical fact. The whole concept of the sacred grove is alien to Christianity and to the ethos of the West. For nearly two millennia Christian missionaries have been chopping down sacred groves, which are idolatrous because they assume spirit in nature.

What we do about ecology depends on our ideas of the man-nature relationship. More science and more technology are not going to get us out of the present ecologic crisis until we find a new religion, or rethink our old one. The beatniks, who are the basic revolutionaries of our time, show a sound instinct in their affinity for Zen Buddhism, which conceives of the man-nature relationship as very nearly the mirror image of the Christian view. Zen, however, is as deeply conditioned by Asian history as Christianity is by the experience of the West, and I am dubious of its viability among us.

Possibly we should ponder the greatest radical in Christian history since Christ: Saint Francis of Assisi. The prime miracle of Saint Francis is the fact that he did not end at the stake, as many of his left-wing followers did. He was so clearly heretical that a General of the Franciscan Order, Saint Bonaventura, a great and perceptive Christian, tried to suppress the early accounts of Franciscanism. The key to an understanding of Francis is his belief in the virtue of humility—not merely for the individual but for man as a species. Francis tried to depose man from his monarchy over creation and set up a democracy of all God's creatures. With him the ant is no longer simply a homily for the lazy, flames a sign of the thrust of the soul toward union with God; now they are Brother Ant and Sister Fire, praising the Creator in their own ways as Brother Man does in his.

Later commentators have said that Francis preached to the birds as a rebuke to men who would not listen. The records do not read so: he urged the little birds to praise God, and in spiritual ecstasy they flapped their wings and chirped rejoicing. Legends of saints, especially the Irish saints, had long told of their dealings with animals but always, I believe, to show their human dominance over creatures. With Francis it is different. The land around Gubbio in the Apennines was ravaged by a fierce wolf. Saint Francis, says the legend, talked to the wolf and persuaded him of the error of his ways. The wolf repented, died in the odor of sanctity, and was buried in consecrated ground.

What Sir Steven Ruciman calls "the Franciscan doctrine of the animal soul" was quickly stamped out. Quite possibly it was in part inspired, consciously or unconsciously, by the belief in reincarnation held by the Cathar heretics who at that time teemed in Italy and southern France, and who presumably had got it originally from India. It is significant that at just the same moment, about 1200, traces of metempsychosis are found also in western Judaism, in the Provencal *Cabbala*. But Francis held neither to transmigration of souls nor to pantheism. His view of nature and of man rested on a unique sort of panpsychism of all things animate and inanimate, designed for the glorification of their transcendent Creator, who, in the ultimate gesture of cosmic humility, assumed flesh, lay helpless in a manger, and hung dying on a scaffold.

I am not suggesting that many contemporary Americans who are concerned about our ecologic crisis will be either able or willing to counsel with wolves or exhort birds. However, the present increasing disruption of the global environment is the product of a dynamic technology and science which were originating in the Western medieval world against which Saint Francis was rebelling in so original a way. Their growth cannot be understood historically apart from distinctive attitudes toward nature which are deeply grounded in Christian dogma. The fact that most people do not think of these attitudes as Christian is irrelevant. No new set of basic values has been accepted in our society to displace those of Christianity. Hence we shall continue to have a worsening ecologic crisis until we reject the Christian axiom that nature has no reason for existence save to serve man.

The greatest spiritual revolutionary in Western history, Saint Francis, proposed what he thought was an alternative Christian view of nature and man's relation to it; he tried to substitute the idea of the equality of all creatures, including man, for the idea of man's limitless rule of creation. He failed. Both our present science and our present technology are so tinctured with orthodox Christian arrogance toward nature that no solution for our ecologic crisis can be expected from them alone. Since the roots of our trouble are so largely religious, the remedy must also be essentially religious, whether we call it that or not. We must rethink and refeel our nature and destiny. The profoundly religious, but heretical, sense of the primitive Franciscans for the spiritual autonomy of all parts of nature may point a direction. I propose Francis as a patron saint for ecologists.

PART III:
HISTORICAL
CRITIQUES

TAO TE CHING

LAO TZU

The earliest recorded critique of technology is found in an ancient Chinese text, the *Tao Te Ching*. This collection of short sayings and aphorisms has been historically attributed to the venerable sage Lao Tzu, who is said to have been an older contemporary of Confucius—placing him in the early part of the 6th century BC. The *Tao Te Ching* is one of the great classics of ancient China, and the central text in the philosophical system known as Taoism. The title means, roughly, 'the book of the Way (*tao*), and of Virtue (*te*)'. The 'Way', or Tao, is the ultimate creative force of the cosmos, and is something every good Taoist wishes to align himself with. Achieving a life of harmony with the Tao is the path of virtue.

Taoism advocates a simple life, free from the luxuries and complexities of civilization. Luxuries—those things that are 'hard to come by'—are more than unnecessary; they are virtual obstacles to achieving a harmonious life: "Goods hard to come by / serve to hinder [man's] progress" (I.12). The wise man, therefore, works to free himself from desire for artificialities and human contrivances. He seeks the simple life.

Artificial things are a result of man's cleverness and hubris. The way of the Tao is simple and natural, and thus requires no complex analytic thinking, no human constructions or creations. Cleverness leads to innovation, which results in the luxuries and artifices that corrupt society, and obstruct the free and open life:

> Exterminate ingenuity, discard profit / and there will be no more thieves and bandits. (I.19)

> When cleverness emerges / there is great hypocrisy. (I.18)

> Woe to him who willfully innovates. (I.16)

> Exterminate learning and there will no longer be worries. (I.20)

"Learning", or technical knowledge, is singled out as particularly important. It very clearly leads one down the wrong path, away from the Tao:

> In the pursuit of learning one knows more every day;
> In the pursuit of the Tao one does less every day.
> One does less and less until one does nothing at all,
> And when one does nothing at all, there is nothing that is undone. (II.48)

Learning leads to 'tools', to technology, and this is a source of great concern to the well-run society. Tools alter and disrupt the social, and natural, orders. As tools become more highly refined, the state becomes more unstable and more at risk; it enters a darkened time, a condition of 'benightenment':

> The more sharpened tools the people have,
> The more benighted the state. (II.57)

Paraphrasing this stanza, we might say: *'The sharper the tools, the darker the times'*. Here we have a prescient, poetic, ominous vision of technology—from well over two thousand years ago.

Given this warning, the wise man knows when enough is enough; he knows when to *stop*:

> One ought to know that it is time to stop.
> Knowing when to stop, one can be free from danger. (I.32)

> Know when to stop
> And you will meet with no danger.
> You can then endure. (II.44)

We have tools that would have been, for Lao Tzu, inconceivably potent. Clearly our era is one of *very sharp tools*. Not coincidentally, it is also, for many, a very dark time. Can we make the connection? Will we know when to stop? Is it *too late* to stop?

CHUANG TZU

CHUANG TZU

Chuang Tzu (380-301 BC) was a quasi-Taoist philosopher who lived some 100 years after Lao Tzu. His main work, also called 'Chuang Tzu', is another classic of ancient Chinese literature. In this excerpt from Chapter 12, a story is told of a Tao-inspired old man, a gardener, who declines to use technology because of its tendency to corrupt the spirit.

Tzu-kung travelled south to Ch'u, turned back towards Chin, and while passing along the south bank of the Han River saw an old man looking after his vegetable garden. He had dug out a passage down into the well, from which he emerged with a pitcher in his arms to water the earth. Splash, splash! It was costing him a lot of effort, with the poorest of results to show for it.

"Suppose you had a machine," said Tzu-kung, "which in one day would irrigate a hundred fields. You would have plentiful results to show for very little effort. Wouldn't you want one of them?"

The gardener lifted his head and looked at him. "How does it work?"

"The thing which makes it go is a piece of wood chiseled to make it heavier at the back end than at the front. It pulls up the water as though you were plucking it straight out of the well, as fast as bubbles in a boiling pot. It's called a well-sweep."

The gardener gave him a dirty look, and then laughed and said, "I have heard from my teacher that those who possess machines must end up becoming mechanical in their affairs. And those who are mechanical in their affairs must end up becoming mechanical in their hearts and minds. If a mechanical heart is situated in one's chest, then one can't prepare oneself to receive pure simplicity. If one can't be prepared to receive pure simplicity, then the spirit becomes unsettled, and Tao has no place to enter. It's not that I wasn't aware of such machines, but that it would be disgraceful to end up that way."

Tzu-kung, too embarrassed to meet his eye, looked down at the ground without answering.

After a while the gardener said: "What do you do, sir?"

"I am a disciple of Confucius."

"Aren't you those people who make themselves so learned to get to be like the sages, go in for fancy talk so that they can look down on the crowd, while crying a mournful song about how alone you are, simply in order to gain a reputation in the world? You forget about your spiritual essence, cause degeneration to your physical body, and for what? You don't even have the ability to set yourselves straight, let alone set anything straight in the world! Go away, don't interrupt my work."

Tzu-kung was shocked out of countenance, in too much of a dither to pull himself together. He had travelled another ten miles before he recovered.

"That man just now," said a disciple, "what would he be, sir? Why was it that when you saw him you looked so upset, you weren't yourself for the rest of the day?"

"I used to think there was only one person in the world who could really reach me, but that was before I'd come across this man. I'd heard from my Master, Confucius, that in order to deal with affairs and be successful at them, one must use as little effort as possible while achieving many things. That's the Tao of a sage. Now I've found that's not so. In order to grasp the Tao one has to make their virtue whole. One who makes their virtue whole has to make their form whole. One who makes their form whole has to make their spirit whole. One who's made their spirit whole — that's the Tao of the sage. By trusting life to take him along with different people without knowing what's really going on, his capacity for purity is boundless! The rewards and benefits obtained by clever machines must be forgotten in the mind of such a person. Such a man goes only where he himself intends, and does nothing which is not from his own heart. Even if the whole world were to praise him and accept what he had to say, he'd consider it to be frivolous talk and ignore it. If the whole world opposed him and rejected what he had to say, he'd figure he didn't fit in, and not let it bother him. The world's praise or opposition wouldn't harm him, and that's what's called being a person whose virtue is whole! I'm someone who still gets disturbed by other people."

When he was back in Lu he told Confucius about it.

"The gardener is a follower and practitioner of the tradition of the House of Hun-t'un [the primordial Chaos]," said Confucius. "He perceives the oneness of everything, does not know about duality in it; he orders it as inward, does not order it as outward. Someone who by illumination enters into simplicity, by doing nothing reverts to the unhewn, who identifies himself with his nature and protects his spirit, as he roams among the vulgar — is he really so astonishing to you? In any case, when it comes to the tradition of the House of Hun-t'un, how would you and I be adequate to understand it?"

PHAEDRUS

PLATO

In this dialogue, circa 380 BC, Plato, drawing probably from Socrates' own view, gives an interesting critique of writing. This is consistent with the fact that Socrates never, to our knowledge, wrote anything. But it is also ironic, given Plato's status as one of the greatest writers in history.

Socrates. [274c] I heard, then, that at Naucratis, in Egypt, was one of the ancient gods of that country, the one whose sacred bird is called the ibis, and the name of the god himself was Theuth. He it was who invented numbers and arithmetic and geometry and astronomy, also checkers and dice, and, most important of all, *letters*.

Now the king of all Egypt at that time was the holy Thamus, who lived in the great city of the upper region, which the Greeks call the Egyptian Thebes. To him came Theuth to show his inventions [*technas*, arts], saying that they ought to be imparted to the other Egyptians. But Thamus asked what use there was in each, and as Theuth enumerated their uses, expressed praise or blame, according as he approved or disapproved.

The story goes that Thamus said many things to Theuth in praise or blame of the various *technês*, which it would take too long to repeat; but when they came to the *letters*, "This invention, O king," said Theuth, "will make the Egyptians wiser and will improve their memories; for it is an elixir of memory and wisdom that I have discovered."

But Thamus replied, "Most ingenious Theuth, one man has the ability to beget *technês*, but the ability to judge of their usefulness or harmfulness to their users belongs to others; [275a] and now you, who are the father of letters, have been led by your affection to ascribe to them a power the *opposite* of that which they really possess. *For this invention will produce forgetfulness in the minds of those who learn to use it, because they will not practice their memory. Their trust in writing, produced by external characters that are no part of themselves, will discourage the use of their own memory within them. You have invented an elixir not of memory, but of reminding.* And you offer your pupils the *appearance* of wisdom, not true wisdom, for they will read many things without instruction and will therefore seem to know many things, when they are for the most part ignorant and hard to get along with, since they are not wise, but only appear wise."

Phaedrus. Socrates, you easily make up stories of Egypt or any country you please.

Socrates. They used to say, my friend, that the words of the oak in the holy place of Zeus at Dodona were the first prophetic utterances. The people of that time, not being so wise as you young folks, were content in their simplicity to listen to an oak [275c] or a rock, provided only it spoke the truth; but to you, perhaps, it makes a difference who the speaker is and where he comes from, for you do not consider only whether his words are true or not.

Phaedrus. Your rebuke is just; and I think Thamus the Theban is right in what he says about letters.

Socrates. He who thinks, then, that he has left behind him any *technên* of writing, and he who receives it in the belief that anything in writing will be clear and certain, would be an utterly simple person, and in truth ignorant of the prophecy of Thamus, if he thinks written words are of any use except to *remind* him who knows the matter about which they are written.

Phaedrus. Very true.

Socrates. Writing, Phaedrus, has this strange quality, and is very like painting; for the creatures of painting stand like living beings, but if one asks them a question, they preserve a solemn silence. And so it is with written words. You might think they spoke as if they had intelligence, but if you question them, wishing to know about their sayings, they always say only one and the same thing. And every word, when once it is written, is bandied about, alike among those who understand and those who have no interest in it, and it knows not to whom to speak or not to speak; when ill-treated or unjustly reviled, it always needs its father to help it; for it has no power to protect or help itself.

Phaedrus. You are quite right about that, too.

Socrates. [276a] Now tell me; is there not another kind of speech [*logôn*], or word, which shows itself to be the legitimate brother of this bastard one, both in the manner of its begetting and in its better and more powerful nature?

Phaedrus. What is this word and how is it begotten, as you say?

Socrates. The word which is written with intelligence in the mind of the learner, which is able to defend itself and knows to whom it should speak, and before whom to be silent.

Phaedrus. You mean the living and breathing *logôn* of him who knows, of which the written word may justly be called the *image*.

Socrates. Exactly. Now tell me this. Would a sensible husbandman, who has seeds which he cares for and which he wishes to bear fruit, plant them with serious purpose in the heat of summer in some garden of Adonis, and delight in seeing them appear in beauty in eight days, or would he do that sort of thing, when he did it at all, only in play and for amusement? Would he not, when he was in earnest, follow the rules of husbandry, plant his seeds in fitting ground, and be pleased when those which he had sowed reached their perfection in the eighth month?

Phaedrus. Yes, Socrates, he would, as you say, act in that way when in earnest and in the other way only for amusement.

Socrates. And shall we suppose that he who has knowledge of the just and the good and beautiful has less sense about his seeds than the husbandman?

Phaedrus. By no means.

Socrates. Then he will not, when in earnest, write them in ink, sowing them through a pen with *logôn* that cannot defend themselves by argument and cannot teach the truth effectively.

Phaedrus. No, at least, probably not.

Socrates. No. The gardens of letters he will, it seems, plant for amusement, and will write, when he writes, to treasure up reminders for himself, when he comes to the forgetfulness of old age, and for others who follow the same path, and he will be pleased when he sees them putting forth tender leaves. When others engage in other amusements, refreshing themselves with banquets and kindred entertainments, he will pass the time in such pleasures as I have suggested.

Phaedrus. A noble pastime, Socrates, and a contrast to those base pleasures, the pastime of the man who can find amusement in discourse, telling stories about justice, and the other subjects of which you speak.

Socrates. Yes, Phaedrus, so it is; but, in my opinion, serious discourse about them is far nobler, when one employs the *dialectic* method and plants and sows in a fitting soul intelligent words which are able to help themselves and him [277a] who planted them, which are not fruitless, but yield seed from which there spring up in other minds other words capable of continuing the process for ever, and which make their possessor happy, to the farthest possible limit of human happiness.

Phaedrus. Yes, that is far nobler.

MINING THE EARTH'S WOMB

CAROLYN MERCHANT

Merchant (1936-) is an ecofeminist philosopher and writer. The following essay dates from 1983.

The domination of the earth through technology and the corresponding rise of the image of the world as *Machin ex Deo*, were features of the Scientific Revolution of the sixteenth and seventeenth centuries. During this period, the two ideas of mechanism and the domination of nature came to be core concepts and controlling images of our modern world. An organically oriented mentality prevalent from ancient times to the Renaissance, in which the female principle played a significant positive role, was gradually undermined and replaced by a technological mindset that used female principles in an exploitative manner. As Western culture became increasingly mechanized during the 1600s, a female nurturing earth and virgin earth spirit were subdued by the machine.

The change in controlling imagery was directly related to changes in human attitudes and behavior toward the earth. Whereas the older nurturing earth image can be viewed as a cultural constraint restricting the types of socially and morally sanctioned human actions allowable with respect to the earth, the new images of mastery and domination functioned as cultural sanctions for the denudation of nature. Society needed these new images as it continued the processes of commercialism and industrialization, which depended on activities directly altering the earth—mining, drainage, deforestation, and assarting (grubbing up stumps to clear fields). The new activities utilized new technologies—lift and force pumps, cranes, windmills, geared wheels, flap valves, chains, pistons, treadmills, under—and overshot watermills, fulling mills, flywheels, bellows, excavators, bucket chains, rollers, geared and wheeled bridges, cranks, elaborate block and tackle systems, worm, spur, crown, and lantern gears, cams and eccentrics, ratchets, wrenches, presses, and screws in magnificent variation and combination.

These technological and commercial changes did not take place quickly; they developed gradually over the ancient and medieval eras, as did the accompanying environmental deterioration: Slowly, over many centuries, early Mediterranean and Greek civilization had mined and quarried the mountainsides, altered the forested landscape, and overgrazed the hills. Nevertheless, technologies were low level, people considered themselves parts of a finite cosmos, and animism and fertility cults that treated nature as sacred were numerous. Roman civilization was more pragmatic, secular, and commercial and its environmental impact more intense. Yet Roman writers such as Ovid, Seneca, Pliny, and, the Stoic philosophers openly deplored mining as an abuse of their mother, the earth. With the disintegration of feudalism and the expansion of Europeans into new worlds and markets, commercial society began to have an accelerated impact on the natural environment. By the sixteenth and seventeenth centuries, the tension between technological development in the world of action and the controlling organic images in the world of the mind had become too great. The old structures were incompatible with the new activities.

Both the nurturing and domination metaphors had existed in philosophy, religion, and literature—the idea of dominion over the earth in Greek philosophy and Christian religion; that of the nurturing earth, in Greek and other pagan philosophies. But, as the economy became modernized and the Scientific Revolution proceeded, the dominion metaphor spread beyond the religious sphere and assumed ascendancy in the social and political spheres as well. These two competing images and their normative associations can be found in sixteenth-century literature, art, philosophy, and science.

The image of the earth as a living organism and nurturing mother had served as a cultural constraint restricting the actions of human beings. One does not readily slay a mother, dig into her entrails for gold, or mutilate her body, although commercial mining would soon require that. As long as the earth was considered to be alive and sensitive, it could be considered a breach of human ethical behavior to carry out destructive acts against it. For most traditional cultures, minerals and metals ripened in the uterus of the Earth Mother, mines were compared to her vagina, and metallurgy was the human hastening of the birth of the living metal in the artificial womb of the furnace—an abortion of the metals' natural growth cycle before its time. Miners offered propitiation to the deities of the soil and subterranean world, performed ceremonial sacrifices, and observed strict cleanliness, sexual abstinence, and fasting before violating the sacredness of the living earth by sinking a mine. Smiths assumed an awesome responsibility in precipitating the metal's birth through smelting, fusing, and beating it with hammer and anvil; they were often accorded the status of shaman in tribal rituals and their tools were thought to hold special powers.

The Renaissance image of the nurturing earth still carried with it subtle ethical controls and restraints. Such imagery found in a culture's literature can play a normative role within the culture. Controlling images operate as ethical restraints or as ethical sanctions—as subtle "oughts" or "ought-nots." Thus, as the descriptive metaphors and images of nature change, a behavioral restraint can be changed into a sanction. Such a change in the image and description of nature was occurring during the course of the Scientific Revolution.

It is important to recognize the normative import of descriptive statements about nature. Contemporary philosophers of language have critically reassessed the earlier positivist distinction between the "is" of science and the "ought" of society, arguing that descriptions and norms are not opposed to one another by linguistic separation into separate "is" and "ought" statements, but are contained within each other. Descriptive statements about the world can presuppose the normative; they are then ethic-laden. A statement's normative function lies in the use itself as description. The norms may be tacit assumptions hidden within the descriptions in such a way as to act as invisible restraints or moral ought-nots. The writer or culture may not be conscious of the ethical import yet may act in accordance with its dictates. The hidden norms may become conscious or explicit when an alternative or contradiction presents itself. Because language contains a culture within itself, when language changes, a culture is also changing in important ways. By examining changes in descriptions of nature, we can then perceive something of the changes in cultural values. To be aware of the interconnectedness of descriptive and normative statements is to be able to evaluate changes in the latter by observing changes in the former.

Not only did the image of nature as nurturing mother contain ethical implications but the organic framework itself, as a conceptual system, also carried with it an associated value system. Contemporary philosophers have argued that a given normative theory is linked with certain conceptual frameworks and not with others. The framework contains within itself certain dimensions of structural and normative variation, while denying others belonging to an alternative or rival framework.

We cannot accept a framework of explanation and yet reject its associated value judgments, because the connections to the values associated with the structure are not fortuitous. New commercial and technological innovations, however, can upset and undermine an established conceptual structure. New human and social needs can threaten associated normative constraints, thereby demanding new ones.

While the organic framework was for many centuries sufficiently integrative to override commercial development and technological innovation, the acceleration of such changes throughout Western Europe during the sixteenth and seventeenth centuries began to undermine the organic unity of the cosmos and society. Because the needs and purposes of society as a whole were changing with the commercial revolution, the values associated with the organic view of nature were no longer applicable; hence, the plausibility of the conceptual framework itself was slowly, but continuously, being threatened.

The Geocosm: The Earth as a Nurturing Mother

Not only was nature in a generalized sense seen as female, but also the earth, or geocosm, was universally viewed as a nurturing mother—sensitive, alive, and responsive to human action. The changes in imagery and attitudes relating to the earth were of enormous significance as the mechanization of nature proceeded. The nurturing earth would lose its function as a normative restraint as it changed to a dead, inanimate, physical system.

The macrocosm theory likened the cosmos to the human body, soul, and spirit with male and female reproductive components. Similarly, the geocosm theory compared the earth to the living human body, with breath, blood, sweat, and elimination systems.

For the Stoics, who flourished in Athens during the third century BC, after the death of Aristotle, and in Rome through the first century AD, the world itself was an intelligent organism; God and matter were synonymous. Matter was dynamic, composed of two forces: expansion and condensation—the former directed outward, the latter inward. The tension between them was the inherent force generating all substances, properties, and living forms in the cosmos and the geocosm.

Zeno of Citium (ca. 304 BC) and M. Tullius Cicero (106-43 BC) held that the world reasons, has sensation, and generates living rational beings: "The world is a living and wise being, since it produces living and wise beings" (Cicero). Every part of the universe and the earth was created for the benefit and support of another part. The earth generated and gave stability to plants, plants supported animals, and animals in turn served human beings; conversely, human skill helped to preserve these organisms. The Universe itself was created for the sake of rational beings — gods and men — but God's foresight insured the safety and preservation of all things. Humankind was given hands to transform the earth's resources and dominion over them: timber was to be used for houses and ships, soil for crops, iron for plows, and gold and silver for ornaments. Each part and imperfection existed for the sake and ultimate perfection of the whole.

The living character of the world organism meant not only that the stars and planets were alive, but that the earth too was pervaded by a force giving life and motion to the living beings on it. Lucius Seneca (4 BC – 65 AD), a Roman Stoic, stated that the earth's breath nourished both the growths on its surface and the heavenly bodies above by its daily exhalations:

> How could she nourish all the different roots that sink into the soil in one place and another, had she not an abundant supply of the breath of life?...all these [heavenly bodies] draw their nourishment from materials of earth...and are sustained...by nothing else than the breath of the earth... Now the earth would be unable to nourish so many bodies...unless it were full of breath, which it exhales from every part of it day and night.

The earth's springs were akin to the human blood system; its other various fluids were likened to the mucus, saliva, sweat, and other forms of lubrication in the human body, the earth being organized "...much after the plan of our bodies, in which there are both veins and arteries, the former blood vessels, the latter air vessels... So exactly alike is the resemblance to our bodies in nature's formation of the earth, that our ancestors have spoken of veins (= springs) of water." Just as the human body contained blood, marrow, mucus, saliva, tears, and lubricating fluids, so in the earth there were various fluids. Liquids that turned hard became metals, such as gold and silver, other fluids turned into stones and veins of sulfur. Like the human body, the earth gave forth sweat: "There is often a gathering of thin, scattered moisture like dew, which from many points flows into one spot. The dowsers call it sweat, because a kind of drop is either squeezed out by the pressure of the ground or raised by the heat."

Leonardo da Vinci (1452-1519) elaborated the Greek analogy between the waters of the earth and the ebb and flow of human blood through the veins and heart:

> The water runs from the rivers to the sea and from the sea to the rivers, always making the same circuit. The water is thrust from the

utmost depth of the sea to the high summits of the mountains, where, finding the veins cut, it precipitates itself and returns to the sea below, mounts once more by the branching veins and then falls back, thus going and coming between high and low, sometimes inside, sometimes outside. It acts like the blood of animals which is always moving, starting from the sea of the heart and mounting to the summit of the head.

The earth's venous system was filled with metals and minerals. Its veins, veinlets, seams, and canals coursed through the entire earth, particularly in the mountains. Its humors flowed from the veinlets into the larger veins. The earth, like the human, even had its own elimination system. The tendency for both to break wind caused earthquakes in the case of the former and another type of quake in the latter:

> The material cause of earthquakes…is no doubt great abundance of wind, or store of gross and dry vapors, and spirits, fast shut up, and as a man would say, emprisoned in the caves, and dungeons of the earth; which wind, or vapors, seeking to be set at liberty, and to get them home to their natural lodgings, in a great fume, violently rush out, and as it were, break prison; which forcible eruption, and strong breath, causeth an earthquake (Gabriel Harvey).

Its bowels were full of channels, fire chambers, glory holes, and fissures through which fire and heat were emitted, some in the form of fiery volcanic exhalations, others as hot water springs. The most commonly used analogy, however, was between the female's reproductive and nurturing capacity and the mother earth's ability to give birth to stones and metals within its womb through its marriage with the sun.

In his *De Rerum Natura* of 1565, the Italian philosopher Bernardino Telesio referred to the marriage of the two great male and female powers: "We can see that the sky and the earth are not merely large parts of the world universe, but are of primary—even principal rank… They are like mother and father to all the others." The earth and the sun served as mother and father to the whole of creation: all things are "made of earth by the sun and that in the constitution of all things the earth and the sun enter respectively as mother and father." According to Giordano Bruno (1548-1600), every human being was "a citizen and servant of the world, a child of Father Sun and Mother Earth."

A widely held alchemical belief was the growth of the baser metals into gold in womblike matrices in the earth. The appearance of silver in lead ores or gold in silvery assays was evidence that this transformation was underway. Just as the child grew in the warmth of the female womb, so the growth of metals was fostered through the agency of heat, some places within the earth's crust being hotter and therefore hastening the maturation process. "Given to gold, silver, and the other metals [was] the vegetative powers whereby they could also reproduce themselves. For, since it was impossible for God to make anything that was not perfect, he gave to all created things, with their being, the power of multiplication." The sun acting on the earth nurtured not only the plants and animals but also "the metals, the broken sulfuric, bituminous, or nitrogenous rocks;...as well as the plants and animals—if these are not made of earth by the sun, one cannot imagine of what else or by what other agent they could be made" (Telesio).

The earth's womb was the matrix or mother not only of metals but also of living things. Paracelsus compared the earth to a female whose womb nurtured all life.

> Woman is like the earth and all the elements and in this sense she may be considered a matrix; she is the tree which grows in the earth and the child is like the fruit born of the tree... Woman is the image of the tree. Just as the earth, its fruits, and the elements are created for the sake of the tree and in order to sustain it, so the members of woman, all her qualities, and her whole nature exist for the sake of her matrix, her womb...
>
> And yet woman in her own way is also a field of the earth and not at all different from it. She replaces it, so to speak: she is the field and the garden mold in which the child is sown and planted.

The earth in the Paracelsian philosophy was the mother or matrix giving birth to plants, animals, and men.

The image of the earth as a nurse, which had appeared in the ancient world in Plato's *Timaeus* and the Emerald Tablet of Hermes Trismegistus, was a popular Renaissance metaphor. According to sixteenth-century alchemist Basil Valentine, all things grew in the womb of the earth, which was alive, and vital, and the nurse of all life:

> The quickening power of the earth produces all things that grow forth from it, and he who says that the earth has no life makes a statement flatly contradicted by facts. What is dead cannot produce life and growth, seeing that it is devoid of the quickening spirit... This spirit is the life and soul that dwell in the earth, and are nourished by heavenly and sidereal influences... This spirit is itself fed by the stars and is thereby rendered capable of imparting nutriment to all things that grow and of nursing them as a mother does her child while it is yet in the womb... If the earth were deserted by this spirit it would be dead.

In general, the Renaissance view was that all things were permeated by life, there being no adequate method by which to designate the inanimate from the animate. It was difficult to differentiate between living and nonliving things because of the resemblance in structures. Like plants and animals, minerals and gems were filled with small pores, tubelets, cavities, and streaks through which they seemed to nourish themselves. Crystalline salts were compared to plant forms, but criteria by which to differentiate the living from the nonliving could not successfully be formulated. This was due not only to the vitalistic framework of the period but to striking similarities between them. Minerals were thought to possess a lesser degree of the vegetative soul, because they had the capacity for medicinal action and often took the form of various parts of plants. By virtue of the vegetative soul, minerals and stones grew in the human body, in animal bodies, within trees, in the air and water, and on the earth's surface in the open country.

Popular Renaissance literature was filled with hundreds of images associating nature, matter, and the earth with the female sex. The earth was alive and considered to be a beneficent, receptive, nurturing female. For most writers, there was a mingling of traditions based on ancient sources. In general, the pervasive animism of nature created a relationship of immediacy with the human being.

An I-thou relationship in which nature was considered to be a person-writ-large was sufficiently prevalent that the ancient tendency to treat it as another human still existed. Such vitalistic imagery was thus so widely accepted by the Renaissance mind that it could effectively function as a restraining ethic.

In much the same way, the cultural belief-systems of many American Indian tribes had for centuries subtly guided group behavior toward nature. Smohalla of the Columbia Basin Tribes voiced the Indian objections to European attitudes in the mid-1800s:

> You ask me to plow the ground! Shall I take a knife and tear my mother's breast? Then when I die she will not take me to her bosom to rest.

> You ask me to dig for stone! Shall I dig under her skin for her bones? Then when I die I cannot enter her body to be born again.

> You ask me to cut grass and make hay and sell it, and be rich like white men! But how dare I cut off my mother's hair?

In the 1960s, the Native American became a symbol in the ecology movement's search for alternatives to Western exploitative attitudes. The Indian animistic belief-system and reverence for the earth as a mother were contrasted with the Judeo-Christian heritage of dominion over nature and with capitalist practices resulting in the "tragedy of the commons" (exploitation of resources available for any person's or nation's use). But as will be seen, European culture was more complex and varied than this judgment allows. It ignores the Renaissance philosophy of the nurturing earth as well as those philosophies and social movements resistant to mainstream economic change.

Normative Constraints against the Mining of Mother Earth

If sixteenth-century descriptive statements and imagery can function as an ethical constraint and if the earth was widely viewed as a nurturing mother, did such imagery actually function as a norm against improper use of the earth? Evidence that this was indeed the case can be drawn from theories of the origins of metals and the debates about mining prevalent during the sixteenth century.

What ethical ideas were held by ancient and early modern writers on the extraction of the metals from the bowels of the living earth? The Roman compiler Pliny (23 – 79 AD), in his *Natural History*, had specifically warned against mining the depth of Mother Earth, speculating that earthquakes were an expression of her indignation at being violated in this manner:

> We trace out all the veins of the earth, and yet…are astonished that it should occasionally cleave asunder or tremble: as though, forsooth, these signs could be any other than expressions of the indignation felt by our sacred parent! We penetrate into her entrails, and seek for treasures…as though each spot we tread upon were not sufficiently bounteous and fertile for us!

He went on to argue that the earth had concealed from view that which she did not wish to be disturbed, that her resources might not be exhausted by human avarice:

> For it is upon her surface, in fact, that she has presented us with these substances, equally with the cereals, bounteous and ever ready, as she is, in supplying us with all things for our benefit! It is what is concealed from our view, what is sunk far beneath her surface, objects, in fact, of no rapid formation, that urge us to our ruin, that send us to the very depth of hell…, when will be the end of thus exhausting the earth, and to what point will avarice finally penetrate!

Here, then, is a striking example of the restraining force of the beneficent mother image—the living earth in her wisdom has ordained against the mining of metals by concealing them in the depths of her womb.

While mining gold led to avarice, extracting iron was the source of human cruelty in the form of war, murder, and robbery. Its use should be limited to agriculture and those activities that contributed to the "honors of more civilized life":

> For by the aid of iron we lay open the ground, we plant trees, we prepare our vineyard trees, and we force our vines each year to resume their youthful state, by cutting away their decayed branches. It is by the aid of iron that we construct houses, cleave rocks, and perform so many other useful offices of life. But it is with iron also that wars, murders, and robberies are effected,… not only hand to hand, but…by the aid of missiles and winged weapons, now launched from engines, now hurled by the human arm, and now furnished with feathery wings. Let us therefore acquit nature of a charge that here belongs to man himself.

In past history, Pliny stated, there had been instances in which laws were passed to prohibit the retention of weapons and to ensure that iron was used solely for innocent purposes, such as the cultivation of fields.

In the *Metamorphoses* (7 AD), the Roman poet Ovid wrote of the violence done to the earth during the age of iron, when evil was let loose in the form of trickery, slyness, plotting, swindling, and violence, as men dug into the earth's entrails for iron and gold:

> The rich earth
> Was asked for more; they dug into her vitals.
> Pried out the wealth a kinder lord had hidden
> In stygian shadow, all that precious metal,
> The root of evil. They found the guilt of iron.
> And gold, more guilty still. And War came forth.

The violation of Mother Earth resulted in new forms of monsters, born of the blood of her slaughter:

> Jove struck them down
> With thunderbolts, and the bulk of those huge bodies
> Lay on the earth, and bled, and Mother earth,
> Made pregnant by that blood, brought forth new bodies,
> And gave them, to recall her older offspring,
> The forms of men. And this new stock was also
> Contemptuous of gods, and murder-hungry
> And violent. You would know they were sons of blood.

Seneca also deplored the activity of mining, although, unlike Pliny and Ovid, he did not consider it a new vice, but one that had been handed down from ancient times. "What necessity caused man, whose head points to the stars, to stoop, below, burying him in mines and plunging him in the very bowels of innermost earth to root up gold?" Not only did mining remove the earth's treasures, but it created "a sight to make [the] hair stand on end—huge rivers and vast reservoirs of sluggish waters." The defiling of the earth's waters was even then a noteworthy consequence of the quest for metals.

These ancient strictures against mining were still operative during the early years of the commercial revolution when mining activities, which had lapsed after the fall of Rome, were once again revived. Ultimately, such constraints would have to be defeated by proponents of the new mercantilist philosophy.

An allegorical tale, reputedly sent to Paul Schneevogel, a professor at Leipzig about 1490-1495, expressed opposition to mining encroachments into the farmlands of Lichtenstat in Saxony, Germany, an area where the new mining activities were developing rapidly. In the following allegorical vision of an old hermit of Lichtenstat, Mother Earth is dressed in a tattered green robe and seated on the right hand of Jupiter who is represented in a court case by "glib-tongued Mercury" who charges a miner with matricide. Testimony is presented by several of nature's deities:

> Bacchus complained that his vines were uprooted and fed to the
> flames and his most sacred places desecrated. Ceres stated that her

fields were devastated; Pluto that the blows of the miners resound like thunder through the depths of the earth, so that he could hardly reside in his own kingdom; the Naiad, that the subterranean waters were diverted and her fountains dried up; Charon that the volume of the underground waters had been so diminished that he was unable to float his boat on Acheron and carry the souls across to Pluto's realm, and the Fauns protested that the charcoal burners had destroyed whole forests to obtain fuel to smelt the miner's ores.

In his defense, the miner argued that the earth was not a real mother, but a wicked stepmother who hides and conceals the metals in her inner parts instead of making them available for human use.

In the old hermit's tale, we have a fascinating example of the relationship between images and values. The older view of nature as a kindly mother is challenged by the growing interests of the mining industry in Saxony, Bohemia, and the Harz Mountains, regions of newly found prosperity. The miner, representing these newer commercial activities, transforms the image of the nurturing mother into that of a step-mother who wickedly conceals her bounty from the deserving and needy children.

Henry Cornelius Agrippa's polemic *The Vanity of Arts and Sciences* (1530) reiterated some of the moral strictures against mining found in the ancient treatises, quoting the passage from Ovid portraying miners digging into the bowels of the earth in order to extract gold and iron. "These men," he declared, "have made the very ground more hurtful and pestiferous, by how much they are more rash and venturous than they that hazard themselves in the deep to dive for pearls." Mining thus despoiled the earth's surface, infecting it, as it were, with an epidemic disease.

If mining were to be freed of such strictures and sanctioned as a commercial activity, the ancient arguments would have to be refuted. This task was taken up by Georg Agricola (1494-1555), who wrote the first "modern" treatise on mining. His *De Re Metallica* ("On Metals," 1556) marshaled the arguments of the detractors of mining in order to refute them and thereby promote the activity itself.

According to Agricola, people who argued against the mining of the earth for metals did so on the basis that nature herself did not wish to be discovered what she herself had concealed:

> The earth does not conceal and remove from our eyes those things which are useful and necessary to mankind, but, on the contrary, like a beneficent and kindly mother she yields in large abundance from her bounty and brings into the light of day the herbs, vegetables, grains, and fruits, and trees. The minerals, on the other hand, she buries far beneath in the depth of the ground, therefore they should not be sought.

This argument, taken directly from Pliny, reveals the normative force of the image of the earth as a nurturing mother.

A second argument of the detractors, reminiscent of Seneca and Agrippa, and based on Renaissance "ecological" concerns was the disruption of the natural environment and the pollutive effects of mining.

> But, besides this, the strongest argument of the detractors [of mining] is that the fields are devastated by mining operations, for which reason formerly Italians were warned by law that no one should dig the earth for metals and so injure their very fertile fields, their vineyards, and their olive groves. Also they argue that the woods and groves are cut down, for there is need of wood for timbers, machines, and the smelting of metals. And when the woods and groves are felled, then are exterminated the beasts and birds, many of which furnish a pleasant and agreeable food for man. Further, when the ores are washed, the water which has been used poisons the brooks and streams, and either destroys the fish or drives them away. Therefore the inhabitants of these regions, on account of the devastation of their fields, woods, groves, brooks, and rivers, find great difficulty in procuring the necessaries of life, and by reason of the destruction of the timber they are forced to greater expense in erecting buildings. Thus it is said, it is clear to all that there is greater detriment from mining than the value of the metals which the mining produces.

Agricola may have been alluding to laws passed by the Florentines between 1420 and 1485, preventing people from dumping lime into rivers upstream from the city for the purpose of "poisoning or catching fish," as it caused severe problems for those living downstream. The laws were enacted both to preserve the trout, "a truly noble and impressive fish" and to provide Florence with "a copious and abundant supply of such fish."

Such ecological consciousness, however, suffered because of the failure of law enforcement, as well as because of the continuing progress of mining activities. Agricola, in his response to the detractors of mining, pointed out the congruences in the need to catch fish and to construct metal tools for the wellbeing of the human race. His effort can be interpreted as an attempt to liberate the activity of mining from the constraints imposed by the organic framework and the nurturing earth image, so that new values could sanction and hasten its development.

To the argument that the woods were cut down and the price of timber therefore raised, Agricola responded that most mines occurred in unproductive, gloomy areas. Where the trees were removed from more productive sites, fertile fields could be created, the profits from which would reimburse the local inhabitants for their losses in timber supplies. Where the birds and animals had been destroyed by mining operations, the profits could be used to purchase "birds without number" and "edible beasts and fish elsewhere" and refurbish the area.

The vices associated with the metals—anger, cruelty, discord, passion for power, avarice, and lust—should be attributed instead to human conduct: "It is not the metals which are to be blamed, but the evil passions of men which become inflamed and ignited; or it is due to the blind and impious desires of their minds." Agricola's arguments are a conscious attempt to separate the older normative constraints from the image of the metals themselves so that new values can then surround them.

Edmund Spenser's treatment of Mother Earth in the *Faerie Queene* (1595) was representative of the concurrent conflict of attitudes about mining the earth. Spenser entered fully into the sixteenth-century debates about the wisdom of mining, the two greatest sins against the earth being, according to him, avarice and lust. The arguments associating mining with avarice had appeared in the ancient texts of Pliny, Ovid, and Seneca, while during Spenser's lifetime the sermons of Johannes Mathesius, entitled *Beregpostilla, oder Sarepta* (1578), inveighed against the moral consequences of human greed for the wealth created by mining for metals.

In Spenser's poem, Guyon presents the arguments against mining taken from Ovid and Agricola, while the description of Mammon's forge is drawn from the illustrations to the *De Re Metallica*. Gold and silver pollute the spirit and debase human values just as the mining operation itself pollutes the "purest streams" of the earth's womb:

> Then gan a cursed hand the quiet wombe
> Of his great Grandmother with steele to wound,
> And the hid treasures in her sacred tombe
> With Sacrilege to dig. Therein he found
> Fountaines of gold and silver to abound,
> Of which the matter of his huge desire
> And pompous pride eftsoones he did compound.

The earth in Spenser's poem is passive and docile, allowing all manner of assault, violence, ill-treatment, rape by lust, and despoilment by greed. No longer a nurturer, she indiscriminately, as in Ovid's verse, supplies flesh to all life and lacking in judgment brings forth monsters and evil creatures. Her offspring fall and bite her in their own death throes. The new mining activities have altered the earth from a bountiful mother to a passive receptor of human rape.

John Milton's *Paradise Lost* (1667) continues the Ovidian image, as Mammon leads "bands of pioneers with Spade and Pickaxe" in the wounding of the living female earth:

> …By him first
> Men also, and by his suggestion taught.
> Ransack'd the Center, and with impious hands
> Rifl'd the bowels of their mother Earth
> For Treasures better hid. Soon had his crew
> Op'nd into the Hill a spacious wound
> And dig'd out ribs of Gold.

Not only did mining encourage the mortal sin of avarice, it was compared by Spenser to the second great sin, human lust. Digging into the matrices and pockets of earth for metals was like mining the female flesh for pleasure. The sixteenth- and seventeenth-century imagination perceived a direct correlation between mining and digging into the nooks and crannies of a woman's body. Both mining and sex represent for Spenser the return to animality and earthly slime. In the *Faerie Queene*, lust is the basest of all human sins. The spilling of human blood, in the rush to rape the earth of her gold, taints and muddies the once fertile fields.

The sonnets of the poet and divine John Donne (1573-1631) also played up the popular identity of mining with human lust. The poem "Love's Alchemie" begins with the sexual image, "Some that have deeper digged loves Myne than I / say where his centrique happiness doth lie." The Platonic lover, searching for the ideal or "centrique" experience of love, begins by digging for it within the female flesh, an act as debasing to the human being as the mining of metals is to the female earth. Happiness is not to be obtained by avarice for gold and silver, nor can the alchemical elixir be produced from base metals. Nor does ideal love result from an ascent up the hierarchical ladder from base sexual love to the love of Poetry, music, and art to the highest Platonic love of the good, virtue; and God. The same equation appears in Elegie XVIII, "Love's Progress":

> Search every sphaere
> And firmament, our Cupid is not there;
> He's an infernal god and under ground,
> With Pluto dwells, where gold and fire abound:
> Men to such Gods, their sacrificing Coles,
> Did not in Altars lay, but pits and holes,
> Although we see Celestial bodies move
> Above the earth, the earth we Till and love:
> So we her ayres contemplate, words and heart
> And Virtues; but we love the Centrique part.

Lust and love of the body do not lead to the celestial love of higher ideals; rather, physical love is associated with the pits and holes of the female body, just as the love of gold depends on the mining of Pluto's caverns within the female earth, "the earth we till and love." Love of the sexual "centrique" part of the female will not lead to the aery spiritual love of virtue. The fatal association of monetary revenue with human avarice, lust, and the female mine is driven home again in the last lines of the poem:

> Rich Nature hath in women wisely made
> Two purses, and their mouths aversely laid:
> They then, which to the lower tribute owe,
> That way which that Exchequer looks, must go.

Avarice and greed after money corrupted the soul, just as lust after female flesh corrupted the body.

The comparison of the female mine with the new American sources of gold, silver, and precious metals appears again in Elegie XIX, "Going to Bed." Here, however, Donne turns the image upside down and uses it to extol the virtues of the mistress.

> License my roaving hands, and let them go,
> Before, behind, between, above, below.
> O my America! my new-found-land,
> My kingdome, safeliest when with one man man'd
> My Myne of precious stones, My Emperie,
> How blest am I in this discovering thee!

In these lines, the comparison functions as a sanction—the search for precious gems and metals, like the sexual exploration of nature or the female, can benefit a kingdom or a man.

Moral restraints were thus clearly affiliated with the Renaissance image of the female earth and were strengthened by associations with greed, avarice, and lust. But the analogies were double-edged. If the new values connected with mining were positive, and mining was viewed as a means to improve the human condition, as they were by Agricola, then the comparison could be turned upside down. Sanctioning mining sanctioned the rape or technological exploration of the earth. The organic framework, in which the Mother-Earth image was a moral restraint against mining, was literally being undermined by the new commercial activity.

In the seventeenth century, Francis Bacon carried the new ethic a step further through metaphors that compared miners and smiths to scientists and technologists penetrating nature and shaping her on the anvil. Bacon's new man of science must not think that the "inquisition of nature is in any part interdicted or forbidden." Nature must be "bound into service" and made a "slave," put "in constraint" and "molded" by the mechanical arts. The "searchers and spies of nature" are to discover her plots and secrets.

This method, so readily applicable when nature is denoted by the female gender, degraded and made possible the exploitation of the natural environment. Nature's womb harbored secrets that through technology could be wrested from her grasp for use in the improvement of the human condition:

> There is therefore much ground for hoping that there are still laid up in the womb of nature many secrets of excellent use having no affinity or parallelism with anything that is now known...only by the method which we are now treating can they be speedily and suddenly and simultaneously presented and anticipated.

The final step was to recover and sanction man's dominion over nature. Due to the Fall from the Garden of Eden (caused by the temptation of a woman), the human race lost its "dominion over creation." Before the Fall, there was no need for power or dominion, because Adam and Eve had been made sovereign over all other creatures. In this state of dominion, mankind was "like unto God." While some, accepting God's punishment, had obeyed the medieval strictures against searching too deeply into God's secrets, Bacon turned the constraints into sanctions. Only by "digging further and further into the mine of natural knowledge" could mankind recover that lost dominion. In this way, "the narrow limits of man's dominion over the universe" could be stretched "to their promised bounds."

Although a female's inquisitiveness may have caused man's fall from his god-given dominion, the relentless interrogation of another female, nature, could be used to regain it. As he argued in *The Masculine Birth of Time*, "I am come in very truth leading to you nature with all her children to bind her to your service and make her your slave." "We have no right," he asserted, "to expect nature to come to us." Instead, "Nature must be taken by the forelock, being bald behind." Delay and subtle argument "permit one only to clutch at nature, never to lay hold of her and capture her."

Nature existed in three states—at liberty, in error, or in bondage:

> She is either free and follows her ordinary course of development as in the heavens, in the animal and vegetable creation, and in the general array of the universe; or she is driven out of her ordinary course by the perverseness, insolence, and forwardness of matter and vi6lence of impediments, as in the case of monsters; or lastly, she is put in constraint, molded, and made as it were new by art and the hand of man; as in things artificial.

The first instance was the view of nature as immanent self-development, the nature naturing herself of the Aristotelians. This was the organic view of nature as a living, growing, self-actualizing being. The second state was necessary to explain the malfunctions and monstrosities that frequently appeared and that could not have been caused by God or another higher power acting on his instruction. Since monstrosities could not be explained by the action of form or spirit, they had to be the result of matter acting perversely. Matter in Plato's *Timaeus* was recalcitrant and had to be forcefully shaped by the demiurge. Bacon frequently described matter in female imagery, as a "common harlot." "Matter is not devoid of an appetite and inclination to dissolve the world and fall back into the old chaos." It therefore must be "restrained and kept in order by the prevailing concord of things." "The vexations of art are certainly as the bonds and handcuffs of Proteus, which betray the ultimate struggles and efforts of matter."

The third instance was the case of art (*techne*)—man operating on nature to create something new and artificial. Here, "nature takes orders from man and works under his authority." Miners and smiths should become the model for the new class of natural philosophers who would interrogate and alter nature. They had developed the two most important methods of wresting nature's secrets from her, "the one searching into the bowels of nature, the other shaping nature as on an anvil." "Why should we not divide natural philosophy into two parts, the mine and the furnace?" For "the truth of nature lies hid in certain deep mines and caves," within the earth's bosom. Bacon, like some of the practically minded alchemists, would "advise the studious to sell their books and build furnaces" and, "forsaking Minerva and the Muses as barren virgins, to rely upon Vulcan."

The new method of interrogation was not through abstract notions, but through the instruction of the understanding "that it may in very truth dissect nature." The instruments of the mind supply suggestions, those of the hand give motion and aid the work. "By art and the hand of man," nature can then be "forced out of her natural state and squeezed and molded." In this way, "human knowledge and human power meet as one."

Here, in bold sexual imagery, is the key feature of the modern experimental method—constraint of nature in the laboratory, dissection by hand and mind, and the penetration of hidden secrets—language still used today in praising a scientist's "hard facts," "penetrating mind," or the "thrust of his argument." The constraints against mining the earth have been turned into sanctions in language that legitimates the exploitation and "rape" of nature for human good.

Scientific method, combined with mechanical technology, would create a "new organon," a new system of investigation, that unified knowledge with material power. The technological discoveries of printing, gunpowder, and the magnet in the fields of learning, warfare, and navigation "help us to think about the secrets still locked in nature's bosom." "They do not, like the old, merely exert a gentle guidance over nature's course; they have the power to conquer and subdue her, to shake her to her foundations." Under the mechanical arts, "nature betrays her secrets more fully...than when in enjoyment of her natural liberty."

Mechanics, which gave man power over nature, consisted in motion; that is, in "the uniting or disuniting of natural bodies." Most useful were the arts that altered the materials of things—"agriculture, cookery, chemistry, dying, the manufacture of glass, enamel, sugar, gunpowder, artificial fires, paper, and the like." But in performing these operations, one was constrained to operate within the chain of causal connections; nature could "not be commanded except by being obeyed." Only by the study, interpretation, and observation of nature could these possibilities be uncovered; only by acting as the interpreter of nature could knowledge be turned into power. Of the three grades of human ambition, the most wholesome and noble was "to endeavor to establish and extend the power and dominion of the human race itself over the universe." In this way, "the human race [could] recover that right over nature which belongs to it by divine bequest."

By the close of the seventeenth century, a new science of mechanics in combination with the Baconian ideal of technological mastery over Nature had helped to create the modern worldview. The core of female principles that had for centuries subtly guided human behavior toward the earth had given way to a new ethic of exploitation. The nurturing earth mother was subdued by science and technology.

OF THE INCERTAINTY AND VANITY OF THE WORLDLY ARTS AND SCIENCES

HENRY (HEINRICH) CORNELIUS AGRIPPA

Agrippa (1486-1535) was a German theologian, philosopher, and alchemist. The following essay, from 1530, was one of the first modern critiques of technology.

Of the Sciences in General

It is an old Opinion, and the concurring and unanimous judgment almost of all Philosophers, whereby they uphold, that every Science addeth so much of a sublime Nature to Man himself, according to the Capacity and Worth of every Person, as many times enables them to Translate themselves beyond the Limits of Humanity, even to the Celestial Seats of the Blessed. From hence have proceeded those various and Innumerable Encomiums [praises] of the Sciences, whereby every one hath endeavored, in accurate, as well as long Orations, to prefer, and as it were to extol beyond the Heavens themselves, those Arts and Mysteries, wherein, with continual Labor, he hath exercised the strength and vigor of his Ingenuity or Invention.

But I, persuaded by reasons of another nature, do verily believe, that *there is nothing more pernicious, nothing more destructive to the well-being of Men, or to the Salvation of our Souls, than the Arts and Sciences themselves*. And therefore quite contrary to what has been hitherto practiced, my Opinion is, That these Arts and Sciences are so far from being to be extolled with such high applauses and Panegyrics [praises], that they are rather for the most part to be dispraised and vilified: And that indeed there is none which does not merit just cause of Reproof and Censure; nor any one which of it self deserves any praise or commendation, unless what it may borrow from the Ingenuity and Virtue of the first possessor.

However, I would have you take this Opinion of mine in that modest Construction, which may imagine, that I neither go about to reprehend those who are of a contrary judgment; or that I intend to arrogate any thing singly singular to my self, above others: Therefore I shall entreat you to suspend your Censure of me, differing in this one thing from all others; so long as you find me laying, an auspicious Foundation of proof, not upon Vulgar Arguments drawn from the Superficies and outside of things, but upon the most firm reasons deduced from the most hidden bowels of secret Knowledge; and this not in the sharp style of Demosthenes or Chrysippus, which may not so well beseem a Professor of Christianity, but would rather shew me to be a vain pursuer of flattery and ostentation, while I endeavor to varnish my Speech with the Fucus's of Eloquence.

For to speak Properly, not Rhetorically, to intend the truth of the Matter, not the ornament of Language, is the duty of one Professing Sacred Literature. For the fear of Truth is not in the Tongue, but in the Heart. Neither is it of importance what Language we use in the Relation of Truth, seeing that falsehood only wants Eloquence, and the trappings of Words, whereby to insinuate into the minds of Men, but the language of Truth, as Euripides writeth, is plain and simple; not seeking the graces of Art, or painted Flourishes. Therefore if this great Work of ours, undertaken without any Flowers of Eloquence (which in the series of our Discourse we have not so much slighted as condemned) do prove offensive to your more delicate ears; we entreat you to bear it with the same patience, as once one of the Roman Emperors made use of, when he stood still with his whole Army to hear the tittle-tattle of an impertinent Woman: and with the same humor that King Archelaus was wont to hear Persons that were Hoarse; and of an unpleasant Utterance; that thereby afterwards he might take the more delight in the pleasing sounds of Eloquent Rhetoricians, and Tuneful Voices. Remember that saying of Theophrastus, "That the most Illiterate were able to speak in the preference of the most Elegant Persons, while they spake nothing but Truth and Reason."

And now that I may no longer keep ye in suspense, through what Tracts and Byways I have as it were hinted out this Opinion of mine, it is time that I declare unto ye. But first I must admonish ye, That all Sciences are as well evil as good, and that they bring us no other advantage to excel as Deities, more than what the Serpent promised of old, when he said, Ye shall be as Gods, knowing good and evil. Let him therefore glory in this Serpent, who boasts himself in knowledge; which we read the Heresy of the Ophites not a little unbeseemingly to have done, who Worshiped a Serpent among the rest of their Superstitions, as being the Creature that first introduced the knowledge of Virtue into Paradise.

To this agrees that Platonic Fable which feigns, That one Theuth being offended with Mankind, was the first raiser of that Devil, called the Sciences; not less hurtful than profitable: as Thamus King of Egypt wisely discourses writing of the Inventors of Arts & Letters. Hence most Grammarians Expound and Interpret the word Demons, as much as to say Artists.

But leaving these Fables to their Poets and Philosophers, suppose there were no other Inventors of Arts than Men themselves, yet were they the Sons of the worst Generation, even the Sons of Cain, of whom it is truly said, The sons of this world are wiser than the sons of light in this generation. If men be therefore the Inventors of Arts, it is not said, Every man is a Liar, neither is there one that doth good?

But grant on the other side, that there may be some good men; yet follows it not, that the Sciences themselves have anything of virtue, anything of truth in them, but what they reap and borrow from the Inventors and possessors thereof: For if they light upon any evil Person, they are hurtful; as a perverse Grammarian, an Ostentatious Poet, a lying Historian, a flattering Rhetorician, a litigious Logician, a turbulent Sophister, a loquacious Lullist, a Lotterist Arithmetician, a lascivious Musician, a shameless Dancing master, a boasting Geometrician, a wandering Cosmographer, a pernicious Architect, a Pirate-Navigator, a fallacious Astrologer, a wicket Magician, a perfidious Cabalist, a dreaming Naturalist, a Wonder-feigning Metaphysician, a morose Ethic, a treacherous Politician, a tyrannical Prince, an oppressing Magistrate, a seditious People, a Schismatical Priest, a superstitious Monk, a prodigal Husband, a bargain-breaking Merchant, a pilling Customer, a slothful Husbandman, a careless Shepherd, an envious Fisherman, a bawling Hunter, a plundering Soldier, an exacting Landlord, a murderous Physician, a poisoning Apothecary, a glutton-Cook, a deceitful Alchemist, a juggling Lawyer, a perfidious Notary, a Bribe-taking Judge, and a heretical and seducing Divine.

So that there is nothing more ominous than Art and Knowledge guarded with impiety, seeing that every man becomes a ready Inventor, a learned Author of evil things. If it light upon a person that is not so evil as foolish, there is nothing more insolent or Dogmatical, having besides its own headstrong obstinacy, the authority of Learning, and the weapons of Argument to defend its own fury; which other fools wanting, are more tame and quietly mad: As Plato saith of the Rhetorician, "That the more simple and illiterate he is, the more he will take upon him to declaim; will imitate all things, and think himself not unworthy of any undertaking." *So that there is nothing more deadly than to be, as it were, rationally mad.*

But if good and just men be the possessors of Knowledge, then Arts and Sciences may probably become useful to the public Weal, though they render their possessors nothing more happy. For it is not, as Porphyrius and Iamblicus report, That Happiness consists in the multitude of Arts, or heaps of Words. For should that be true, they that were most laden with Sciences, would be most happy; and those that wanted them, would on the other side be altogether unhappy; and hence it would come to pass, That Philosophers would be more happy than Divines.

For true Beatitude consists not in the knowledge of good Things, but in good Life; not in Understanding, but in living Understandingly. Neither is it great Learning, but Good Will, that joins Men to God. Nor do outward Arts avail to Happiness, only as Conditional means, not the Causes of completing our Happiness, unless afflicted with a Life answerable to the nature of those good things we possess. Therefore saith Cicero in his Oration for Archias, "Experience tells us, That Nature without Learning is more diligent in the pursuit of Praise and Virtue, than Learning without natural Inclination."

NOVUM ORGANUM

FRANCIS BACON

Bacon (1561-1626) was an English philosopher and one of the first major thinkers of the Renaissance. He was influenced by the literal Biblical view in which man is given dominion over the Earth. Bacon took it as humanity's mission to examine, explore, and manipulate the natural world to achieve human ends. But we cannot do this without the proper *knowledge* of nature; knowledge gives us the power to act as we choose in the world. His most famous work is the *Novum Organum* (1620), from which the following short passages are taken.

Book I.1: Man is Nature's agent and interpreter.

I.3: Human *knowledge* and human *power* come to the same thing, because ignorance of cause frustrates effect. For *Nature is conquered only by obedience*; and that which in thought is a cause, is like a rule in practice.

I.129: And it would not be irrelevant to distinguish three kinds and degrees of human ambition. [The first is personal ambition; the second is the ambition to conquer other nations.] But if anyone attempts to renew and extend the power and empire of the human race itself *over the universe of things* [i.e. nature], his ambition is without a doubt both more sensible and more majestic than the others'. And the empire of man over things lies solely in the arts [i.e. technology] and sciences. ... Let man recover the *right over nature* which belongs to him by God's gift...

DISCOURSE ON THE ARTS AND SCIENCES

JEAN-JACQUES ROUSSEAU

Rousseau (1712-1778) was a prominent Swiss philosopher, writer, and social theorist. This essay, his first major work, was written for the Dijon Prize Competition in 1750; it won first prize.

The question before me is: "Whether the Restoration of the arts and sciences has had the effect of purifying or corrupting morals." Which side am I to take? That, gentlemen, which becomes an honest man, who is sensible of his own ignorance, and thinks himself none the worse for it.

I feel the difficulty of treating this subject fittingly, before the tribunal which is to judge of what I advance. How can I presume to belittle the sciences before one of the most learned assemblies in Europe, to commend ignorance in a famous Academy, and reconcile my contempt for study with the respect due to the truly learned?

I was aware of these inconsistencies, but not discouraged by them. It is not science, I said to myself, that I am attacking; it is virtue that I am defending, and that before virtuous men—and goodness is ever dearer to the good than learning to the learned.

What then have I to fear? The sagacity of the assembly before which I am pleading? That, I acknowledge, is to be feared; but rather on account of faults of construction than of the views I hold. Just sovereigns have never hesitated to decide against themselves in doubtful cases; and indeed the most advantageous situation in which a just claim can be, is that of being laid before a just and enlightened arbitrator, who is judge in his own case.

To this motive, which encouraged me, I may add another which finally decided me. And this is, that as I have upheld the cause of truth to the best of my natural abilities, whatever my apparent success, there is one reward which cannot fail me. That reward I shall find in the bottom of my heart.

The First Part

It is a noble and beautiful spectacle to see man raising himself, so to speak, from nothing by his own exertions; dissipating, by the light of reason, all the thick clouds in which he was by nature enveloped; mounting above himself; soaring in thought even to the celestial regions; like the sun, encompassing with giant strides the vast extent of the universe; and, what is still grander and more wonderful, going back into himself, there to study man and get to know his own nature, his duties and his end. All these miracles we have seen renewed within the last few generations.

Europe had relapsed into the barbarism of the earliest ages; the inhabitants of this part of the world, which is at present so highly enlightened, were plunged, some centuries ago, in a state still worse than ignorance. A scientific jargon, more despicable than mere ignorance, had usurped the name of knowledge, and opposed an almost invincible obstacle to its restoration.

Things had come to such a pass, that it required a *complete revolution* to bring men back to common sense. This came at last from the quarter from which it was least to be expected. It was the stupid Mussulman, the eternal scourge of letters, who was the immediate cause of their revival among us. The fall of the throne of Constantine brought to Italy the relics of ancient Greece; and with these precious spoils France in turn was enriched. The sciences soon followed literature, and the art of thinking joined that of writing: an order which may seem strange, but is perhaps only too natural. The world now began to perceive the principal advantage of an intercourse with the Muses, that of rendering mankind more sociable by inspiring them with the desire to please one another with performances worthy of their mutual approbation.

The mind, as well as the body, has its needs: those of the body are the basis of society, those of the mind its ornaments.

So long as government and law provide for the security and well-being of men in their common life, the arts, literature, and the sciences, less despotic though perhaps more powerful, fling garlands of flowers over the chains which weigh them down. They stifle in men's breasts that *sense of original liberty*, for which they seem to have been born; *cause them to love their own slavery*, and so make of them what is called a civilized people.

Necessity raised up thrones; the arts and sciences have made them strong. Powers of the earth, cherish all talents and protect those who cultivate them[1]. Civilized peoples, cultivate such pursuits: to them, happy slaves, you owe that delicacy and exquisiteness of taste, which is so much your boast, that sweetness of disposition and urbanity of manners which make intercourse so easy and agreeable among you—in a word, the appearance of all the virtues, without being in possession of one of them.

It was for this sort of accomplishment, which is by so much the more captivating, as it seems less affected, that Athens and Rome were so much distinguished in the boasted times of their splendour and magnificence: and it is doubtless in the same respect that our own age and nation will excel all periods and peoples. An air of philosophy without pedantry; an address at once natural and engaging, distant equally from Teutonic clumsiness and Italian pantomime; these are the effects of a taste acquired by liberal studies and improved by conversation with the world. What happiness would it be for those who live among us, if our external appearance were always a true mirror of our hearts; if decorum were but virtue; if the maxims we professed were the rules of our conduct; and if real philosophy were inseparable from the title of a philosopher! But so many good qualities too seldom go together; virtue rarely appears in so much pomp and state.

Richness of apparel may proclaim the man of fortune, and elegance the man of taste; but true health and manliness are known by different signs. It is under the homespun of the labourer, and not beneath the gilt and tinsel of the courtier, that we should look for strength and vigour of body.

External ornaments are no less foreign to virtue, which is the strength and activity of the mind. The honest man is an athlete, who loves to wrestle stark naked; he scorns all those vile trappings, which prevent the exertion of his strength, and were, for the most part, invented only to conceal some deformity.

Before art had *moulded our behaviour*, and taught our passions to speak an artificial language, our morals were rude but natural; and the different ways in which we behaved proclaimed at the first glance the difference of our dispositions. Human nature was not at bottom better then than now; but men found their security in the ease with which they could see through one another, and this advantage, of which we no longer feel the value, prevented their having many vices.

In our day, now that more subtle study and a more refined taste have reduced the art of pleasing to a system, there prevails in modern manners a servile and deceptive conformity; so that one would think every mind had been cast in the same mould. Politeness requires this thing; decorum that; ceremony has its forms, and fashion its laws, and these we must always follow, never the promptings of our own nature.

We no longer dare seem what we really are, but lie under a perpetual restraint; in the meantime the herd of men, which we call society, all act under the same circumstances exactly alike, unless very particular and powerful motives prevent them. Thus we never know with whom we have to deal; and even to know our friends we must wait for some critical and pressing occasion; that is, till it is too late; for it is on those very occasions that such knowledge is of use to us.

What a train of vices must attend this uncertainty! Sincere friendship, real esteem, and perfect confidence are banished from among men. Jealousy, suspicion, fear, coldness, reserve, hate, and fraud lie constantly concealed under that uniform and deceitful veil of politeness; that boasted candour and urbanity, for which we are indebted to the light and leading of this age. We shall no longer take in vain by our oaths the name of our Creator; but we shall insult Him with our blasphemies, and our scrupulous ears will take no offence. We have grown too modest to brag of our own deserts; but we do not scruple to decry those of others. We do not grossly outrage even our enemies, but artfully calumniate them. Our hatred of other nations diminishes, but patriotism dies with it. Ignorance is held in contempt; but a dangerous scepticism has succeeded it. Some vices indeed are condemned and others grown dishonourable; but we have still many that are honoured with the names of virtues, and it is become necessary that we should either have, or at least pretend to have them. Let who will extol the moderation of our modern sages, I see nothing in it but a refinement of intemperance as unworthy of my commendation as their *artificial* simplicity.

Such is the purity to which our morals have attained; this is the virtue we have made our own. Let the arts and sciences claim the share they have had in this salutary work. I shall add but one reflection more; suppose an inhabitant of some distant country should endeavour to form an idea of European morals from the state of the sciences, the perfection of the arts, the propriety of our public entertainments, the politeness of our behaviour, the affability of our conversation, our constant professions of benevolence, and from those tumultuous assemblies of people of all ranks, who seem, from morning till night, to have no other care than to oblige one another. Such a stranger, I maintain, would arrive at a totally false view of our morality.

Where there is no effect, it is idle to look for a cause: but here the effect is *certain* and the depravity *actual*; *our minds have been corrupted in proportion as the arts and sciences have improved.* Will it be said, that this is a misfortune peculiar to the present age? No, gentlemen, the evils resulting from our vain curiosity are as old as the world. The daily ebb and flow of the tides are not more regularly influenced by the moon than the morals of a people by the progress of the arts and sciences. As their light has risen above our horizon, virtue has taken flight, and the same phenomenon has been constantly observed in all times and places.

Take Egypt, the first school of mankind, that ancient country, famous for its fertility under a brazen sky; the spot from which Sesostris once set out to conquer the world. Egypt became the mother of philosophy and the fine arts; soon she was conquered by Cambyses [King of Persia, ca. 525 BC], and then successively by the Greeks, the Romans, the Arabs, and finally the Turks.

Take Greece, once peopled by heroes, who twice vanquished Asia. Letters, as yet in their infancy, had not corrupted the disposition of its inhabitants; but the progress of the sciences soon produced a dissoluteness of manners and the imposition of the Macedonian yoke [by King Philip, ca. 338 BC]: from which time Greece, always learned, always voluptuous, and always a slave, has experienced amid all its revolutions no more than a change of masters. Not all the eloquence of Demosthenes could breathe life into a body which luxury and the arts had once enervated.

It was not till the days of Ennius and Terence [ca. 200 BC] that Rome, founded by a shepherd, and made illustrious by peasants, began to degenerate. But after the appearance of an Ovid, a Catullus, a Martial, and the rest of those numerous obscene authors, whose very names are enough to put modesty to the blush, Rome, once the shrine of virtue, became the theatre of vice, a scorn among the nations, and an object of derision even to barbarians. Thus the capital of the world at length submitted to the yoke of slavery it had imposed on others, and the very day of its fall was the eve of that on which it conferred on one of its citizens the title of Arbiter of Good Taste.

What shall I say of that metropolis of the Eastern Empire, which, by its situation, seemed destined to be the capital of the world; that refuge of the arts and sciences, when they were banished from the rest of Europe, more perhaps by wisdom than barbarism? The most profligate debaucheries, the most abandoned villainies, the most atrocious crimes, plots, murders, and assassinations form the warp and woof of the history of Constantinople. Such is the pure source from which have flowed to us the floods of knowledge on which the present age so prides itself.

But wherefore should we seek, in past ages, for proofs of a truth, of which the present affords us ample evidence? There is in Asia a vast empire, where learning is held in honour, and leads to the highest dignities in the State. If the sciences improved our morals, if they inspired us with courage and taught us to lay down our lives for the good of our country, the Chinese should be wise, free, and invincible. But, if there be no vice they do not practise, no crime with which they are not familiar; if the sagacity of their ministers, the supposed wisdom of their laws, and the multitude of inhabitants who people that vast empire, have alike failed to preserve them from the yoke of the rude and ignorant Tartars, of what use were their men of science and literature? What advantage has that country reaped from the honours bestowed on its learned men? Can it be that of being peopled by a race of scoundrels and slaves?

Contrast with these instances the morals of those few nations which, being preserved from the contagion of useless knowledge, have by their virtues become happy in themselves and afforded an example to the rest of the world. Such were the first inhabitants of Persia, a nation so singular that virtue was taught among them in the same manner as the sciences are with us. They very easily subdued Asia, and possess the exclusive glory of having had the history of their political institutions regarded as a philosophical romance. Such were the Scythians, of whom such wonderful eulogies have come down to us. Such were the Germans, whose simplicity, innocence, and virtue afforded a most delightful contrast to the pen of an historian [Tacitus, ca. 100 AD], weary of describing the baseness and villainies of an enlightened, opulent, and voluptuous nation. Such had been even Rome in the days of its poverty and ignorance. And such has shown itself to be, even in our own times, that rustic nation [Switzerland], whose justly renowned courage not even adversity could conquer, and whose fidelity no example could corrupt.

It is not through stupidity that the people have preferred other activities to those of the mind. They were not ignorant that in other countries there were men who spent their time in disputing idly about the sovereign good, and about vice and virtue. They knew that these useless thinkers were lavish in their own praises, and stigmatized other nations contemptuously as barbarians. But they noted the morals of these people, and so learnt what to think of their learning.

Can it be forgotten that, in the very heart of Greece, there arose a city as famous for the happy ignorance of its inhabitants, as for the wisdom of its laws; a republic of demigods rather than of men, so greatly superior their virtues seemed to those of mere humanity? Sparta, eternal proof of the vanity of science, while the vices, under the conduct of the fine arts, were being introduced into Athens, even while its tyrant [Pisistratus, ca. 550 BC] was carefully collecting together the works of the prince of poets, was driving from her walls artists and the arts, the learned and their learning!

The difference was seen in the outcome. Athens became the seat of politeness and taste, the country of orators and philosophers. The elegance of its buildings equaled that of its language; on every side might be seen marble and canvas, animated by the hands of the most skilled artists. From Athens we derive those astonishing performances, which will serve as models to every corrupt age. The picture of Lacedaemon [Sparta] is not so highly coloured. There, the neighbouring nations used to say, "men were born virtuous, their native air seeming to inspire them with virtue." But its inhabitants have left us nothing but the memory of their heroic actions: monuments that should not count for less in our eyes than the most curious relics of Athenian marble.

It is true that, among the Athenians, there were some few wise men who withstood the general torrent, and preserved their integrity even in the company of the Muses. But hear the judgment which the principal, and most unhappy of them, passed on the artists and learned men of his day.

"I have considered the poets," says he, "and I look upon them as people whose talents impose both on themselves and on others; they give themselves out for wise men, and are taken for such; but in reality they are anything sooner than that."

"From the poets," continues Socrates, "I turned to the artists. Nobody was more ignorant of the arts than myself; nobody was more fully persuaded that the artists were possessed of amazing knowledge. I soon discovered, however, that they were in as bad a way as the poets, and that both had fallen into the same misconception. Because the most skilful of them excel others in their particular jobs, they think themselves wiser than all the rest of mankind. This arrogance spoilt all their skill in my eyes, so that, putting myself in the place of the oracle, and asking myself whether I would rather be what I am or what they are, know what they know, or know that I know nothing, I very readily answered, for myself and the god, that I had rather remain as I am."

"None of us, neither the sophists, nor the poets, nor the orators, nor the artists, nor I, know what is the nature of the true, the good, or the beautiful. But there is this difference between us; that, though none of these people know anything, they all think they know something; whereas for my part, if I know nothing, I am at least in no doubt of my ignorance. So the superiority of wisdom, imputed to me by the oracle, is reduced merely to my being fully convinced that I am ignorant of what I do not know."

Thus we find Socrates, the wisest of men in the judgment of the gods, and the most learned of all the Athenians in the opinion of all Greece, speaking in praise of ignorance. Were he alive now, there is little reason to think that our modern scholars and artists would induce him to change his mind. No, gentlemen, that honest man would still persist in despising our vain sciences. He would lend no aid to swell the flood of books that flows from every quarter: he would leave to us, as he did to his disciples, only the example and memory of his virtues; that is the noblest method of instructing mankind.

Socrates had begun at Athens, and the elder Cato proceeded at Rome [ca. 200 BC], to inveigh against those seductive and subtle Greeks, who corrupted the virtue and destroyed the courage of their fellow-citizens: culture, however, prevailed. Rome was filled with philosophers and orators, military discipline was neglected, agriculture was held in contempt, men formed sects, and forgot their country. To the sacred names of liberty, disinterestedness, and obedience to law, succeeded those of Epicurus, Zeno, and Arcesilaus. It was even a saying among their own philosophers that since learned men appeared among them, honest men had been in eclipse [to quote Seneca]. Before that time the Romans were satisfied with the practice of virtue; they were undone when they began to study it.

What would the great soul of Fabricius [ca. 200 BC] have felt, if it had been his misfortune to be called back to life, when he saw the pomp and magnificence of that Rome, which his army had saved from ruin, and his honourable name made more illustrious than all its conquests. "Ye gods!" he would have said, "what has become of those thatched roofs and rustic hearths, which were formerly the habitations of temperance and virtue? What fatal splendour has succeeded the ancient Roman simplicity? What is this foreign language, this effeminacy of manners? What is the meaning of these statues, paintings, and buildings? Fools, what have you done? You, the lords of the earth, have made yourselves the slaves of the frivolous nations you have subdued. You are governed by rhetoricians, and it has been only to enrich architects, painters, sculptors, and stage-players that you have watered Greece and Asia with your blood. Even the spoils of Carthage are the prize of a flute- player. Romans! Romans! make haste to demolish those amphitheatres, break to pieces those statues, burn those paintings; drive from among you those slaves who keep you in subjection, and whose fatal arts are corrupting your morals. Let other hands make themselves illustrious by such vain talents; the only talent worthy of Rome is that of conquering the world and making virtue its ruler. When Cyneas took the Roman senate for an assembly of kings, he was not struck by either useless pomp or studied elegance. He heard there none of that futile eloquence, which is now the study and the charm of frivolous orators. What then was the majesty that Cyneas beheld? Fellow-citizens, he saw the noblest sight that ever existed under heaven, a sight which not all your riches or your arts can show; an assembly of two hundred virtuous men, worthy to command in Rome, and to govern the world."

But let pass the distance of time and place, and let us see what has happened in our own time and country; or rather let us banish odious descriptions that might offend our delicacy, and spare ourselves the pains of repeating the same things under different names. It was not for nothing that I invoked the Manes of Fabricius; for what have I put into his mouth that might not have come with as much propriety from Louis the Twelfth or Henry the Fourth? It is true that in France Socrates would not have drunk the hemlock, but he would have drunk of a potion infinitely more bitter, of insult, mockery, and contempt a hundred times worse than death.

Thus it is that luxury, profligacy, and slavery have been, in all ages, the scourge of the efforts of our pride to emerge from that happy state of ignorance, in which the wisdom of providence had placed us. That thick veil with which it has covered all its operations seems to be a sufficient proof that it never designed us for such fruitless researches. But is there, indeed, one lesson it has taught us, by which we have rightly profited, or which we have neglected with impunity? Let men learn for once that nature would have preserved them from science, as a mother snatches a dangerous weapon from the hands of her child. Let them know that all the secrets she hides are so many evils from which she protects them, and that the very difficulty they find in acquiring knowledge is not the least of her bounty towards them. Men are perverse; but they would have been far worse, if they had had the misfortune to be born learned.

How humiliating are these reflections to humanity, and how mortified by them our pride should be! What! it will be asked, is uprightness the child of ignorance? Is virtue inconsistent with learning? What consequences might not be drawn from such suppositions? But to reconcile these apparent contradictions, we need only examine closely the emptiness and vanity of those pompous titles, which are so liberally bestowed on human knowledge, and which so blind our judgment. Let us consider, therefore, the arts and sciences in themselves. Let us see what must result from their advancement, and let us not hesitate to admit the truth of all those points on which our arguments coincide with the inductions we can make from history.

The Second Part

An ancient tradition passed out of Egypt into Greece, that some god, who was an enemy to the repose of mankind, was the inventor of the sciences [recall Plato's *Phaedrus*]. What must the Egyptians, among whom the sciences first arose, have thought of them? And they beheld, near at hand, the sources from which they sprang.

In fact, whether we turn to the annals of the world, or eke out with philosophical investigations the uncertain chronicles of history, we shall not find for human knowledge an origin answering to the idea we are pleased to entertain of it at present. Astronomy was born of superstition, eloquence of ambition, hatred, falsehood, and flattery; geometry of avarice; physics of an idle curiosity; and even moral philosophy of human pride. Thus the arts and sciences owe their birth to our vices; we should be less doubtful of their advantages, if they had sprung from our virtues.

Their evil origin is, indeed, but too plainly reproduced in their objects. What would become of the arts, were they not cherished by luxury? If men were not unjust, of what use were jurisprudence? What would become of history, if there were no tyrants, wars, or conspiracies? In a word who would pass his life in barren speculations, if everybody, attentive only to the obligations of humanity and the necessities of nature, spent his whole life in serving his country, obliging his friends, and relieving the unhappy? Are we then made to live and die on the brink of that well at the bottom of which Truth lies hid? This reflection alone is, in my opinion, enough to discourage at first setting out every man who seriously endeavours to instruct himself by the study of philosophy.

What a variety of dangers surrounds us! What a number of wrong paths present themselves in the investigation of the sciences! Through how many errors, more perilous than truth itself is useful, must we not pass to arrive at it? The disadvantages we lie under are evident; for falsehood is capable of an infinite variety of combinations; but the truth has only one manner of being. Besides, where is the man who sincerely desires to find it? Or even admitting his good will, by what characteristic marks is he sure of knowing it? Amid the infinite diversity of opinions where is the criterion[2] by which we may certainly judge of it? Again, what is still more difficult, should we even be fortunate enough to discover it, who among us will know how to make right use of it?

If our sciences are futile in the objects they propose, they are no less dangerous in the effects they produce. Being the effect of idleness, they generate idleness in their turn; and an irreparable loss of time is the first prejudice which they must necessarily cause to society. To live without doing some good is a great evil as well in the political as in the moral world; and hence every useless citizen should be regarded as a pernicious person. Tell me then, illustrious philosophers, of whom we learn the ratios in which attraction acts *in vacuo*; and in the revolution of the planets, the relations of spaces traversed in equal times; by whom we are taught what curves have conjugate points, points of inflexion, and cusps; how the soul and body correspond, like two clocks, without actual communication; what planets may be inhabited; and what insects reproduce in an extraordinary manner. Answer me, I say, you from whom we receive all this sublime information, whether we should have been less numerous, worse governed, less formidable, less flourishing, or more perverse, supposing you had taught us none of all these fine things.

Reconsider therefore the importance of your productions; and, since the labours of the most enlightened of our learned men and the best of our citizens are of so little utility, tell us what we ought to think of that numerous herd of obscure writers and useless litterateurs, who devour without any return the substance of the State.

Useless, do I say? Would God they were! Society would be more peaceful, and morals less corrupt. But these vain and futile declaimers go forth on all sides, armed with their fatal paradoxes, to sap the foundations of our faith, and nullify virtue. They smile contemptuously at such old names as patriotism and religion, and consecrate their talents and philosophy to the destruction and defamation of all that men hold sacred. Not that they bear any real hatred to virtue or dogma; they are the enemies of public opinion alone; to bring them to the foot of the altar, it would be enough to banish them to a land of atheists. What extravagancies will not the rage of singularity induce men to commit!

The waste of time is certainly a great evil; but still greater evils attend upon literature and the arts. One is luxury, produced like them by indolence and vanity. Luxury is seldom unattended by the arts and sciences; and they are always attended by luxury. I know that our philosophy, fertile in paradoxes, pretends, in contradiction to the experience of all ages, that luxury contributes to the splendour of States. But, without insisting on the necessity of sumptuary laws, can it be denied that rectitude of morals is essential to the duration of empires, and that luxury is diametrically opposed to such rectitude? Let it be admitted that luxury is a certain indication of wealth; that it even serves, if you will, to increase such wealth; what conclusion is to be drawn from this paradox, so worthy of the times? And what will become of virtue if riches are to be acquired at any cost? The politicians of the ancient world were always talking of morals and virtue; ours speak of nothing but commerce and money. One of them will tell you that in such a country a man is worth just as much as he will sell for at Algiers: another, pursuing the same mode of calculation, finds that in some countries a man is worth nothing, and in others still less than nothing; they value men as they do droves of oxen. According to them, a man is worth no more to the State than the amount he consumes; and thus a Sybarite would be worth at least thirty Lacedaemonians. Let these writers tell me, however, which of the two republics, Sybaris or Sparta, was subdued by a handful of peasants, and which became the terror of Asia.

The monarchy of Cyrus was conquered by thirty thousand men, led by a prince poorer than the meanest of Persian Satraps [Alexander the Great, ca. 330 BC]: in like manner the Scythians, the poorest of all nations, were able to resist the most powerful monarchs of the universe. When two famous republics contended for the empire of the world, the one rich and the other poor, the former was subdued by the latter. The Roman empire in its turn, after having engulfed all the riches of the universe, fell a prey to peoples who knew not even what riches were. The Franks conquered the Gauls, and the Saxons England [ca. 450 AD], without any other treasures than their bravery and their poverty. A band of poor mountaineers [i.e. Swiss peasants] whose whole cupidity was confined to the possession of a few sheep-skins, having first given a check to the arrogance of Austria, went on to crush the opulent and formidable house of Burgundy, which at that time made the potentates of Europe tremble. In short, all the power and wisdom of the heir of Charles the Fifth, backed by all the treasures of the Indies, broke before a few herring-fishers [when the Dutch defeated Spain, ca. 1570]. Let our politicians condescend to lay aside their calculations for a moment, to reflect on these examples; let them learn for once that money, though it buys everything else, cannot buy morals and citizens. What then is the precise point in dispute about luxury? It is to know which is most advantageous to empires, that their existence should be brilliant and momentary, or virtuous and lasting. I say brilliant, but with what lustre? A taste for ostentation never prevails in the same minds as a taste for honesty. No, it is impossible that understandings, degraded by a multitude of futile cares, should ever rise to what is truly great and noble; even if they had the strength, they would want the courage.

Every artist loves applause. The praise of his contemporaries is the most valuable part of his recompense. What then will he do to obtain it, if he have the misfortune to be born among a people, and at a time, when learning is in vogue, and the superficiality of youth is in a position to lead the fashion; when men have sacrificed their taste to those who tyrannize over their liberty, and one sex dare not approve anything but what is proportionate to the pusillanimity of the other; when the greatest masterpieces of dramatic poetry are condemned, and the noblest of musical productions neglected? This is what he will do. He will lower his genius to the level of the age, and will rather submit to compose mediocre works, that will be admired during his lifetime, than labour at sublime achievements which will not be admired till long after he is dead. Let the famous Voltaire tell us how many nervous and masculine beauties he has sacrificed to our false delicacy, and how much that is great and noble, that spirit of gallantry, which delights in what is frivolous and petty, has cost him.

It is thus that the dissolution of morals, the necessary consequence of luxury, brings with it in its turn the corruption of taste. Further, if by chance there be found among men of average ability, an individual with enough strength of mind to refuse to comply with the spirit of the age, and to debase himself by puerile productions, his lot will be hard. He will die in indigence and oblivion. This is not so much a prediction as a fact already confirmed by experience! Yes, Carle and Pierre [two contemporary painters], the time is already come when your pencils, destined to increase the majesty of our temples by sublime and holy images, must fall from your hands, or else be prostituted to adorn the panels of a coach with lascivious paintings. And you, inimitable Pigalle [sculptor], rival of Phidias and Praxiteles, whose chisel the ancients would have employed to carve them gods, whose images almost excuse their idolatry in our eyes; even your hand must condescend to fashion the belly of an ape, or else remain idle.

We cannot reflect on the morality of mankind without contemplating with pleasure the picture of the simplicity which prevailed in the earliest times. This image may be justly compared to a beautiful coast, adorned only by the hands of nature; towards which our eyes are constantly turned, and which we see receding with regret. While men were innocent and virtuous and loved to have the gods for witnesses of their actions, they dwelt together in the same huts; but when they became vicious, they grew tired of such inconvenient onlookers, and banished them to magnificent temples. Finally, they expelled their deities even from these, in order to dwell there themselves; or at least the temples of the gods were no longer more magnificent than the palaces of the citizens. This was the height of degeneracy; nor could vice ever be carried to greater lengths than when it was seen, supported, as it were, at the doors of the great, on columns of marble, and graven on Corinthian capitals.

As the conveniences of life increase, as the arts are brought to perfection, and luxury spreads, true courage flags, the virtues disappear; and all this is the effect of the sciences and of those acts which are exercised in the privacy of men's dwellings. When the Goths ravaged Greece, the libraries only escaped the flames owing to an opinion that was set on foot among them, that it was best to leave the enemy with a possession so calculated to divert their attention from military exercises, and keep them engaged in indolent and sedentary occupations.

Charles the Eighth found himself master of Tuscany and the kingdom of Naples, almost without drawing sword; and all his court attributed this unexpected success to the fact that the princes and nobles of Italy applied themselves with greater earnestness to the cultivation of their understandings than to active and martial pursuits. In fact, says the sensible person who records these characteristics, experience plainly tells us that in military matters and all that resemble them, application to the sciences tends rather to make men effeminate and cowardly than resolute and vigorous.

The Romans confessed that military virtue was extinguished among them, in proportion as they became connoisseurs in the arts of the painter, the engraver, and the goldsmith, and began to cultivate the fine arts. Indeed, as if this famous country was to be for ever an example to other nations, the rise of the Medici and the revival of letters has once more destroyed, this time perhaps for ever, the martial reputation which Italy seemed a few centuries ago to have recovered.

The ancient republics of Greece, with that wisdom which was so conspicuous in most of their institutions, forbade their citizens to pursue all those inactive and sedentary occupations, which by enervating and corrupting the body diminish also the vigour of the mind. With what courage, in fact, can it be thought that hunger and thirst, fatigues, dangers, and death, can be faced by men whom the smallest want overwhelms and the slightest difficulty repels? With what resolution can soldiers support the excessive toils of war, when they are entirely unaccustomed to them? With what spirits can they make forced marches under officers who have not even the strength to travel on horseback? It is no answer to cite the reputed valour of all the modern warriors who are so scientifically trained. I hear much of their bravery in a day's battle; but I am told nothing of how they support excessive fatigue, how they stand the severity of the seasons and the inclemency of the weather. A little sunshine or snow, or the want of a few superfluities, is enough to cripple and destroy one of our finest armies in a few days. Intrepid warriors! permit me for once to tell you the truth, which you seldom hear. Of your bravery I am fully satisfied. I have no doubt that you would have triumphed with Hannibal at Cannae, and at Trasimene: that you would have passed the Rubicon with Caesar, and enabled him to enslave his country; but you never would have been able to cross the Alps with the former, or with the latter to subdue your own ancestors, the Gauls.

A war does not always depend on the events of battle: there is in generalship an art superior to that of gaining victories. A man may behave with great intrepidity under fire, and yet be a very bad officer. Even in the common soldier, a little more strength and vigour would perhaps be more useful than so much courage, which after all is no protection from death. And what does it matter to the State whether its troops perish by cold and fever, or by the sword of the enemy?

If the cultivation of the sciences is prejudicial to military qualities, it is still more so to moral qualities. Even from our infancy an absurd system of education serves to adorn our wit and corrupt our judgment. We see, on every side, huge institutions, where our youth are educated at great expense, and instructed in everything but their duty. Your children will be ignorant of their own language, when they can talk others which are not spoken anywhere. They will be able to compose verses which they can hardly understand; and, without being capable of distinguishing truth from error, they will possess the art of making them unrecognizable by specious arguments. But magnanimity, equity, temperance, humanity, and courage will be words of which they know not the meaning. The dear name of country will never strike on their ears; and if they ever hear speak of God, it will be less to fear than to be frightened of Him. I would as soon, said a wise man, that my pupil had spent his time in the tennis court as in this manner; for there his body at least would have got exercise.

I well know that children ought to be kept employed, and that idleness is for them the danger most to be feared. But what should they be taught? This is undoubtedly an important question. Let them be taught what they are to practice when they come to be men; not what they ought to forget.

Our gardens are adorned with statues and our galleries with pictures. What would you imagine these masterpieces of art, thus exhibited to public admiration, represent? The great men who have defended their country, or the still greater men who have enriched it by their virtues? Far from it. They are the images of every perversion of heart and mind, carefully selected from ancient mythology, and presented to the early curiosity of our children, doubtless that they may have before their eyes the representations of vicious actions, even before they are able to read.

Whence arise all those abuses, unless it be from that fatal inequality introduced among men by the difference of talents and the cheapening of virtue? This is the most evident effect of all our studies, and the most dangerous of all their consequences. The question is no longer whether a man is honest, but whether he is clever. We do not ask whether a book is useful, but whether it is well written. Rewards are lavished on wit and ingenuity, while virtue is left unhonoured. There are a thousand prizes for fine discourses, and none for good actions. I should be glad, however, to know whether the honour attaching to the best discourse that ever wins the prize in this Academy is comparable with the merit of having founded the prize.

A wise man does not go in chase of fortune; but he is by no means insensible to glory, and when he sees it so ill distributed, his virtue, which might have been animated by a little emulation, and turned to the advantage of society, droops and dies away in obscurity and indigence. It is for this reason that the agreeable arts must in time everywhere be preferred to the useful; and this truth has been but too much confirmed since the revival of the arts and sciences. We have physicists, geometricians, chemists, astronomers, poets, musicians, and painters in plenty; but we have no longer a citizen among us; or if there be found a few scattered over our abandoned countryside, they are left to perish there unnoticed and neglected. Such is the condition to which we are reduced, and such are our feelings towards those who give us our daily bread, and our children milk.

I confess, however, that the evil is not so great as it might have become. The eternal providence, in placing salutary simples beside noxious plants, and making poisonous animals contain their own antidote, has taught the sovereigns of the earth, who are its ministers, to imitate its wisdom. It is by following this example that the truly great monarch [King Louis XIV, ca. 1670], to whose glory every age will add new lustre, drew from the very bosom of the arts and sciences the very fountains of a thousand lapses from rectitude, those famous societies, which, while they are depositaries of the dangerous trust of human knowledge, are yet the sacred guardians of morals, by the attention they pay to their maintenance among themselves in all their purity, and by the demands which they make on every member whom they admit.

These wise institutions, confirmed by his august successor and imitated by all the kings of Europe, will serve at least to restrain men of letters, who, all aspiring to the honour of being admitted into these Academies, will keep watch over themselves, and endeavour to make themselves worthy of such honour by useful performances and irreproachable morals. Those Academies also, which, in proposing prizes for literary merit, make choice of such subjects as are calculated to arouse the love of virtue in the hearts of citizens, prove that it prevails in themselves, and must give men the rare and real pleasure of finding learned societies devoting themselves to the enlightenment of mankind, not only by agreeable exercises of the intellect, but also by useful instructions.

An objection which may be made is, in fact, only an additional proof of my argument. So much precaution proves but too evidently the need for it. We never seek remedies for evils that do not exist. Why, indeed, must these bear all the marks of ordinary remedies, on account of their inefficacy? The numerous establishments in favour of the learned are only adapted to make men mistake the objects of the sciences, and turn men's attention to the cultivation of them. One would be inclined to think, from the precautions everywhere taken, that we are overstocked with husbandmen, and are afraid of a shortage of philosophers. I will not venture here to enter into a comparison between agriculture and philosophy, as they would not bear it. I shall only ask: What is philosophy? What is contained in the writings of the most celebrated philosophers? What are the lessons of these friends of wisdom. To hear them, should we not take them for so many mountebanks, exhibiting themselves in public, and crying out, *Here, Here, come to me, I am the only true doctor*? One of them teaches that there is no such thing as matter, but that everything exists only in representation. Another declares that there is no other substance than matter, and no other God than the world itself. A third tells you that there are no such things as virtue and vice, and that moral good and evil are chimeras; while a fourth informs you that men are only beasts of prey, and may conscientiously devour one another. Why, my great philosophers, do you not reserve these wise and profitable lessons for your friends and children? You would soon reap the benefit of them, nor should we be under the apprehension of our own becoming your disciples.

Such are the wonderful men, whom their contemporaries held in the highest esteem during their lives, and to whom immortality has been attributed since their decease. Such are the wise maxims we have received from them, and which are transmitted, from age to age, to our descendants. Paganism, though given over to all the extravagances of human reason, has left nothing to compare with the shameful monuments which have been prepared by the art of printing, during the reign of the gospel. The impious writings of Leucippus and Diagoras perished with their authors. The world, in their days, was ignorant of the art of immortalizing the errors and extravagances of the human mind. But thanks to the art of printing and the use we make of it, the pernicious reflections of Hobbes and Spinoza will last forever. Go, famous writings, of which the ignorance and rusticity of our forefathers would have been incapable. Go to our descendants, along with those still more pernicious works which reek of the corrupted manners of the present age! Let them together convey to posterity a faithful history of the progress and advantages of our arts and sciences. If they are read, they will leave not a doubt about the question we are now discussing, and unless mankind should then be still more foolish than we, they will lift up their hands to Heaven and exclaim in bitterness of heart: "Almighty God! Thou who holdest in Thy hand the minds of men, deliver us from the fatal arts and sciences of our forefathers; give us back ignorance, innocence, and poverty, which alone can make us happy and are precious in Thy sight."

But if the progress of the arts and sciences had added nothing to our real happiness; if it has corrupted our morals, and if that corruption has vitiated our taste, what are we to think of the herd of text-book authors, who have removed those impediments which nature purposely laid in the way to the Temple of the Muses, in order to guard its approach and try the powers of those who might be tempted to seek knowledge? What are we to think of those compilers who have indiscreetly broken open the door of the sciences, and introduced into their sanctuary a populace unworthy to approach it, when it was greatly to be wished that all who should be found incapable of making a considerable progress in the career of learning should have been repulsed at the entrance, and thereby cast upon those arts which are useful to society. A man who will be all his life a bad versifier, or a third-rate geometrician, might have made nevertheless an excellent clothier. Those whom nature intended for her disciples have not needed masters. Bacon, Descartes, and Newton, those teachers of mankind, had themselves no teachers. What guide indeed could have taken them so far as their sublime genius directed them? Ordinary masters would only have cramped their intelligence, by confining it within the narrow limits of their own capacity. It was from the obstacles they met with at first that they learned to exert themselves, and bestirred themselves to traverse the vast field which they covered. If it be proper to allow some men to apply themselves to the study of the arts and sciences, it is only those who feel themselves able to walk alone in their footsteps and to outstrip them. It belongs only to these few to raise monuments to the glory of the human understanding.

But if we are desirous that nothing should be above their genius, nothing should be beyond their hopes. This is the only encouragement they require. The soul insensibly adapts itself to the objects on which it is employed, and thus it is that great occasions produce great men. The greatest orator in the world was Consul of Rome, and perhaps the greatest of philosophers Lord Chancellor of England. Can it be conceived that, if the former had only been a professor at some University, and the latter a pensioner of some Academy, their works would not have suffered from their situation. Let not princes disdain to admit into their councils those who are most capable of giving them good advice. Let them renounce the old prejudice, which was invented by the pride of the great, that the art of governing mankind is more difficult than that of instructing them; as if it was easier to induce men to do good voluntarily than to compel them to it by force. Let the learned of the first rank find an honourable refuge in their courts; let them there enjoy the only recompense worthy of them, that of promoting by their influence the happiness of the peoples they have enlightened by their wisdom. It is by this means only that we are likely to see what virtue, science and authority can do, when animated by the noblest emulation, and working unanimously for the happiness of mankind.

But so long as power alone is on one side, and knowledge and understanding alone on the other, the learned will seldom make great objects their study, princes will still more rarely do great actions, and the peoples will continue to be, as they are, mean, corrupt, and miserable.

As for us, ordinary men, on whom Heaven has not been pleased to bestow such great talents; as we are not destined to reap such glory, let us remain in our obscurity. Let us not covet a reputation we should never attain, and which, in the present state of things, would never make up to us for the trouble it would have cost us, even if we were fully qualified to obtain it. Why should we build our happiness on the opinions of others, when we can find it in our own hearts? Let us leave to others the task of instructing mankind in their duty, and confine ourselves to the discharge of our own. We have no occasion for greater knowledge than this.

Virtue! sublime science of simple minds, are such industry and preparation needed if we are to know you? Are not your principles graven on every heart? Need we do more, to learn your laws, than examine ourselves and listen to the voice of conscience, when the passions are silent?

This is the true philosophy, with which we must learn to be content, without envying the fame of those celebrated men, whose names are immortal in the republic of letters. Let us, instead of envying them, endeavour to make, between them and us, that honourable distinction which was formerly seen to exist between two great peoples, that the one knew how to speak, and the other how to act, aright.

THE END

Notes

[1] Sovereigns always see with pleasure a taste for the arts of amusement and superfluity, which do not result in the exportation of bullion, increase among their subjects. They very well know that, besides nourishing that littleness of mind which is proper to slavery, *the increase of artificial wants only binds so many more chains upon the people*. Alexander, wishing to keep the *Ichthyophagi* [fish-eaters] in a state of dependence, compelled them to give up fishing, and subsist on the customary food of civilized nations. The American savages [i.e. Indians], who go naked, and live entirely on the products of the chase, have been always impossible to subdue. What yoke, indeed, can be imposed on men who stand in need of nothing?

[2] The less we know, the more we think we know. The Peripatetics [Aristotle and his followers] doubted of nothing. Did not Descartes construct the universe with cubes and vortices? And is there in all Europe one single physicist who does not boldly explain the inexplicable mysteries of electricity, which will, perhaps, be for ever the

THE LUDDITES

Definition: A "Luddite" was originally a person who violently opposed the use of labor-saving (and thus unemployment-inducing) machinery, but has now expanded to refer to anyone who opposes advanced technology in general. Sometimes one sees the term 'neo-Luddite'.

The term comes from a possibly mythical Briton named Ned Ludd. Ludd was supposedly a dimwitted lad who accidentally destroyed two English 'stocking frames' (devices to make socks and stockings) sometime in the late 1700s. Word spread that perhaps he deliberately destroyed them in order to protect the higher quality, higher paid work of the skilled weavers. Thereafter, tradesmen who violently opposed such machinery claimed to be acting under the orders of 'General Ned Ludd'.

The Luddite attacks occurred primarily in the early 1800's, but as early as 1721 the English government had acted to defend new technology; they passed the first law making 'machine-breaking' a capital offense, i.e. punishable by death. The first violent acts occurred in 1779, when a Bill restricting the frame-knitting industry failed to pass, whereon 300 frames were smashed and thrown into the streets. But more serious actions would not occur until some 30 years later.

In the early months of 1811 the first threatening letters from 'General Ned Ludd and the Army of Redressers', were sent to employers in Nottingham. Workers, upset by wage reductions and the use of unapprenticed workmen, began to break into factories at night to destroy the new machines that the employers were using. In a three-week period, over two hundred stocking frames were destroyed. By March, 1811, several attacks were taking place every night and the Nottingham authorities had to enroll 400 special constables to protect the factories. To help catch the culprits, the Prince Regent offered £50 to anyone "giving information on any person or persons wickedly breaking the frames."

Luddism gradually spread to Yorkshire, Lancashire, Leicestershire and Derbyshire. In Yorkshire, the croppers (a small and highly skilled group of cloth finishers) turned their anger on the new shearing frame that they feared would put them out of work. In February and March, 1812, factories were attacked in Huddersfield, Halifax, Wakefield and Leeds.

In February 1812 the government of Spencer Perceval proposed that frame-breaking, specifically, should become a capital offence. Despite a passionate speech by Lord Byron [the poet – see following] in the House of Lords, Parliament passed the Frame Breaking Act that enabled people convicted of machine-breaking to be sentenced to death. As a further precaution, the government ordered 12,000 troops into the areas where the Luddites were active.

One of the most serious Luddite attacks took place at Rawfolds Mill in Yorkshire, on April 11, 1812. William Cartwright, the owner of Rawfolds Mill, had been using cloth-finishing machinery since 1811. Local croppers began losing their jobs, and after a public meeting they decided to try to destroy the machinery. Cartwright was suspecting trouble and arranged for the mill to be protected by armed guards. The Luddites failed to gain entry, and two of their men were killed in the attempt.

Seven days later the Luddites killed William Horsfall, another large mill-owner in the area. The authorities rounded up over a hundred suspects. Of these, sixty-four were indicted. Ultimately, 14 were hung for the Rawfolds Mill attack, and 3 executed for the murder of Horsfall.

Throughout 1812 there were attacks on Lancashire cotton mills. Local handloom weavers objected to the introduction of *steam-powered* looms. On 20th March, 1812 the Stockport warehouse of William Radcliffe, one of the first manufacturers to use the power-loom, was attacked.

Wheat prices soared in 1812. Unable to feed their families, workers became desperate. There were food riots in several cities. On 20th April several *thousand* men attacked Burton's Mill near Manchester. Emanuel Burton, who knew that his policy of buying power-looms had upset local handloom weavers, had recruited armed guards; 3 members of the crowd were killed by musket-fire. The following day the men returned and, after failing to break-in to the mill, they proceeded to Burton's house and burned it to the ground. The military arrived and another 7 men were killed.

Three days later, Wray & Duncroff's Mill near Manchester was set on fire. William Hulton, the High Sheriff of Lancashire, arrested twelve men suspected of taking part in the attack; 4 were executed, including a 12-year-old boy.

In the summer of 1812, eight men in Lancashire were sentenced to death and 13 exiled to Australia for attacks on cotton mills. Another 15 were executed at York.

In 1816 there was a revival of violence, following a bad harvest and a downturn in trade. On June 28 the Luddites attacked Heathcote & Boden's mill in Loughborough, smashing 53 frames. Troops were brought in, and 6 men executed and 3 exiled. This was followed by further sporadic outbreaks of violence, but by 1817 the Luddite movement had ceased to be active in Britain.

SONG OF THE LUDDITES

LORD BYRON (1816)

As the Liberty lads o'er the sea
Bought their freedom, and cheaply with blood,
So we, boys, we
Will die fighting, or live free,
And down with all kings but King Ludd!

When the web that we weave is complete,
And the shuttle exchanged for the sword,
We will fling the winding sheet
O'er the despot at our feet,
And dye it deep in the gore he has poured.

Though black as his heart its hue,
Since his veins are corrupted to mud,
Yet this is the dew
Which the tree shall renew
Of Liberty, planted by Ludd!

SIGNS OF THE TIMES

THOMAS CARLYLE

Carlyle (1795-1881) was a Scottish philosopher and essayist. This article was originally published in the *Edinburgh Review* in 1829

Were we required to characterise this age of ours by any single epithet, we should be tempted to call it, not an Heroical, Devotional, Philosophical, or Moral Age, but, above all others, the Mechanical Age. It is the Age of Machinery, in every outward and inward sense of that word; the age which, with its whole undivided might, forwards, teaches and practises the great art of adapting means to ends. Nothing is now done directly, or by hand; all is by rule and calculated contrivance. For the simplest operation, some helps and accompaniments, some cunning abbreviating process is in readiness. Our old modes of exertion are all discredited, and thrown aside. On every hand, the living artisan is driven from his workshop, to make room for a speedier, inanimate one. The shuttle drops from the fingers of the weaver, and falls into iron fingers that ply it faster. The sailor furls his sail, and lays down his oar; and bids a strong, unwearied servant, on vaporous wings, bear him through the waters. Men have crossed oceans by steam; the Birmingham Fire-king ship has visited the fabulous East; and the genius of the Cape, were there any [poet] Camoens now to sing it, has again been alarmed, and with far stranger thunders than Gama's.

There is no end to machinery. Even the horse is stripped of his harness, and finds a fleet fire-horse invoked in his stead. Nay, we have an artist that hatches chickens by steam; the very brood-hen is to be superseded! For all earthly, and for some unearthly purposes, we have machines and mechanic furtherances; for mincing our cabbages; for casting us into magnetic sleep. We remove mountains, and make seas our smooth highways; nothing can resist us. We war with rude Nature; and, by our resistless engines, come off always victorious, and loaded with spoils.

What wonderful accessions have thus been made, and are still making, to the physical power of mankind; how much better fed, clothed, lodged and, in all outward respects, accommodated men now are, or might be, by a given quantity of labour, is a grateful reflection which forces itself on every one. What changes, too, this addition of power is introducing into the Social System; how wealth has more and more increased, and at the same time gathered itself more and more into masses, strangely altering the old relations, and increasing the distance between the rich and the poor, will be a question for Political Economists, and a much more complex and important one than any they have yet engaged with.

But leaving these matters for the present, let us observe how the mechanical genius of our time has diffused itself into quite other provinces. Not the external and physical alone is now managed by machinery, but the internal and spiritual also. Here too nothing follows its spontaneous course, nothing is left to be accomplished by old natural methods. Everything has its cunningly devised implements, its preestablished apparatus; it is not done by hand, but by machinery. Thus we have machines for Education: Lancastrian machines; Hamiltonian machines; monitors, maps and emblems. Instruction, that mysterious communing of Wisdom with Ignorance, is no longer an indefinable tentative process, requiring a study of individual aptitudes, and a perpetual variation of means and methods, to attain the same end; but a secure, universal, straightforward business, to be conducted in the gross, by proper mechanism, with such intellect as comes to hand. Then, we have Religious machines, of all imaginable varieties; the Bible-Society, professing a far higher and heavenly structure, is found, on inquiry, to be altogether an earthly contrivance: supported by collection of moneys, by fomenting of vanities, by puffing, intrigue and chicane; a machine for converting the Heathen. It is the same in all other departments. Has any man, or any society of men, a truth to speak, a piece of spiritual work to do; they can nowise proceed at once and with the mere natural organs, but must first call a public meeting, appoint committees, issue prospectuses, eat a public dinner; in a word, construct or borrow machinery, wherewith to speak it and do it. Without machinery, they were hopeless, helpless; a colony of Hindu weavers squatting in the heart of Lancashire. Mark, too, how every machine must have its moving power, in some of the great currents of society; every little sect among us, Unitarians, Utilitarians, Anabaptists, Phrenologists, must have its Periodical, its monthly or quarterly Magazine;— hanging out, like its windmill, into the *popularis aura*, to grind meal for the society.

With individuals, in like manner, natural strength avails little. No individual now hopes to accomplish the poorest enterprise single-handed and without mechanical aids; he must make interest with some existing corporation, and till his field with their oxen. In these days, more emphatically than ever, "to live, signifies to unite with a party, or to make one."

Philosophy, Science, Art, Literature, all depend on machinery. No Newton, by silent meditation, now discovers the system of the world from the falling of an apple; but some quite other than Newton stands in his Museum, his Scientific Institution, and behind whole batteries of retorts, digesters, and galvanic piles imperatively "interrogates Nature," who however, shows no haste to answer. In defect of Raphaels, and Michelangelos, and Mozarts, we have Royal Academies of Painting, Sculpture, Music; whereby the languishing spirits of Art may be strengthened, as by the more generous diet of a Public Kitchen. Literature, too, has its Paternoster-row mechanism, its Trade-dinners, its Editorial conclaves, and huge subterranean, puffing bellows; so that books are not only printed, but, in a great measure, written and sold, by machinery.

National culture, spiritual benefit of all sorts, is under the same management. No Queen Christina, in these times, needs to send for her Descartes; no King Frederick for his Voltaire, and painfully nourish him with pensions and flattery: any sovereign of taste, who wishes to enlighten his people, has only to impose a new tax, and with the proceeds establish Philosophic Institutes. Hence the Royal and Imperial Societies, the Bibliothèques, Glyptothèques, Technothèques, which front us in all capital cities; like so many well-finished hives, to which it is expected the stray agencies of Wisdom will swarm of their own accord, and hive and make honey. In like manner, among ourselves, when it is thought that religion is declining, we have only to vote half-a-million's worth of bricks and mortar, and build new churches. In Ireland it seems they have gone still farther, having actually established a "Penny-a-week Purgatory-Society"! Thus does the Genius of Mechanism stand by to help us in all difficulties and emergencies, and with his iron back bears all our burdens.

《《——》》

These things, which we state lightly enough here, are yet of deep import, and indicate a mighty change in our whole manner of existence. For the same habit regulates not our modes of action alone, but our modes of thought and feeling. Men are grown mechanical in head and in heart, as well as in hand. They have lost faith in individual endeavour, and in natural force, of any kind. Not for internal perfection, but for external combinations and arrangements, for institutions, constitutions, — for Mechanism of one sort or other, do they hope and struggle. Their whole efforts, attachments, opinions, turn on mechanism, and are of a mechanical character.

We may trace this tendency in all the great manifestations of our time; in its intellectual aspect, the studies it most favours and its manner of conducting them; in its practical aspects, its politics, arts, religion, morals; in the whole sources, and throughout the whole currents, of its spiritual, no less than its material activity.

Consider, for example, the state of Science generally, in Europe, at this period. It is admitted, on all sides, that the Metaphysical and Moral Sciences are falling into decay, while the Physical are engrossing, every day, more respect and attention. In most of the European nations there is now no such thing as a Science of Mind; only more or less advancement in the general science, or the special sciences, of matter. The French were the first to desert Metaphysics; and though they have lately affected to revive their school, it has yet no signs of vitality. The land of Malebranche, Pascal, Descartes and Fenelon, has now only its [Victor] Cousins and [Abel-Francois] Villemains; while, in the department of Physics, it reckons far other names. Among ourselves, the Philosophy of Mind, after a rickety infancy, which never reached the vigour of manhood, fell suddenly into decay, languished and finally died out, with its last amiable cultivator, Professor [Dugald] Stewart. In no nation but Germany has any decisive effort been made in psychological science; not to speak of any decisive result. The science of the age, in short, is physical, chemical, physiological; in all shapes mechanical.

Our favourite Mathematics, the highly prized exponent of all these other sciences, has also become more and more mechanical. Excellence in what is called its higher departments depends less on natural genius than on acquired expertness in wielding its machinery. Without undervaluing the wonderful results which a Lagrange or Laplace educes by means of it, we may remark, that their calculus, differential and integral, is little else than a more cunningly-constructed arithmetical mill; where the factors, being put in, are, as it were, ground into the true product, under cover, and without other effort on our part than steady turning of the handle. We have more Mathematics than ever; but less Mathesis. Archimedes and Plato could not have read the *Mécanique Céleste*; but neither would the whole French Institute see aught in that saying, "God geometrises!" but a sentimental rodomontade.

Nay, our whole Metaphysics itself, from Locke's time downward, has been physical; not a spiritual philosophy, but a material one. The singular estimation in which his *Essay* was so long held as a scientific work (an estimation grounded, indeed, on the estimable character of the man) will one day be thought a curious indication of the spirit of these times. His whole doctrine is mechanical, in its aim and origin, in its method and its results. It is not a philosophy of the mind: it is a mere discussion concerning the origin of our consciousness, or ideas, or whatever else they are called; a genetic history of what we see in the mind. The grand secrets of Necessity and Freewill, of the Mind's vital or non-vital dependence on Matter, of our mysterious relations to Time and Space, to God, to the Universe, are not, in the faintest degree touched on in these inquiries; and seem not to have the smallest connexion with them.

This condition of the two great departments of knowledge—the outward, cultivated exclusively on mechanical principles; the inward, finally abandoned, because, cultivated on such principles, it is found to yield no result—sufficiently indicates the intellectual bias of our time, its all-pervading disposition towards that line of inquiry. In fact, an inward persuasion has long been diffusing itself, and now and then even comes to utterance, That, except the external, there are no true sciences; that to the inward world (if there be any) our only conceivable road is through the Outward; that, in short, what cannot be investigated and understood mechanically, cannot be investigated and understood at all. We advert the more particularly to these intellectual propensities, as to prominent symptoms of our age, because Opinion is at all times doubly related to Action, first as cause, then as effect; and the speculative tendency of any age will therefore give us, on the whole, the best indications of its practical tendency.

Nowhere, for example, is the deep, almost exclusive faith we have in Mechanism more visible than in the Politics of this time. Civil government does by its nature include much that is mechanical, and must be treated accordingly. We term it indeed, in ordinary language, the Machine of Society, and talk of it as the grand working wheel from which all private machines must derive, or to which they must adapt, their movements. Considered merely as a metaphor, all this is well enough; but here, as in so many other cases, the "foam hardens itself into a shell," and the shadow we have wantonly evoked stands terrible before us and will not depart at our bidding. Government includes much also that is not mechanical, and cannot be treated mechanically; of which latter truth, as appears to us, the political speculations and exertions of our time are taking less and less cognisance.

Nay, in the very outset, we might note the mighty interest taken in *mere political arrangements*, as itself the sign of a mechanical age. The whole discontent of Europe takes this direction. The deep, strong cry of all civilised nations, — a cry which, every one now sees, must and will be answered, is: Give us a reform of Government! A good structure of legislation, a proper check upon the executive, a wise arrangement of the judiciary, is all that is wanting for human happiness. The Philosopher of this age is not a Socrates, a Plato, a Hooker, or Taylor, who inculcates on men the necessity and infinite worth of moral goodness, the great truth that our happiness depends on the mind which is within us, and not on the circumstances which are without us; but a [Adam] Smith, a [John] De Lolme, a [Jeremy] Bentham, who chiefly inculcates the reverse of this, — that our happiness depends entirely on external circumstances; nay, that the strength and dignity of the mind within us is itself the creature and consequence of these. Were the laws, the government, in good order, all were well with us; the rest would care for itself! Dissentients from this opinion, expressed or implied, are now rarely to be met with; widely and angrily as men differ in its application, the principle is admitted by all.

«««—»»»

To us who live in the midst of all this, and see continually the faith, hope and, practice of every one founded on Mechanism of one kind or other, it is apt to seem quite natural, and as if it could never have been otherwise. Nevertheless, if we recollect or reflect a little, we shall find both that it has been, and might again be otherwise. The domain of Mechanism—meaning thereby political, ecclesiastical or other outward establishments—was once considered as embracing but a limited portion of man's interests, and by no means the highest portion.

To speak a little pedantically, there is a science of *Dynamics* in man's fortunes and nature, as well as of *Mechanics*. There is a science which treats of, and practically addresses, the primary, unmodified forces and energies of man, the mysterious springs of Love, and Fear, and Wonder, of Enthusiasm, Poetry, Religion, all which have a truly vital and *infinite* character; as well as a science which practically addresses the finite, modified developments of these, when they take the shape of immediate 'motives,' as hope of reward, or as fear of punishment.

Now it is certain, that in former times the wise men, the enlightened lovers of their kind, who appeared generally as Moralists, Poets or Priests, did, without neglecting the Mechanical province, deal chiefly with the Dynamical; applying themselves chiefly to regulate, increase and purify the inward primary powers of man; and fancying that herein lay the main difficulty, and the best service they could undertake. But a wide difference is manifest in our age. For the wise men who now appear as Political Philosophers, deal exclusively with the Mechanical province; and occupying themselves in counting-up and estimating men's motives, strive by curious checking and balancing, and other adjustments of Profit and Loss, to guide them to their true advantage: while, unfortunately, those same 'motives' are so innumerable, and so variable in every individual, that no really useful conclusion can ever be drawn from their enumeration. But though Mechanism, wisely contrived, has done much for man in a social and moral point of view, we cannot be persuaded that it has ever been the chief source of his worth or happiness.

Consider the great elements of human enjoyment, the attainments and possessions that exalt man's life to its present height, and see what part of these he owes to institutions, to Mechanism of any kind; and what to the instinctive, unbounded force, which Nature herself lent him, and still continues to him. Shall we say, for example, that Science and Art are indebted principally to the founders of Schools and Universities? Did not Science originate rather, and gain advancement, in the obscure closets of the Roger Bacons, Keplers, Newtons; in the workshops of the Fausts and the Watts; wherever, and in what guise soever Nature, from the first times downwards, had sent a gifted spirit upon the earth? Again, were Homer and Shakespeare members of any beneficed guild, or made Poets by means of it? Were Painting and Sculpture created by forethought, brought into the world by institutions for that end? No; Science and Art have, from first to last, been the free gift of Nature; an unsolicited, unexpected gift; often even a fatal one. These things rose up, as it were, by sponta- neous growth, in the free soil and sunshine of Nature. They were not planted or grafted, nor even greatly multiplied or improved by the culture or manuring of insti- tutions. Generally speaking, they have derived only partial help from these; often enough have suffered damage. They made constitutions for themselves. They origi- nated in the Dynamical nature of man, not in his Mechanical nature.

These and the like facts are so familiar, the truths which they preach so obvious, and have in all past times been so universally believed and acted on, that we should almost feel ashamed for repeating them; were it not that, on every hand, the memory of them seems to have passed away, or at best died into a faint tradition, of no value as a practical principle. To judge by the loud clamour of our Constitution-builders, Statists, Economists, directors, creators, reformers of Public Societies; in a word, all manner of Mechanists, from the Cartwright up to the Code-maker; and by the nearly total silence of all Preachers and Teachers who should give a voice to Poetry, Religion and Morality, we might fancy either that man's Dynamical nature was, to all spiritual intents, extinct, or else so perfected that nothing more was to be made of it by the old means; and henceforth only in his Mechanical contrivances did any hope exist for him.

To define the limits of these two departments of man's activity, which work into one another, and by means of one another, so intricately and inseparably, were by its nature an impossible attempt. Their relative importance, even to the wisest mind, will vary in different times, according to the special wants and dispositions of those times. Meanwhile, it seems clear enough that only in the right coordination of the two, and the vigorous forwarding of *both*, does our true line of action lie. Undue cultivation of the inward or Dynamical province leads to idle, visionary, impracticable courses, and, especially in rude eras, to Superstition and Fanaticism, with their long train of baleful and well-known evils. Undue cultivation of the outward, again, though less immedi- ately prejudicial, and even for the time productive of many palpable benefits, must, in the long-run, by destroying Moral Force, which is the parent of all other Force, prove not less certainly, and perhaps still more hopelessly, pernicious. This, we take it, is the grand characteristic of our age. By our skill in Mechanism, it has come to pass, that in the management of external things we excel all other ages; while in whatever respects the pure moral nature, in true dignity of soul and character, we are perhaps inferior to most civilised ages.

«《——》»

In fact, if we look deeper, we shall find that this faith in Mechanism has now struck its roots down into man's most intimate, primary sources of conviction; and is thence sending up, over his whole life and activity, innumerable stems, — fruit-bearing and poison-bearing. The truth is, men have lost their belief in the Invisible, and believe, and hope, and work only in the Visible; or, to speak it in other words: This is not a Religious age. Only the material, the immediately practical, not the divine and spiritual, is important to us. The infinite, absolute character of Virtue has passed into a finite, conditional one; it is no longer a worship of the Beautiful and Good; but a calculation of the Profitable. Worship, indeed, in any sense, is not recognised among us, or is mechanically explained into Fear of pain, or Hope of pleasure. Our true Deity is Mechanism. It has subdued external Nature for us, and we think it will do all other things. We are Giants in physical power: in a deeper than metaphorical sense, we are Titans, that strive, by heaping mountain on mountain, to conquer Heaven also.

The strong Mechanical character, so visible in the spiritual pursuits and methods of this age, may be traced much farther into the condition and prevailing disposition of our spiritual nature itself. Consider, for example, the general fashion of Intellect in this era. Intellect, the power man has of knowing and believing, is now nearly synonymous with Logic, or the mere power of arranging and communicating. Its implement is not Meditation, but Argument. 'Cause and effect' is almost the only category under which we look at, and work with, all Nature. Our first question with regard to any object is not, What is it? but, How is it? We are no longer instinctively driven to apprehend, and lay to heart, what is Good and Lovely, but rather to inquire, as onlookers, how it is produced, whence it comes, whither it goes. Our favourite Philosophers have no love and no hatred; they stand among us not to do, nor to create anything, but as a sort of Logic mills, to grind out the true causes and effects of all that is done and created. To the eye of a Smith, a Hume or a [Benjamin] Constant, all is well that works quietly. An Order of Ignatius Loyola, a Presbyterianism of John Knox, a Wickliffe or a Henry the Eighth, are simply so many mechanical phenomena, caused or causing.

An intellectual dapperling of these times boasts chiefly of his irresistible perspicacity, his "dwelling in the daylight of truth," and so forth; which, on examination, turns out to be a dwelling in the rush-light of "closet logic," and a deep unconsciousness that there is any other light to dwell in or any other objects to survey with it. Wonder, indeed, is, on all hands, dying out: it is the sign of uncultivation to wonder. Speak to any small man of a high, majestic Reformation, of a high majestic Luther; and forthwith he sets about "accounting" for it; how the "circumstances of the time" called for such a character, and found him, we suppose, standing girt and road-ready, to do its errand; how the "circumstances of the time" created, fashioned, floated him quietly along into the result; how, in short, this small man, had he been there, could have performed the like himself! For it is the "force of circumstances" that does everything; the force of one man can do nothing. Now all this is grounded on little more than a metaphor. We figure Society as a "Machine," and that mind is opposed to mind, as body is to body; whereby two, or at most ten, little minds must be stronger than one great mind. Notable absurdity! For the plain truth, very plain, we think is, that minds are opposed to minds in quite a different way; and *one* man that has a higher Wisdom, a hitherto unknown spiritual Truth in him, is stronger, not than ten men that have it not, or than ten thousand, but than *all* men that have it not; and stands among them with a quite ethereal, angelic power, as with a sword out of Heaven's own armory, sky-tempered, which no buckler, and no tower of brass, will finally withstand.

To what extent theological Unbelief, we mean intellectual dissent from the Church, in its view of Holy Writ, prevails at this day, would be a highly important, were it not, under any circumstances, an almost impossible inquiry. But the Unbelief, which is of a still more fundamental character, every man may see prevailing, with scarcely any but the faintest contradiction, all around him; even in the Pulpit itself. Religion in most countries, more or less in every country, is no longer what it was, and should be, — a thousand-voiced psalm from the heart of Man to his invisible Father, the fountain of all Goodness, Beauty, Truth, and revealed in every revelation of these; but for the most part, a wise prudential feeling grounded on mere calculation; a matter, as all others now are, of Expediency and Utility; whereby some smaller quantum of earthly enjoyment may be exchanged for a far larger quantum of celestial enjoyment. Thus Religion too is Profit, a working for wages; not Reverence, but vulgar Hope or Fear. Many, we know, very many we hope, are still religious in a far different sense; were it not so, our case were too desperate: but to witness that such is the temper of the times, we take any calm observant man, who agrees or disagrees in our feeling on the matter, and ask him whether our view of it is not in general well-founded.

Literature, too, if we consider it, gives similar testimony. At no former era has Literature, the printed communication of Thought, been of such importance, as it is now. We often hear that the Church is in danger; and truly so it is, — in a danger it seems not to know of: for, with its tithes in the most perfect safety, its functions are becoming more and more superseded. The true Church of England, at this moment, lies in the Editors of its Newspapers. These preach to the people daily, weekly; admonishing kings themselves; advising peace or war, with an authority which only the first Reformers, and a long-past class of Popes, were possessed of; inflicting moral censure; imparting moral encouragement, consolation, edification; in all ways dili-gently "administering the Discipline of the Church." It may be said too, that in private disposition the new Preachers somewhat resemble the Mendicant Friars of old times: outwardly full of holy zeal; inwardly not without stratagem, and hunger for terrestrial things. But omitting this class, and the boundless host of watery personages who pipe, as they are able, on so many scrannel straws, let us look at the higher regions of Literature, where, if any where, the pure melodies of Poesy and Wisdom should be heard. Of natural talent there is no deficiency: one or two richly-endowed individuals even give us a superiority in this respect. But what is the song they sing? Is it a tone of the Memnon Statue [of ancient Egypt], breathing music as the light first touches it? A "liquid wisdom," disclosing to our sense the deep, infinite harmonies of Nature and man's soul? Alas, no! It is not a matin or vesper hymn to the Spirit of Beauty, but a fierce clashing of cymbals, and shouting of multitudes, as children pass through the fire to Moloch! Poetry itself has no eye for the Invisible.

Beauty is no longer the god it worships, but some brute image of Strength; which we may call an idol, for true Strength is one and the same with Beauty, and its worship also is a hymn. The meek, silent Light can mould, create and purify all Nature; but the loud Whirlwind, the sign and product of Disunion, of Weakness, passes on, and is forgotten. How widely this veneration for the physically Strongest has spread itself through Literature, any one may judge who reads either criticism or poem. We praise a work, not as "true," but as "strong"; our highest praise is that it has "affected" us, has "terrified" us. All this, it has been well observed, is the "maximum of the Barbarous," the symptom, not of vigorous refinement, but of luxurious corruption. It speaks much, too, for men's indestructible love of truth, that nothing of this kind will abide with them; that even the talent of a Byron cannot permanently seduce us into idol worship; that he too, with all his wild siren charming, already begins to be disregarded and forgotten.

Again, with respect to our Moral condition: here also he who runs may read that the same physical, mechanical influences are everywhere busy. For the "superior morality," of which we hear so much, we too would desire to be thankful: at the same time, it were but blindness to deny that this "superior morality" is properly rather an "inferior criminality," produced not by greater love of Virtue, but by greater perfection of Police; and of that far subtler and stronger Police, called Public Opinion. This last watches over us with its Argus eyes more keenly than ever; but the "inward eye" seems heavy with sleep. Of any belief in invisible, divine things, we find as few traces in our Morality as elsewhere. It is by tangible, material considerations that we are guided, not by inward and spiritual. Self-denial, the parent of all virtue, in any true sense of that word, has perhaps seldom been rarer: so rare is it, that the most, even in their abstract speculations, regard its existence as a chimera. Virtue is Pleasure, is Profit; no celestial, but an earthly thing. Virtuous men, Philanthropists, Martyrs are happy accidents; their "taste" lies the right way! In all senses, we worship and follow after Power; which may be called a physical pursuit.

No man now loves Truth, as Truth must be loved, with an infinite love; but only with a finite love, as it were *par amours*. Nay, properly speaking, he does not *believe* and know it, but only "*thinks*" it, and that "there is every probability!" He preaches it aloud, and rushes courageously forth with it, — if there is a multitude huzzaing at his back; yet ever keeps looking over his shoulder, and the instant the huzzaing languishes, he too stops short. In fact, what morality we have takes the shape of Ambition, or "Honour"; beyond money and money's worth, our only rational blessedness is Popularity. It were but a fool's trick to die for conscience. Only for "character," by duel, or in case of extremity, by suicide, is the wise man bound to die. By arguing on the "force of circumstances," we have argued away all force from ourselves; and stand leashed together, uniform in dress and movement, like the rowers of some boundless galley. This and that may be right and true; *but* we must not do it. Wonderful "Force of Public Opinion"! We must act and walk in all points as it prescribes; follow the traffic it bids us, realise the sum of money, the degree of "influence" it expects of us, or we shall be lightly esteemed; certain mouthfuls of articulate wind will be blown at us, and this what mortal courage can front? Thus, while civil liberty is more and more secured to us, our moral liberty is all but lost. Practically considered, our creed is Fatalism; and, free in hand and foot, we are shackled in heart and soul with far straiter than feudal chains. Truly may we say, with the Philosopher [Novalis, aka Georg von Hardenberg], "the deep meaning of the Laws of Mechanism lies heavy on us"; and in the closet, in the marketplace, in the temple, by the social hearth, encumbers the whole movements of our mind, and over our noblest faculties is spreading a nightmare sleep.

«««—»»»

These dark features, we are aware, belong more or less to other ages, as well as to ours. This faith in Mechanism, in the all-importance of physical things, is in every age the common refuge of Weakness and blind Discontent; of all who believe, as many will ever do, that man's true good lies without him, not within. We are aware also, that, as applied to ourselves in all their aggravation, they form but half a picture; that in the whole picture there are bright lights as well as gloomy shadows. If we here dwell chiefly on the latter, let us not be blamed: it is in general more profitable to reckon up our defects than to boast of our attainments.

Neither, with all these evils more or less clearly before us, have we at any time despaired of the fortunes of society. Despair, or even despondency, in that respect, appears to us, in all cases, a groundless feeling. We have a faith in the imperishable dignity of man; in the high vocation to which, throughout this his earthly history, he has been appointed. However it may be with individual nations, whatever melancholic speculators may assert, it seems a well-ascertained fact, that in all times, reckoning even from those of the [ancient Greek tribes] Heraclides and Pelasgi, the happiness and greatness of mankind at large have been continually progressive. Doubtless this age also is advancing. Its very unrest, its ceaseless activity, its discontent contains matter of promise. Knowledge, education are opening the eyes of the humblest; are increasing the number of thinking minds without limit. This is as it should be; for not in turning back, not in resisting, but only in resolutely struggling forward, does our life consist.

Nay, after all, our spiritual maladies are but of Opinion; we are but fettered by chains of our own forging, and which ourselves also can rend asunder. This deep, paralysed subjection to physical objects comes not from Nature, but from our own unwise mode of viewing Nature. Neither can we understand that man wants, at this hour, any faculty of heart, soul or body, that ever belonged to him. 'He, who has been born, has been a First Man'; has had lying before his young eyes, and as yet unhardened into scientific shapes, a world as plastic, infinite, divine, as lay before the eyes of Adam himself.

If Mechanism, like some glass bell, encircles and imprisons us; if the soul looks forth on a fair heavenly country which it cannot reach, and pines, and in its scanty atmosphere is ready to perish, — yet the bell is but of glass, 'one bold stroke to break the bell in pieces, and thou art delivered!' Not the invisible world is wanting, for it dwells in man's soul, and this last is still here. Are the solemn temples, in which the Divinity was once visibly revealed among us, crumbling away? We can repair them, we can rebuild them. The wisdom, the heroic worth of our forefathers, which we have lost, we can recover. That admiration of old nobleness, which now so often shows itself as a faint *dilettantism*, will one day become a generous emulation, and man may again be all that he has been, and more than he has been.

Nor are these the mere daydreams of fancy; they are clear possibilities; nay, in this time they are even assuming the character of hopes. Indications we do see in other countries and in our own, signs infinitely cheering to us, that Mechanism is not always to be our hard taskmaster, but one day to be our pliant, all-ministering servant; that a new and brighter spiritual era is slowly evolving itself for all man. But on these things our present course forbids us to enter.

Meanwhile, that great outward changes are in progress can be doubtful to no one. The time is sick and out of joint. Many things have reached their height; and it is a wise adage that tells us, "the darkest hour is nearest the dawn." Wherever we can gather indication of the public thought, whether from printed books' as in France or Germany, or from Carbonari rebellions and other political tumults, as in Spain, Portugal, Italy, and Greece, the voice it utters is the same. The thinking minds of all nations call for change.

There is a deep-lying struggle in the whole fabric of society; a boundless grinding collision of the New with the Old. The French Revolution, as is now visible enough, was not the parent of this mighty movement, but its offspring. Those two hostile influences, which always exist in human things, and on the constant intercommunion of which depends their health and safety, had lain in separate masses, accumulating through generations, and France was the scene of their fiercest explosion; but the final issue was not unfolded in that country: nay, it is not yet anywhere unfolded. Political freedom is hitherto the object of these efforts; but they will not and cannot stop there. It is towards a higher freedom than mere freedom from oppression by his fellow-mortal, that man dimly aims. Of this higher, heavenly freedom, which is "man's reasonable service," all his noble institutions, his faithful endeavours and loftiest attainments, are but the body, and more and more approximated emblem.

On the whole, as this wondrous planet, Earth, is journeying with its fellows through infinite Space, so are the wondrous destinies embarked on it journeying through infinite Time, under a higher guidance than ours. For the present, as our astronomy informs us, its path lies towards *Hercules*, the constellation of *Physical Power*: but that is not our most pressing concern. Go where it will, the deep HEAVEN will be around it. Therein let us have hope and sure faith.

To reform a world, to reform a nation, no wise man will undertake; and all but foolish men know, that the only solid, though a far slower reformation, is what each begins and perfects on *himself.*

WALDEN

HENRY DAVID THOREAU

Thoreau (1817-1862) was an American writer, poet, philosopher, and naturalist. *Walden* chronicles his two-year stay (1845-1847) at his handmade cabin in the woods near Concord, Massachusetts. *(Editor's comments in italics.)*

Chapter 1 – "Economy"

Thoreau believed that, for the most part, 'civilized' society was demeaning and counterproductive to human wellbeing. In it, most men work as common laborers, either on the farms or in the employment of another who seeks to profit from their work. This life of hard labor is a waste of a man's life; it vastly diminishes his ability to enjoy his own freedom:

Most men, even in this comparatively free country, through mere ignorance and mistake, are so occupied with the factitious cares and superfluously coarse labors of life that its finer fruits cannot be plucked by them. Their fingers, from excessive toil, are too clumsy and tremble too much for that. Actually, the laboring man has not leisure for a true integrity day by day; he cannot afford to sustain the manliest relations to men; his labor would be depreciated in the market. He has no time to be anything but a machine.

The 'benefits' of modern society are in fact barriers to the good life:

Most of the luxuries, and many of the so-called comforts of life, are not only not indispensable, but positive hindrances to the elevation of mankind. With respect to luxuries and comforts, the wisest have ever lived a more simple and meager life than the poor. The ancient philosophers, Chinese, Hindu, Persian, and Greek, were a class than which none has been poorer in outward riches, none so rich in inward. We know not much about them. It is remarkable that we know so much of them as we do.

The same is true of the more modern reformers and benefactors of their race. None can be an impartial or wise observer of human life but from the vantage ground of what we should call voluntary poverty. Of a life of luxury the fruit is luxury, whether in agriculture, or commerce, or literature, or art. There are nowadays professors of philosophy, but not philosophers. Yet it is admirable to profess, because it was once admirable to live. To be a philosopher is not merely to have subtle thoughts, nor even to found a school, but so to love wisdom as to live according to its dictates, a life of simplicity, independence, magnanimity, and trust. It is to solve some of the problems of life, not only theoretically, but practically.

The success of great scholars and thinkers is [merely] a courtier-like success, not kingly, not manly. They make shift to live merely by conformity, practically as their fathers did, and are in no sense the progenitors of a noble race of men. But why do men degenerate ever? What makes families run out? What is the nature of the luxury which enervates and destroys nations? Are we sure that there is none of it in our own lives? The philosopher is in advance of his age even in the outward form of his life. He is not fed, sheltered, clothed, warmed, like his contemporaries. How can a man be a philosopher and not maintain his vital heat by better methods than other men?

One of the most important industries of the mid-19th century was textiles, and the manufacture of clothing. It was also notorious—especially in England—for terrible working conditions, long hours, and abuse of women and children. The factory system was in fact one of the first clear demonstrations of the dangerous nature of industrial technology.

I cannot believe that our factory system is the best mode by which men may get clothing. The condition of the operatives [in the US] is becoming every day more like that of the English; and it cannot be wondered at, since, as far as I have heard or observed, the principal object is, not that mankind may be well and honestly clad, but, unquestionably, that corporations may be enriched. In the long run men hit only what they aim at. Therefore, though they should fail immediately, they had better aim at something high.

A centerpiece of the American Dream was, and is, to own your own home. Modern society supposedly offers the best housing ever obtained by the common man—and yet he typically must go into debt for many years to own such a house. Years of debt equate to years of servitude. The pre-technological man, on the other hand, has a perfectly fine abode with no enslaving debt:

In the savage state [such as with the American Indians] every family owns a shelter as good as the best, and sufficient for its coarser and simpler wants; but I think that I speak within bounds when I say that, though the birds of the air have their nests, and the foxes their holes, and the savages their wigwams, in modern civilized society not more than one half the families own a shelter. In the large towns and cities, where civilization especially prevails, the number of those who own a shelter is a very small fraction of the whole. ...

But, answers one, by merely paying this tax, the poor civilized man secures an abode which is a *palace* compared with the savage's. An annual rent of from twenty-five to a hundred dollars (these are the country rates) entitles him to the benefit of the improvements of centuries, spacious apartments, clean paint and paper, Rumford fireplace, back plastering, Venetian blinds, copper pump, spring lock, a commodious cellar, and many other things. But how happens it that he who is said to enjoy these things is so commonly a *poor* civilized man, while the savage, who has them not, is *rich* as a savage?

If it is asserted that civilization is a real advance in the condition of man—and I think that it is, though only the wise improve their advantages—it must be shown that it has produced better dwellings without making them more costly; and the cost of a thing is the amount of what I will call 'life' which is required to be exchanged for it, immediately or in the long run. An average house in this neighborhood costs perhaps eight hundred dollars, and to lay up this sum will take from ten to fifteen years of the laborer's life, even if he is not encumbered with a family...so that he must have spent more than half his life commonly before his wigwam will be earned.... Would the savage have been wise to exchange his wigwam for a palace on these terms?

Thus there is an "important distinction between the civilized man and the savage": primitive society respects the individual, whereas modern society
[makes] the life of a civilized people an *institution*, in which the life of the individual is to a great extent absorbed, in order to preserve and perfect that of the race. But I wish to show at what a sacrifice this advantage is at present obtained, and to suggest that we may possibly so live as to secure all the advantage without suffering any of the disadvantage.

The very simplicity and nakedness of man's life in the primitive ages imply this advantage, at least, that they left him still but a sojourner in nature. When he was refreshed with food and sleep, he contemplated his journey again. He dwelt, as it were, in a tent in this world, and was either threading the valleys, or crossing the plains, or climbing the mountain-tops. But lo! men have become the tools of their tools. The man who [once] independently plucked the fruits when he was hungry is [now] become a farmer; and he who stood under a tree for shelter, a housekeeper. We now no longer camp as for a night, but have settled down on earth and forgotten heaven. ... We have built for this world a family mansion, and for the next a family *tomb*. The best works of art are the expression of man's struggle to free himself from this condition, but the effect of our art is merely to make this low state comfortable and that higher state to be forgotten.

We like to think of higher education and college as one of the enhancements of modern, technological society, but it is far from clear that humanity is any better off for such things:
As with our colleges, so with a hundred "modern improvements"; there is an illusion about them; there is not always a positive advance. ... Our inventions are wont to be pretty toys, which distract our attention from serious things. They are but improved means to an unimproved end, an end which it was already but too easy to arrive at; as railroads lead to Boston or New York. We are in great haste to construct a magnetic telegraph from Maine to Texas; but Maine and Texas, it may be, have nothing important to communicate. ... As if the main object were to talk fast and not to talk sensibly. We are eager to tunnel under the Atlantic and bring the Old World some weeks nearer to the New; but perchance the first news that will leak through into the broad, flapping American ear will be that the Princess Adelaide has the whooping cough. After all, the man whose horse trots a mile in a minute does not carry the most important messages...

One says to me, "I wonder that you do not [save] money; you love to travel; you might take the [railroad] cars and go to Fitchburg today and see the country." But I am wiser than that. I have learned that the swiftest traveller is he that goes afoot. I say to my friend, Suppose we try [to see] who will get there first. The distance is thirty miles; the [train] fare ninety cents. That is almost a day's wages. ... Well, I start now on foot, and get there before night; I have travelled at that rate by the week together. You will in the meanwhile have earned your fare, and arrive there some time tomorrow, or possibly this evening, if you are lucky enough to get a job in season. Instead of going to Fitchburg, you will be working here the greater part of the day. And so, if the railroad reached round the world, I think that I should keep ahead of you...

Chapter 2 – "Where I Lived, and What I Lived For"

I went to the woods because I wished to live deliberately, to front only the essential facts of life, and see if I could not learn what it had to teach, and not, when I came to die, discover that I had not lived. I did not wish to live what was not life, living is so dear; nor did I wish to practise resignation, unless it was quite necessary. I wanted to live deep and suck out all the marrow of life, to live so sturdily and Spartan-like as to put to rout all that was not life, to cut a broad swath and shave close, to drive life into a corner, and reduce it to its lowest terms, and, if it proved to be mean, why then to get the whole and genuine meanness of it, and publish its meanness to the world; or if it were sublime, to know it by experience, and be able to give a true account of it in my next excursion. For most men, it appears to me, are in a strange uncertainty about it, whether it is of the devil or of God, and have somewhat hastily concluded that it is the chief end of man here to "glorify God and enjoy him forever."

Still we live meanly, like ants; though the fable tells us that we were long ago changed into men; like pygmies we fight with cranes; it is error upon error, and clout upon clout, and our best virtue has for its occasion a superfluous and evitable wretchedness. Our life is frittered away by detail. An honest man has hardly need to count more than his ten fingers, or in extreme cases he may add his ten toes, and lump the rest.

Simplicity, simplicity, simplicity! I say, let your affairs be as two or three, and not a hundred or a thousand; instead of a million count half a dozen, and keep your accounts on your thumb-nail. In the midst of this chopping sea of civilized life, such are the clouds and storms and quicksands and thousand-and-one items to be allowed for, that a man has to live, if he would not founder and go to the bottom and not make his port at all, by dead reckoning, and he must be a great calculator indeed who succeeds.

Simplify, simplify. Instead of three meals a day, if it be necessary eat but one; instead of a hundred dishes, five; and reduce other things in proportion. ... Our nation itself, with all its so-called internal improvements, which, by the way are all external and superficial, is just such an unwieldy and overgrown establishment, cluttered with furniture and tripped up by its own traps, ruined by luxury and heedless expense, by want of calculation and a worthy aim, as the million households in the land; and the only cure for it, as for them, is in a rigid economy, a stern and more than Spartan simplicity of life and elevation of purpose.

It lives too fast. Men think that it is essential that the Nation have commerce, and export ice, and talk through a telegraph, and ride thirty miles an hour, without a doubt, whether they do or not; but whether we should live like baboons or like men, is a little uncertain.

If we do not get out sleepers [i.e. the large wood timbers that support railroad tracks], and forge rails, and devote days and nights to the work, but go to tinkering upon our lives to improve them, who will build railroads? And if railroads are not built, how shall we get to heaven in season? But if we stay at home and mind our business, who will want railroads? We do not ride on the railroad; it rides upon us.

Did you ever think what those sleepers are that underlie the railroad? Each one is a man, an Irishman, or a Yankee man. The rails are laid on them, and they are covered with sand, and the cars run smoothly over them. They are sound sleepers, I assure you. And every few years a new lot is laid down and run over; so that, if some have the pleasure of riding on a rail, others have the misfortune to be ridden upon. And when they run over a man that is walking in his sleep, a supernumerary sleeper in the wrong position, and wake him up, they suddenly stop the cars, and make a hue and cry about it, as if this were an exception. I am glad to know that it takes a gang of men for every five miles to keep the sleepers down and level in their beds as it is, for this is a sign that they may sometime get up again.

WALKING (1851-1862)

My desire for knowledge is intermittent; but my desire to bathe my head in atmospheres unknown to my feet is perennial and constant. The highest that we can attain to is not Knowledge, but Sympathy with Intelligence. I do not know that this higher knowledge amounts to anything more definite than a novel and grand surprise on a sudden revelation of the *insufficiency* of all that we called Knowledge before—a discovery that there are more things in heaven and earth than are dreamed of in our philosophy. It is the lighting up of the mist by the sun. Man cannot know in any higher sense than this, any more than he can look serenely and with impunity in the face of the sun...

There is something servile in the habit of seeking after a law [of nature] which we may obey. We may study the laws of matter at and for our convenience, but a successful life knows no law. It is an unfortunate discovery certainly, that of a law which binds us where we did not know before that we were bound. Live free, Child of the Mist—and with respect to knowledge we are all children of the mist. The man who takes the liberty to live is superior to all the laws both of heaven and earth...

DARWIN AMONG
THE MACHINES

SAMUEL BUTLER

Butler (1835-1902) was an English novelist, philosopher, and social critic. Two of his anti-technology essays are reprinted here, the first from 1863 and the second from 1865. These themes ultimately became part of Butler's most famous novel, *Erewhon* (1872).

There are few things of which the present generation is more justly proud than of the wonderful improvements which are daily taking place in all sorts of mechanical appliances. And indeed it is matter for great congratulation on many grounds. It is unnecessary to mention these here, for they are sufficiently obvious; our present business lies with considerations which may somewhat tend to humble our pride and to make us think seriously of the future prospects of the human race. If we revert to the earliest primordial types of mechanical life, to the lever, the wedge, the inclined plane, the screw and the pulley, or (for analogy would lead us one step further) to that one primordial type from which all the mechanical kingdom has been developed, we mean to the lever itself, and if we then examine the machinery of the *Great Eastern*, we find ourselves almost awestruck at the vast development of the mechanical world, at the gigantic strides with which it has advanced in comparison with the slow progress of the animal and vegetable kingdom. We shall find it impossible to refrain from asking ourselves what the end of this mighty movement is to be. In what direction is it tending? What will be its upshot? To give a few imperfect hints towards a solution of these questions is the object of the present letter.

We have used the words "mechanical life," "the mechanical kingdom," "the mechanical world" and so forth, and we have done so advisedly, for as the vegetable kingdom was slowly developed from the mineral, and as in like manner the animal supervened upon the vegetable, so now in these last few ages an entirely new kingdom has sprung up, of which we as yet have only seen what will one day be considered the antediluvian prototypes of the race.

We regret deeply that our knowledge both of natural history and of machinery is too small to enable us to undertake the gigantic task of classifying machines into the genera and sub-genera, species, varieties and sub-varieties, and so forth, of tracing the connecting links between machines of widely different characters, of pointing out how subservience to the use of man has played that part among machines which natural selection has performed in the animal and vegetable kingdoms, of pointing out rudimentary organs which exist in some few machines, feebly developed and perfectly useless, yet serving to mark descent from some ancestral type which has either perished or been modified into some new phase of mechanical existence. We can only point out this field for investigation; it must be followed by others whose education and talents have been of a much higher order than any which we can lay claim to.

Some few hints we have determined to venture upon, though we do so with the profoundest diffidence. Firstly, we would remark that as some of the lowest of the vertebrata attained a far greater size than has descended to their more highly organised living representatives, so a diminution in the size of machines has often attended their development and progress. Take the watch for instance. Examine the beautiful structure of the little animal, watch the intelligent play of the minute members which compose it; yet this little creature is but a development of the cumbrous clocks of the thirteenth century— it is no deterioration from them. The day may come when clocks, which certainly at the present day are not diminishing in bulk, may be entirely superseded by the universal use of watches, in which case clocks will become extinct like the earlier saurians, while the watch (whose tendency has for some years been rather to decrease in size than the contrary) will remain the only existing type of an extinct race.

The views of machinery which we are thus feebly indicating will suggest the solution of one of the greatest and most mysterious questions of the day. We refer to the question: What sort of creature man's next successor in the supremacy of the earth is likely to be. We have often heard this debated; but it appears to us that we are ourselves creating our own successors; we are daily adding to the beauty and delicacy of their physical organisation; we are daily giving them greater power and supplying by all sorts of ingenious contrivances that self-regulating, self-acting power which will be to them what intellect has been to the human race. In the course of ages we shall find ourselves the inferior race. Inferior in power, inferior in that moral quality of self-control, we shall look up to them as the acme of all that the best and wisest man can ever dare to aim at. No evil passions, no jealousy, no avarice, no impure desires will disturb the serene might of those glorious creatures. Sin, shame, and sorrow will have no place among them. Their minds will be in a state of perpetual calm, the contentment of a spirit that knows no wants, is disturbed by no regrets. Ambition will never torture them. Ingratitude will never cause them the uneasiness of a moment. The guilty conscience, the hope deferred, the pains of exile, the insolence of office, and the spurns that patient merit of the unworthy takes—these will be entirely unknown to them. If they want "feeding" (by the use of which very word we betray our recognition of them as living organism) they will be attended by patient slaves whose business and interest it will be to see that they shall want for nothing. If they are out of order they will be promptly attended to by physicians who are thoroughly acquainted with their constitutions; if they die, for even these glorious animals will not be exempt from that necessary and universal consummation, they will immediately enter into a new phase of existence, for what machine dies entirely in every part at one and the same instant?

We take it that when the state of things shall have arrived which we have been above attempting to describe, man will have become to the machine what the horse and the dog are to man. He will continue to exist, nay even to improve, and will be probably better off in his state of domestication under the beneficent rule of the machines than he is in his present wild state. We treat our horses, dogs, cattle, and sheep, on the whole, with great kindness; we give them whatever experience teaches us to be best for them, and there can be no doubt that our use of meat has added to the happiness of the lower animals far more than it has detracted from it; in like manner it is reasonable to suppose that the machines will treat us kindly, for their existence is as dependent upon ours as ours is upon the lower animals. They cannot kill us and eat us as we do sheep; they will not only require our services in the parturition of their young (which branch of their economy will remain always in our hands), but also in feeding them, in setting them right when they are sick, and burying their dead or working up their corpses into new machines.

It is obvious that if all the animals in Great Britain save man alone were to die, and if at the same time all intercourse with foreign countries were by some sudden catastrophe to be rendered perfectly impossible, it is obvious that under such circumstances the loss of human life would be something fearful to contemplate—in like manner were mankind to cease, the machines would be as badly off or even worse. The fact is that our interests are inseparable from theirs, and theirs from ours. Each race is dependent upon the other for innumerable benefits, and, until the reproductive organs of the machines have been developed in a manner which we are hardly yet able to conceive, they are entirely dependent upon man for even the continuance of their species. It is true that these organs may be ultimately developed, inasmuch as man's interest lies in that direction; there is nothing which our infatuated race would desire more than to see a fertile union between two steam engines; it is true that machinery is even at this present time employed in begetting machinery, in becoming the parent of machines often after its own kind, but the days of flirtation, courtship, and matrimony appear to be very remote, and indeed can hardly be realised by our feeble and imperfect imagination.

Day by day, however, the machines are gaining ground upon us; day by day we are becoming more subservient to them; more men are daily bound down as slaves to tend them, more men are daily devoting the energies of their whole lives to the development of mechanical life. The upshot is simply a question of time, but that the time will come when the machines will hold the real supremacy over the world and its inhabitants is what no person of a truly philosophic mind can for a moment question.

Our opinion is that war to the death should be instantly proclaimed against them. Every machine of every sort should be destroyed by the well-wisher of his species. Let there be no exceptions made, no quarter shown; let us at once go back to the primeval condition of the race. If it be urged that this is impossible under the present condition of human affairs, this at once proves that the mischief is already done, that our servitude has commenced in good earnest, that we have raised a race of beings whom it is beyond our power to destroy, and that we are not only enslaved but are absolutely acquiescent in our bondage.

For the present we shall leave this subject, which we present gratis to the members of the Philosophical Society. Should they consent to avail themselves of the vast field which we have pointed out, we shall endeavour to labour in it ourselves at some future and indefinite period.

THE MECHANICAL CREATION

SAMUEL BUTLER

Those who are familiar with Mr. Darwin's theory on the origin of species, will be aware that it amounts to something of this nature: "Given life," says Darwin, "no matter how low, and those modes of action which we see around us show how all exiling and extinct species may have come about." Of the origin of life he is as much ignorant (as far as we can see from his book) as we are ourselves, but he shows very clearly that the struggle for existence, following upon descent with modification, results in natural selection, which accumulates divergence and ends in species. The last thing which we should wish to do, would be to throw ridicule on Darwin's magnificent work, but it has set us thinking on our own account, and though we think crudely, yet we feel that we are warranted in expressing the half-shadow, half-substance, of our own views, and in leaving the intelligent reader to draw his own inferences.

It is not at first easy to decide whether we should regard the mechanical kingdom as the commencement of a new phase of life, a phase as distinct from any that have preceded it as the animal from the vegetable kingdom, or as the process by which man's body is at present undergoing modification and improvement. Much has to be said on both sides; it will therefore be our object in this, and an article that will follow it, to point out the inferences which suggest themselves—firstly, if we assume the possibility of an eventual development of mechanical life, far superior to, and widely differing from, any yet known; and secondly, if we regard machines as the extra-corporaneous members of the machinate mammal, man.

It is clear that there was a time when the phases of life, which we now observe in plants and animals, had no existence perceptible by human organs; to all intents and purposes the world was as though they were not. True, the germs were there, but we may safely say that there were once neither plants nor animals upon the face of the earth. Were we permitted to see another world in this condition, without having any knowledge of the transitions which our own planet has undergone, we should almost indignantly deny that plants or animal life could ever come there; were we to see a world with plants only, we should deny that it could ever become peopled with insects, fishes, or anything of the human kind: had we been shown the germs of reason only as visible in the lower animals, we should scoff at the idea of our human intellect being evolved from such rude materials as this; yet those who accept the Darwinian theory will not feel inclined to deny that whatever impulse the animal and vegetable kingdoms have sprung from, has been derived from within the natural influences which operate upon this world, and not from any extra natural source. They will believe that the charges and chances with which countless millions of years have been pregnant, have brought the existing organizations to their present condition without any specially creative effort of an overruling mind.

What shall we think then? That the resources of nature are at an end, and that the animal phase is to be the last which life on this globe is to assume? Or shall we conceive that we are living in the first faint dawning of a new one? Of a life which in another ten or twenty million years shall be to us as we to the vegetable? What has been may be again, and although we grant that hardly any mistake would be more puerile than to individualize and animalize the at present existing machines—or to endow them with human sympathies, yet we can see no a priori objection to the gradual development of a mechanical life, though that life shall be so different from ours that it is only by a severe discipline that we can think of it as life at all.

We despair of condensing our remarks within the limit of an article, yet we must make an attempt. We cannot conceive of a life without the notion of an individual centre of action and consciousness; without an appearance of spontaneity, a reproductive system, and the consumption of some sort or sorts of food. A spade appears to be deficient in all these properties. A spade does not know that it is a spade, the food by which it digs a garden is eaten, not by itself, but by man, its only spontaneity is obedience to the laws of matter, and its reproductive system is provided for by man. When we look at a spade we incline to the extra-corporaneous member theory; we regard it as a process of the forearm—as one of the innumerable ways in which man has modified his own body. Yet when we look at a steam-engine we observe a startling change. It eats its own food for itself; it consumes it by inhaling the very air which we ourselves breathe; it rejects what it cannot digest as man rejects it; it has a very considerable power of self-regulation and adaptability to contingency. It cannot be said to be conscious, but the strides which it has made are made in the direction of consciousness. It is employed in the manufacture of machinery, and though steam engines are as the angels in heaven, with respect to matrimony, yet in their reproduction of machinery we seem to catch a glimpse of the extraordinary vicarious arrangement whereby it is not impossible that the reproductive system of the mechanical world will be always carried on.

It must be borne in mind that we are not thinking so much of what the steam engine is at present, as of what it may become. The steam engine of today is to the mechanical prodigies which are to come as the spade to the steam engine, as the ovum to the human being. All we can see at present is that a new set of organisms has begun to appear—we say begun, for our ideas must be enlarged, and we cannot call ten thousand or even a million years anything but a mere point in the duration of a class of life; it probably is a full million of years since the lever was invented by the gorilla. But what is that? A mere speck of time. Let us assume, however, that the interval between the stick and the steam engine is a million years—and allowing for the increasing ratio at which mechanical progress advances, who will deny that in another million years they may be more alive than man himself?

The interests of man do for the machines what natural selection and the struggle for existence has done for plants and animals. There is as sharp a contention between inventions thus established as though the machines fought among themselves and ate each other up. For if a single new machine is born, which is obviously better than those heretofore in use for the same purpose, it kills the old ones as though a miasma breathed upon them. They may die out faster or slower, and odd ones may linger long, but they are doomed, as the Aborigines of a new country on the approach of European civilization. The old ones may, in a fit of despair, urge on their attendant human beings to oppose the invader; they may break it in pieces, and perhaps secure a short respite for themselves, but their doom is certain if the improvement be bona fide. As the British rat has had to go before his Norwegian conqueror, so had the old steam engines before Watt's; the hand looms and the spinning-wheels are gone; the crossbow is clean forgotten, the stage coach, once the very pride and flower of mechanical chivalry, is now the fast dying remnant of a race. The difference between the conquerors and the conquered in men is often very small: if so, the fight is longer; so is it among the machines, yet it sometimes happens that there is a greater difference between the prowess of an old and of a new machine than there ever is between two races of men; a new machine may sometimes be as much better than an old one as a man is than a pig; this is rare, but it does sometimes happen, and no one can foresee the bounds by which the advance may leap, if a point as yet unseen be once passed.

On the other hand, the development of one species of animal to another slightly higher is a slow and precarious thing; many a good hopeful creature dies young, which had it lived might have changed the destinies of its race; or it may be crossed in love and leave no issue, or marry unadvisedly, or be overcome with its progeny by the jealousy of its fellows, before its stock is large enough to secure permanence. The progress has been very uphill work, by little and little, so that no one can lay his hand upon the change at any time and say "it is here." We are not speaking of animals under domestication, but of those who have been left to themselves. Nature, as she is called (as though man were not a part of nature!) makes a dozen failures for one slight success. She walks very slowly, and puts her improvements to cruelly severe tests. They must pass through the fire for ages before they can take their degree in permanence and be allowed to pass as new and higher species; she is an arbitrary examiner, and plucks many unfairly; but with the machines whatever it may once have been, it is not so now. The tests are fair, and they are at least as certain as man's knowledge of his own interest, and this is becoming somewhat more correct in gross material things. Doubtful cases occur sometimes, as in the gun controversy, yet even here we see that it is not for want of pains if a mistake is made.

What they call nature never took such pains to see that her contending creatures fought fair; yet these two champions for the existence of their race are put to all manner of tests to see which is really, and not merely accidentally, better; in fact, they have at last come to an actual stand-up fight, such as has hardly yet been known in all machinery. They have tried which can smash the other, and the first round not being considered conclusive, they are, or were to have another under the supervision of their respective inventors. We grant that this is an exceptional case; steam engines do not fight with individual steam engines, they are liable to the struggle of race with race, by the competitive examination of champion specimens, and with the fall of the champion the race falls also. They never fight individual with individual; but guns delight to bark and bite naturally, and it is no wonder that they should refuse to fight according to the accepted canons of mechanical warfare; their trade is war, while other machines live by peace, it is fit therefore that guns should fight it out, while the disputes between more peaceable inventions are left for the decision of the ordinary law courts of the country.

We could write volumes on the inventions which seem ply that the machines around us are only the first race new phase of life, but our space is short, and we prefer to occupy the little that remains to us with a few remarks on the probable fate of mankind, if mechanical life should prove ultimately higher than animal. At first sight it seems as though such a consummation were impossible, for since it is man's interest which has been, and is, the sole developer of the machines, how shall it be that a thing so contrary to man's interests as his own inferiority should be suffered to come about without his finding it out and checking it in time?

This question is easily answered. For firstly, man is committed hopelessly to the machines. He cannot stop. If he would continue to marry as early as he does, and bring up his children with a fair prospect of their thriving, he must go on improving the machines; these objects are far dearer to him than the remote subjugation of his race. It will not be in our time, and ten thousand years hence may be left to take care of itself. Secondly, man's interests may not be really opposed by his becoming the lower creature; the interest of the two races may continue in the same direction, notwithstanding the change in their relative situations, and man is not generally sentimental when his material interests are concerned. It is true that here and there some ardent soul may "look upon himself and curse his fate" that he was not born a steam engine, but the insensate mass will readily acquiesce in any arrangement which gives them cheaper comforts without yielding to unreasonable jealousy, merely because the mechanical destinies are more glorious than their own.

The change will be so slow and subtle that man's sense of what is due to himself will never be rudely violated at any given moment; and custom will deaden our senses to the noiseless and imperceptible aggressions of our own creations. Their desires will probably never clash with ours, nor ours with theirs, and we may probably fare as much better under domestication as those creatures have done, towards whom man appears to entertain the most implacable, yet jealously conservative, enmity. Even Jupiter never wooed in a mechanical disguise, neither are the machines likely to want man as a delicacy for the table. They will breed, and beyond a doubt, varieties and sub-varieties of the human race will be developed with a special view to the requirements of certain classes of machinery; we can see the germs of this already in the different aspects of men who attend on different classes of machinery, but they will, as far as we can see, find us always in so many respects serviceable that it would hardly better suit their turn to exterminate us than it would ours to do the like by them.

It will be obvious to anyone that in this article we have done nothing more than make a suggestion, the development of which would be far beyond our own limits. We have proceeded on the assumption that mechanical life is to be distinct from animal, but in a future article we propose to consider it from a different view, and to regard machinery as a component part of the human organism.

DAS KAPITAL

KARL MARX

Marx (1818-1883) was a prominent German philosopher, political theorist, revolutionary, and founder of modern communism. He recognized three basic ways of producing things: (1) *handicraft*—in which a single artisan or craftsman made the complete object; (2) *manufacture*—in which a group of specialists worked by hand, and cooperatively, to produce a relatively large quantity of goods; and (3) *industry*—in which power tools and power machines, owned by wealthy capitalists, were the primary means of achieving high-volume production of goods. Industry, especially in its large-scale form, was the driving force behind the development of new technology in the 19th century. The following material is excerpted from Chapter 15 of *Das Kapital* (1867).

Machinery and Large-Scale Industry

Section 1:

John Stewart Mill says in his *Principles of Political Economy*: "It is questionable if all the mechanical inventions yet made have lightened the day's toil of any human being." That is, however, by no means the aim of the application of machinery under capitalism. Like every other instrument for increasing the productivity of labor, machinery is intended to cheapen commodities and, by shortening the part of the working day in which the worker works for himself, to lengthen the other part, the part he gives to the capitalist for nothing. The machine is a means for producing surplus-value. ...

In large-scale industry, the instruments of labor are the starting point. We have first to investigate, then, how the instruments of labor are converted from tools into machines, or what the difference is between a machine and an implement used in a handicraft. ...

All fully developed machinery consists of three essentially different parts: the motor mechanism, the transmitting mechanism, and finally the tool or working machine... The machine, therefore, is a mechanism that, after being set in motion, performs with its tools the same operations as the worker formerly did with similar tools. ...

The steam engine itself...did not give rise to any industrial revolution. It was, on the contrary, the invention of machines that made a revolution in the form of steam engines necessary. ...

The machine, which is the starting point of the industrial revolution, replaces the worker, who handles a single tool, by a mechanism operating with a number of similar tools and set in motion by a single motive power, whatever the form of that power. Here we have the machine, but in its first role as a simple element in production by machinery. ...

A system of machinery, whether it is based simply on the cooperation of similar machines, as in weaving, or on a combination of different machines, as in spinning, constitutes in itself a vast automaton as soon as it is driven by a self-acting prime mover [such as a steam engine]. ...

An organized system of machines to which motion is communicated by the transmitting mechanism from an automatic center is the most developed form of production by machinery. Here we have, in place of the insolated machine, a mechanical monster whose body fills whole factories, and whose demonic power, at first hidden by the slow and measured motions of its gigantic members, finally bursts forth in the fast and feverish whirl of its countless working organs. ...

The transformation of the mode of production in one sphere of industry necessitates a similar transformation in other spheres. Thus machine-spinning made machine-weaving necessary, and both together made a mechanical and chemical revolution compulsory in bleaching, printing, and dying. So too, on the other hand, the revolution in cotton-spinning called forth the invention of the gin, for separating the seeds from the cotton fiber; it was only by means of this invention that the production of cotton became possible on the enormous scale at present required. But as well as this, the revolution in the modes of production of industry and agriculture made necessary a revolution in the general conditions of the social process of production, i.e. in the means of communication and transport. ...

Large-scale industry therefore had to take over the machine itself, its own characteristic instrument of production, and to produce machines *by means of machines*. It was not till it did this that it could create for itself an adequate technical foundation, and stand on its own feet.

Section 2:

As we have seen, the machine does not drive out the tool. Rather does the tool expand and multiply, changing from a dwarf implement of the human organism to the implement of a mechanism created by man. Capital now sets the worker to work, not with a manual tool, but with a machine which itself handles the tools. ...

Before the labor of women and children under ten years old was forbidden in mines, the capitalist considered the employment of naked women and girls, often in company in men, so far sanctioned by their moral code, that it was only after the passing of the Factory Act that they had recourse to machinery. In England women are still occasionally used instead of horses for hauling barges, because the labor required to produce horses and machines is an accurately known quantity, while that required to retain the women of the surplus population is beneath all calculation. Hence we nowhere find a more shameless squandering of human labor-power for despicable purposes than in England, the land of machinery.

Section 3:

In so far as machinery dispenses with muscular power, it becomes a means for employing workers of slight muscular strength, or whose bodily development is incomplete, but whose limbs are all the more supple. The labor of women and children was therefore the first result of the capitalist application of machinery! That mighty substitute for labor and for workers, the machine, was immediately transformed into a means for increasing the number of wage-laborers by enrolling, under the direct sway of capital, every member of the worker's family, without distinction of age or sex. ...

Machinery, by throwing every member of the family onto the labor-market, spreads the value of the man's labor-power over his whole family. It thus depreciates it. ... Thus we see that machinery, while augmenting the human material that forms capital's most characteristic field of exploitation, at the same time raises the *degree* of that exploitation. ...

Because it is capital, the automatic mechanism [of the industrial factory] is endowed, in the person of the capitalist, with *consciousness* and a *will*. As capital, therefore, it is animated by the drive to reduce to a minimum the resistance offered by man, that obstinate yet elastic natural barrier. This resistance is moreover lessened by the apparently undemanding nature of work at a machine, and the more pliant and docile character of the women and children employed by preference. ...

The capitalist application of machinery on the one hand supplies new and powerful incentives for an unbounded prolongation of the working day, and produces such a revolution in the mode of labor as well as the character of the social working organism that it is able to break all resistance to this tendency. Hence that remarkable phenomenon in the history of modern industry, that machinery sweeps away every moral and natural restriction on the length of the working day. Hence too the economic paradox that the most powerful instrument for reducing labor-time suffers a dialectical inversion and becomes the most unfailing means for turning the whole lifetime of the worker and his family into labor-time at capital's disposal for its own increase in value. ...

The reduction of the working day to twelve hours dates in England from 1832. [The day was reduced to ten hours in 1847.] ... Agitation for a working day of eight hours has now (1867) begun in Lancashire among the factory workers.

Section 4:

The lifelong specialty of handling the same tool now becomes the lifelong specialty of serving the same machine. Machinery is misused in order to transform the worker, from his very childhood, into a part of a specialized machine. In this way, not only are the expenses necessary for his reproduction considerably lessened, but at the same time his helpless dependence upon the factory as a whole, and therefore upon the capitalist, is rendered complete. ...

In handicrafts and manufacture, the worker makes use of a tool; in the factory, the machine makes use of him. ... The special skill of each individual machine-operator, who has been deprived of all significance, vanishes as an infinitesimal quantity in the face of the science, the gigantic natural forces, and the mass of social labor embodied in the system of machinery, which, together with those three forces, constitutes the power of the 'master'.

<u>Section 5</u>:

The character of independence from and estrangement toward the worker, which the capitalist mode of production gives to the conditions of labor and the product of labor, develops into a complete and total antagonism with the advent of machinery. It is therefore when machinery arrives on the scene that the worker for the first time revolts savagely against the instruments of labor. The instrument of labor strikes down the worker.

<u>Section 6</u>:

It is an undoubted fact that machinery is not, as such, responsible for 'setting free' (i.e. throwing out of work) the worker from the means of subsistence. It cheapens and increases production in the branch it seizes on, and at first leaves unaltered the quantity of the means of subsistence produced in other branches. Hence, after the introduction of machinery, society possesses as much of the necessaries of life as before, if not more, for the workers who have been displaced. ... And this is the point relied on by our economic apologists! The contradictions and antagonisms inseparable from the capitalist application of machinery do not exist, they say, because they do not arise out of machinery *as such*, but out of its *capitalist application*!

Therefore, since machinery *in itself* shortens the hours of labor, but when employed by *capital* it lengthens them; since *in itself* it lightens labor, when employed by capital it heightens its intensity; since *in itself* it is a victory of man over the forces of nature but in the hands of capital it makes man the slave of those forces; since *in itself* it increases the wealth of the producers [laborers], but in the hands of capital it makes them into paupers – the bourgeois economist simply states that the contemplation of machinery in itself demonstrates with exactitude that all these evident contradictions are a mere semblance [i.e. appearance], present in everyday reality, but not existing in themselves, and therefore having no theoretical existence either. Thus he manages to avoid racking his brains any more, and in addition implies that his opponent is guilty of the stupidity of contending, not against the capitalist application of machinery, but against *machinery itself*.

No doubt the bourgeois economist is far from denying that 'temporary inconveniences' may result from the capitalist use of machinery. But where is the medal without its reverse side! Any other utilization of machinery than the capitalist one is to him impossible. Exploitation of the worker by the machine is therefore identical for him with exploitation of the machine by the worker. Therefore whoever reveals the real situation with the capitalist employment of machinery does not want machinery to be employed at all, and is an enemy of social progress!

Section 7:

As soon as the factory system has attained a reasonable space to exist in, and reached a definite degree of maturity, ...machinery is itself produced by machinery...; as soon as the general conditions of production appropriate to large-scale industry have been established, the mode of production acquires an elasticity, a capacity for sudden extension by leads and bounds, which comes up against no barriers but those presented by the availability of raw materials... The cheapness of the articles produced by machinery, and the revolution in the means of transport and communication, provide the weapons for the conquest of foreign markets. By ruining handicraft production of finished articles in other countries, machinery forcibly converts them into fields for the production of its raw material. ... By constantly turning workers into 'supernumeraries', large-scale industry, in all countries where it has taken root, spurs on rapid increases in emigration and colonization of foreign lands, which are thereby converted into settlements for growing the raw material of the mother country...

Section 9:

As we have seen, large-scale industry sweeps away by technical means the division of labor characteristic of *manufacture*, under which each man is bound hand and foot for life to a single specialized operation. At the same time, the capitalist form of large-scale industry reproduces the same division of labor in a still more monstrous shape; in the factory proper, by converting the worker into a living appendage of the machine...

Right down to the 18th century, the different trades were called 'mysteries', into whose secrets none but those initiated by their profession and their practical experience could penetrate. Large-scale industry tore aside the veil that concealed from them their own social process of production and turned the various spontaneously divided branches of production into riddles, not only to outsiders but even to the initiated. Its principle, which is to view each process of production in and for itself, and to resolve it into its constituent elements without looking first at the ability of the human hand to perform the new processes, brought into existence the whole of the modern science of technology.

Section 10:

In the sphere of agriculture, large-scale industry has a more revolutionary effect than elsewhere, for the reason that it annihilates the bulwark of the old society, the peasant, and substitutes for him the wage-laborer. ...Capitalist production collects the population together in great centers, and causes the urban population to achieve an evergrowing preponderance. This has two results. On the one hand it concentrates the historical motive power of society; on the other hand, it disturbs the metabolic interaction between man and the earth, i.e. it prevents the return to the soil of its constituent elements consumed by man in the form of food and clothing; hence it hinders the operation of the eternal natural condition for the lasting fertility of the soil. Thus it destroys at the same time the physical health of the urban worker, and the intellectual life of the rural worker. ...The social combination of labor processes appears as an organized *suppression* of [the laborer's] individual vitality, freedom, and autonomy. ... Moreover, all progress in capitalist agriculture is a progress in the art, not only of robbing the worker, but of robbing the soil; all progress in increasing the fertility of the soil for a given time is a progress towards ruining the more long-lasting sources of the fertility. The more a country proceeds from large-scale industry as the background of its development, as in the case of the United States, the more rapid is this process of destruction. Capitalist production, therefore, only develops the techniques and the degree of combination of the social process of production by simultaneously undermining the original sources of all wealth—the soil and the worker.

HUMAN, ALL TOO HUMAN

FRIEDRICH NIETZSCHE

Nietzsche (1844-1900) was one of the greatest and most flamboyant of all German philosophers. A driving motive for him was his disgust with the modern industrial lifestyle that had taken over Europe in the late 1800s. Nietzsche was appalled at the lowly, dehumanized existence of the mass of people, and at their subservience to the demands of industrial production. He wrote primarily against social values in general, but did make a few pointed comments on technology per se. The following short excerpts are taken from his 1878 book.

218

The machine as teacher. — The machine of itself teaches the mutual cooperation of hordes of men in operations where each man has to do only one thing: it provides the model for the party apparatus and the conduct of warfare. On the other hand, it does not teach individual autocracy: it makes of many *one* machine and of every individual an instrument to *one* end. Its most generalized effect is to teach the utility of centralization.

220

Reaction against machine-culture. — The machine, itself a product of the highest intellectual energies, sets in motion in those who serve it almost nothing but the lower, non-intellectual energies. It thereby releases a vast quantity of energy in general that would otherwise be dormant, it is true; but it provides no instigation to enhancement, to improvement, to becoming an artist. It makes men *active* and *uniform*—but in the long run this engenders a counter-effect, a despairing boredom of soul, which teaches them to long for idleness in all its varieties.

278

Premises of the machine age. — The printing press, the machine, the railway, the telegraph are premises whose thousand-year conclusion no one has yet dared to draw.

288

To what extent the machine debases us. — The machine is impersonal, it deprives the piece of work of its pride, of the individual *goodness* and *faultiness* that adheres to all work not done by a machine — that is to say, of its little bit of humanity. In earlier times all purchasing from artisans was a *bestowing of a distinction on individuals*, and the things with which we surrounded ourselves were the insignia of these distinctions: household furniture and clothing thus became symbols of mutual esteem and personal solidarity, whereas we now seem to live in the midst of nothing but an anonymous and impersonal slavery. — We must not purchase the alleviation of work at too high a price.

On the Genealogy of Morals (Part III, sec. 9)
Our entire modern way of life…has the appearance of sheer hubris and godlessness… Our whole attitude toward nature, the way we violate her with the aid of machines and the heedless inventiveness of our technicians and engineers, is hubris…

PART IV:
TWENTIETH CENTURY
ASSESSMENTS AND
CRITIQUES

SCIENCE AND THE MODERN WORLD

ALFRED NORTH WHITEHEAD

Whitehead (1861-1947) was a prominent British mathematician, physicist, and philosopher. He is perhaps most famous for his articulation of *process philosophy*, a conception of reality based on temporal processes rather than fixed things or substances. His technical background made him particularly well-suited to comment on the philosophical meaning of science and technology. The following is from the final chapter of his 1925 book *Science and the Modern World*.

The general conceptions introduced by science into modern thought cannot be separated from the philosophical situation as expressed by Descartes. I mean the assumption of bodies and minds as independent individual substances, each existing in its own right apart from any necessary reference to each other. ... [Descartes] implicitly transformed...individual value, inherent in the very fact of his own reality, into a private [i.e. personal] world of passions, or modes, of independent substances.

The independence ascribed to bodily [physical] substances carried them away from the realm of values altogether. They degenerated into a 'mechanism', entirely valueless... Accordingly, the Cartesian scientific doctrine of bits of matter, bare of intrinsic value, [was established]. ... Science, as equipped by Descartes, gave stability and intellectual status to a point of view which has had *very mixed effects* upon the moral presuppositions of modern communities. Its good effects arose from its efficiency as a method for scientific researches... But in the 19th century, when society was undergoing transformation into the manufacturing system, the bad effects of those doctrines have been very fatal. [Moral worth resides now only in the person himself; thus] the western world is now suffering from the limited moral outlook of the three previous generations.

It is very arguable that the science of political economy, as studied in its first period after the death of Adam Smith, did *more harm than good*. It destroyed many economic fallacies, and taught how to think about the economic revolution then in progress. But it riveted on men a certain set of abstractions which were disastrous in their influence on modern mentality. *It dehumanized industry*. This is only one example of a general danger inherent in modern science. ...

The assumption of the bare valuelessness of mere matter led to a *lack of reverence* in the treatment of natural or artistic beauty. ... In the most advanced industrial countries, art was treated as a frivolity. ... What is wanted [today] is an appreciation of the infinite variety of vivid values achieved by an organism in its proper environment. ...

What I mean is *art* and *aesthetic education*. It is, however, art in such a general sense of the term that I hardly like to call it by that name. Art is a special example. What we want is to draw out habits of aesthetic apprehension. According to the metaphysical doctrine I have been developing, to do so is to increase the depth of individuality. ...

The fertilization of the soul is the reason for the necessity of art. A static value, however serious and important, becomes unendurable by its appalling monotony of endurance. The soul cries aloud for release into change. ... Great art is the arrangement of the environment so as to provide for the soul vivid, but transient, values. ... [But] great art is more than a transient refreshment. It is something which adds to the permanent richness of the soul's self-attainment. ... It transforms the soul into the permanent realization of values extending beyond its former self. .... The importance of a living art, which moves on and yet leaves its permanent mark, can hardly be exaggerated.

In regard to the aesthetic needs of civilized society, the reactions of science have so far been unfortunate. Its materialistic basis has directed attention to *things* as opposed to *values*. ... A creed of competitive business morality was evolved, in some respects curiously high; but entirely devoid of consideration for the value of human life. ... *It may be that civilization will never recover from the bad climate which enveloped the introduction of machinery. ...*

At the present moment a discussion is raging as to the future of civilization in the novel circumstances of rapid scientific and technological progress. ... At the present time, as at other epochs, society is decaying, and there is need for preservative action. Professionals [specialists] are not new to the world. But in the past, professionals have formed unprogressive castes. The point is that professionalism has now been mated with progress. *The world is now faced with a self-evolving system, which it cannot stop.* There are dangers and advantages in this situation. ... It is obvious that the gain in material power affords opportunity for social betterment. But material power is in itself ethically neutral. It can equally well work in the wrong direction. ...

During the last three generations, the exclusive direction of attention to [material power] has been a disaster of the first magnitude. The watchwords of the 19th century have been: struggle for existence, competition, class warfare, commercial antagonism between nations, military warfare. ...

Modern science has imposed on humanity the necessity for wandering. Its progressive thought and its progressive technology make the transition through time, from generation to generation, a true migration into uncharted seas of adventure. The very benefit of wandering is that it is dangerous and needs skill to avert evils. We must expect, therefore, that the future will disclose dangers. ... In the immediate future there will be less security than in the immediate past, less stability. It must be admitted that there is a degree of instability which is inconsistent with civilization. But, on the whole, the great ages have been unstable ages.

164

THE ROAD TO WIGAN PIER

GEORGE ORWELL

George Orwell (1903-1950), the pen name adopted by British writer Eric Blair, was best known for his fictional satires of social and political life—books such as *Animal Farm* (1945) and *1984* (1949). Below is an excerpt from one of his few nonfiction books, written in 1937. Orwell saw world events as a battle between socialism and fascism. In principle he supported the former, but was greatly concerned about socialism's connection with "machine-civilization", industrial collectivism, and over-socialization. In this book he argues for a higher form of socialism, one removed from the undignified and dehumanizing effects of a mechanized society.

Chapter 12

However, there is a much more serious difficulty than the local and temporary objections which I discussed in the last chapter. [...] The first thing to notice is that the idea of Socialism is bound up, more or less inextricably, with the idea of machine-production. Socialism is essentially an *urban* creed. It grew up more or less concurrently with industrialism, it has always had its roots in the town proletariat and the town intellectual, and it is doubtful whether it could ever have arisen in any but an industrial society. Granted industrialism, the idea of Socialism presents itself naturally, because private ownership is only tolerable when every individual (or family or other unit) is at least moderately self-supporting; but the effect of industrialism is to make it impossible for anyone to be self-supporting, even for a moment. Industrialism, once it rises above a fairly low level, *must* lead to some form of collectivism. Not necessarily to Socialism, of course; conceivably it might lead to the Slave-State of which Fascism is a kind of prophecy. And the converse is also true. Machine-production suggests Socialism, but Socialism as a world-system implies machine-production, because it demands certain things not compatible with a primitive way of life. It demands, for instance, constant intercommunication and exchange of goods between all parts of the earth; it demands some degree of centralized control; it demands an approximately equal standard of life for all human beings and probably a certain uniformity of education. We may take it, therefore, that any world in which Socialism was a reality would be at least as highly mechanized as the United States at this moment, probably much more so. In any case, no Socialist would think of denying this. The Socialist world is always pictured as a completely mechanized, immensely organized world, depending on the machine as the civilizations of antiquity depend on the slave.

So far so good, or so bad. Many, perhaps a majority, of thinking people are not in love with machine-civilization, but everyone who is not a fool knows that it is non-sense to talk at this moment about scrapping the machine. But the unfortunate thing is that Socialism, as usually presented, is bound up with the idea of mechanical progress, not merely as a necessary development but as an end in itself, almost as a kind of religion. This idea is implicit in, for instance, most of the propagandist stuff that is written about the rapid mechanical advance in Soviet Russia (the Dneiper dam, tractors, etc., etc.). Karel Capek hits it off well enough in the horrible ending of [his play] *R.U.R.* (1921), when the Robots, having slaughtered the last human being, announce their intention to 'build many houses' (just for the sake of building houses, you see). The kind of person who *most* readily accepts Socialism is also the kind of person who views mechanical progress, *as such*, with enthusiasm. And this is so much the case that Socialists are often unable to grasp that the opposite opinion exists. As a rule the most persuasive argument they can think of is to tell you that the present mechanization of the world is as nothing to what we shall see when Socialism is estab-lished. Where there is one airplane now, in those days there will be fifty! All the work that is now done by hand will then be done by machinery: everything that is now made of leather, wood, or stone will be made of rubber, glass, or steel; there will be no dis-order, no loose ends, no wildernesses, no wild animals, no weeds, no disease, no poverty, no pain—and so on and so forth. The Socialist world is to be above all things an *ordered* world, an *efficient* world. But it is precisely from that vision of the future as a sort of glittering Wells-world that sensitive minds recoil. Please notice that this essentially fat-bellied version of 'progress' is not an integral part of Socialist doctrine; but it has come to be thought of as one, with the result that the temperamental conser-vatism which is latent in all kinds of people is easily mobilized against Socialism.

Every sensitive person has moments when he is suspicious of machinery and to some extent of physical science. But it is important to sort out the various motives, which have differed greatly at different times, for hostility to science and machinery, and to disregard the jealousy of the modern literary gent who hates science because science has stolen literature's thunder. The earliest full-length attack on science and machinery that I am acquainted with is in the third part of *Gulliver's Travels*. But Swift's attack, though brilliant as a *tour de force*, is irrelevant and even silly, because it is written from the standpoint—perhaps this seems a queer thing to say of the author of *Gulliver's Travels*—of a man who lacked imagination. To Swift, science was merely a kind of futile muckraking and the machines were non-sensical contraptions that would never work. His standard was that of practical usefulness, and he lacked the vision to see that an experiment which is not demonstrably useful at the moment may yield results in the future. Elsewhere in the book he names it as the best of all achievements 'to make two blades of grass grow where one grew before'; not seeing, apparently, that this is just what the machine can do. A little later the despised machines began working, physical science increased its scope, and there came the cel-ebrated conflict between religion and science which agitated our grandfathers. That conflict is over and both sides have retreated and claimed a victory, but an anti-scientific bias still lingers in the minds of most religious believers.

All through the nineteenth century protesting voices were raised against science and machinery (see Dickens's *Hard Times*, for instance), but usually for the rather shallow reason that industrialism in its first stages was cruel and ugly. Samuel Butler's attack on the machine in the well-known chapter of *Erewhon* is a different matter. But Butler himself lives in a less desperate age than our own, an age in which it was still possible for a first-rate man to be a dilettante part of the time, and therefore the whole thing appeared to him as a kind of intellectual exercise. He saw clearly enough our abject dependence on the machine, but instead of bothering to work out its consequences he preferred to exaggerate it for the sake of what was not much more than a joke. It is only in our own age, when mechanization has finally triumphed, that we can actually *feel* the tendency of the machine to make a fully human life impossible. There is probably no one capable of thinking and feeling who has not occasionally looked at a gas-pipe chair and reflected that the machine is the enemy of life. As a rule, however, this feeling is instinctive rather than reasoned. People know that in some way or another 'progress' is a swindle, but they reach this conclusion by a kind of mental shorthand; my job here is to supply the logical steps that are usually left out. But first one must ask, what is the function of the machine? Obviously its primary function is to save work, and the type of person to whom machine-civilization is entirely acceptable seldom sees any reason for looking further. Here for instance is a person who claims, or rather screams, that he is thoroughly at home in the modern mechanized world. I am quoting from *World Without Faith*, by Mr John Beevers. This is what he says:

> It is plain lunacy to say that the average £2 to £4 a week man of today is a lower type than an eighteenth-century farm labourer. Or than the labourer or peasant of any exclusively agricultural community now or in the past. It just isn't true. It is so damn silly to cry out about the civilizing effects of work in the fields and farmyards as against that done in a big locomotive works or an automobile factory. Work is a nuisance. We work because we have to and all work is done to provide us with leisure and the means of spending that leisure as enjoyably as possible.

And again:

> Man is going to have time enough and power enough to hunt for his own heaven on earth without worrying about the super-natural one. The earth will be so pleasant a place that the priest and the parson won't be left with much of a tale to tell. Half the stuffing is knocked out of them by one neat blow. (Etc., etc., etc.)

There is a whole chapter to this effect (Chapter 4 of Mr Beevers's book), and it is of some interest as an exhibition of machine-worship in its most completely vulgar, ignorant, and half-baked form. It is the authentic voice of a large section of the modern world. Every aspirin-eater in the outer suburbs would echo it fervently. Notice the shrill wail of anger ('It just isn't troo-o-o!', etc.) with which Mr Beevers meets the suggestion that his grandfather may have been a better man than himself; and the still more horrible suggestion that if we returned to a simpler way of life he might have to toughen his muscles with a job of work. Work, you see, is done 'to provide us with leisure'. Leisure for what? Leisure to become more like Mr Beevers, presumably. Though as a matter of fact, from that line of talk about 'heaven on earth', you can make a fairly good guess at what he would like civilization to be; a sort of Lyons Corner House lasting *in saecula saeculorum* and getting bigger and noisier all the time. And in any book by anyone who feels at home in the machine-world—in any book by H. G. Wells, for instance—you will find passages of the same kind. How often have we not heard it, that glutinously uplifting stuff about 'the machines, our new race of slaves, which will set humanity free', etc., etc., etc. To these people, apparently, the only danger of the machine is its possible use for destructive purposes; as, for instance, aero-planes are used in war. Barring wars and unforeseen disasters, the future is envisaged as an ever more rapid march of mechanical progress; machines to save work, machines to save thought, machines to save pain, hygiene, efficiency, organization, more hygiene, more efficiency, more organization, more machines—until finally you land up in the by now familiar Wellsian Utopia, aptly caricatured by Huxley in *Brave New World*, the paradise of little fat men. Of course in their day-dreams of the future the little fat men are neither fat nor little; they are Men Like Gods. But why should they be?

All mechanical progress is towards greater and greater efficiency; ultimately, therefore, towards a world in which *nothing goes wrong*. But in a world in which nothing went wrong, many of the qualities which Mr Wells regards as 'godlike' would be no more valuable than the animal faculty of moving the ears. The beings in *Men Like Gods* and *The Dream* are represented, for example, as brave, generous, and physically strong. But in a world from which physical danger had been banished—and obviously mechanical progress tends to eliminate danger—would physical courage be likely to survive? *Could* it survive? And why should physical strength survive in a world where there was never the need for physical labour? As for such qualities as loyalty, generosity, etc., in a world where nothing went wrong, they would be not only irrelevant but probably unimaginable.

The truth is that many of the qualities we admire in human beings can only function in opposition to some kind of disaster, pain, or difficulty; but the tendency of mechanical progress is to eliminate disaster, pain, and difficulty. In books like *The Dream* and *Men Like Gods* it is assumed that such qualities as strength, courage, generosity, etc., will be kept alive because they are comely qualities and necessary attributes of a full human being. Presumably, for instance, the inhabitants of Utopia would create artificial dangers in order to exercise their courage, and do dumb-bell exercises to harden muscles which they would never be obliged to use. And here you observe the huge contradiction which is usually present in the idea of progress. The tendency of mechanical progress is to make your environment safe and soft; and yet you are striving to keep yourself brave and hard. You are at the same moment furiously pressing forward and desperately holding back. It is as though a London stockbroker should go to his office in a suit of chain mail and insist on talking medieval Latin. So in the last analysis the champion of progress is also the champion of anachronisms.

Meanwhile I am assuming that the tendency of mechanical progress *is* to make life safe and soft. This may be disputed, because at any given moment the effect of some recent mechanical invention may appear to be the opposite. Take for instance the transition from horses to motor vehicles. At a first glance one might say, considering the enormous toll of road deaths, that the motor-car does not exactly tend to make life safer. Moreover it probably needs as much toughness to be a first-rate dirt-track rider as to be a bronco-buster or to ride in the Grand National. Nevertheless the *tendency* of all machinery is to become safer and easier to handle. The danger of accidents would disappear if we chose to tackle our road-planning problem seriously, as we shall do sooner or later; and meanwhile the motor-car has evolved to a point at which anyone who is not blind or paralytic can drive it after a few lessons. Even now it needs far less nerve and skill to drive a car ordinarily well than to ride a horse ordinarily well; in twenty years' time it may need no nerve or skill at all. Therefore, one must say that, taking society as a whole, the result of the transition from horses to cars has been an increase in human softness.

Presently somebody comes along with another invention, the airplane for instance, which does not at first sight appear to make life safer. The first men who went up in airplanes were superlatively brave, and even today it must need an exceptionally good nerve to be a pilot. But the same tendency as before is at work. The airplane, like the motor-car, will be made foolproof; a million engineers are working, almost unconsciously, in that direction. Finally—this is the objective, though it may never quite be reached—you will get an airplane whose pilot needs no more skill or courage than a baby needs in its perambulator. And all mechanical progress is and must be in this direction. A machine evolves by becoming more efficient, that is, more foolproof; hence the objective of mechanical progress is a foolproof world—which may or may not mean a world inhabited by fools. Mr Wells would probably retort that the world can never become fool-proof, because, however high a standard of efficiency you have reached, there is always some greater difficulty ahead. For example (this is Mr Wells's favourite idea—he has used it in goodness knows how many perorations), when you have got this planet of ours perfectly into trim, you start upon the enormous task of reaching and colonizing another. But this is merely to push the objective further into the future; the objective itself remains the same. Colonize another planet, and the game of mechanical progress begins anew; for the foolproof world you have substituted the foolproof solar system—the foolproof universe. In tying yourself to the ideal of mechanical efficiency, you tie yourself to the ideal of softness. But softness is repulsive; and thus all progress is seen to be a frantic struggle towards an objective which you hope and pray will never be reached. Now and again, but not often, you meet somebody who grasps that what is usually called progress also entails what is usually called degeneracy, and who is nevertheless in favour of progress. Hence the fact that in Mr Shaw's Utopia a statue was erected to Falstaff, as the first man who ever made a speech in favour of cowardice.

But the trouble goes immensely deeper than this. Hitherto I have only pointed out the absurdity of aiming at mechanical progress and also at the preservation of qualities which mechanical progress makes unnecessary. The question one has got to consider is whether there is *any* human activity which would not be maimed by the dominance of the machine.

The function of the machine is to save work. In a fully mechanized world all the dull drudgery will be done by machinery, leaving us free for more interesting pursuits. So expressed, this sounds splendid. It makes one sick to see half a dozen men sweating their guts out to dig a trench for a water-pipe, when some easily devised machine would scoop the earth out in a couple of minutes. Why not let the machine do the work and the men go and do something else. But presently the question arises, what else are they to do? Supposedly they are set free from 'work' in order that they may do something which is not 'work'. But what is work and what is not work? Is it work to dig, to carpenter, to plant trees, to fell trees, to ride, to fish, to hunt, to feed chickens, to play the piano, to take photographs, to build a house, to cook, to sew, to trim hats, to mend motor bicycles? All of these things are work to somebody, and all of them are play to somebody. There are in fact very few activities which cannot be classed either as work or play according as you choose to regard them. The labourer set free from digging may want to spend his leisure, or part of it, in playing the piano, while the professional pianist may be only too glad to get out and dig at the potato patch.

Hence the antithesis between work, as something intolerably tedious, and not-work, as something desirable, is false. The truth is that when a human being is not eating, drinking, sleeping, making love, talking, playing games, or merely lounging about—and these things will not fill up a lifetime—he needs work and usually looks for it, though he may not call it work. Above the level of a third- or fourth-grade moron, life has got to be lived largely in terms of effort. For man is not, as the vulgarer hedonists seem to suppose, a kind of walking stomach; he has also got a hand, an eye, and a brain. Cease to use your hands, and you have lopped off a huge chunk of your consciousness. And now consider again those half-dozen men who were digging the trench for the water-pipe. A machine has set them free from digging, and they are going to amuse themselves with something else—carpentering, for instance. But whatever they want to do, they will find that another machine has set them free from *that*. For in a fully mechanized world there would be no more need to carpenter, to cook, to mend motor bicycles, etc., than there would be to dig. There is scarcely anything, from catching a whale to carving a cherry stone, that could not conceivably be done by machinery. The machine would even encroach upon the activities we now class as 'art'; it is doing so already, via the camera and the radio. Mechanize the world as fully as it might be mechanized, and whichever way you turn there will be some machine cutting you off from the chance of working—that is, of living.

At a first glance this might not seem to matter. Why should you not get on with your 'creative work' and disregard the machines that would do it for you? But it is not so simple as it sounds. Here am I, working eight hours a day in an insurance office; in my spare time I want to do something 'creative', so I choose to do a bit of carpentering—to make myself a table, for instance. Notice that from the very start there is a touch of artificiality about the whole business, for the factories can turn me out a far better table than I can make for myself. But even when I get to work on my table, it is not possible for me to feel towards it as the cabinet-maker of a hundred years ago felt towards his table, still less as Robinson Crusoe felt towards his. For before I start, most of the work has already been done for me by machinery. The tools I use demand the minimum of skill. I can get, for instance, wood-planes which will cut out any molding; the cabinet-maker of a hundred years ago would have had to do the work with chisel and gouge, which demanded real skill of eye and hand. The boards I buy are ready planed and the legs are ready turned by the lathe. I can even go to the wood-shop and buy all the parts of the table ready-made and only needing to be fitted together; my work being reduced to driving in a few pegs and using a piece of sand-paper. And if this is so at present, in the mechanized future it will be enormously more so. With the tools and materials available *then*, there will be no possibility of mistake, hence no room for skill. Making a table will be easier and duller than peeling a potato. In such circumstances it is nonsense to talk of 'creative work'. In any case the arts of the hand (which have got to be transmitted by apprenticeship) would long since have disappeared. Some of them have disappeared already, under the competition of the machine. Look round any country churchyard and see whether you can find a decently-cut tombstone later than 1820. The art, or rather the craft, of stonework has died out so completely that it would take centuries to revive it.

But it may be said, why not retain the machine *and* retain 'creative work'? Why not cultivate anachronisms as a spare-time hobby? Many people have played with this idea; it seems to solve with such beautiful ease the problems set by the machine. The citizen of Utopia, we are told, coming home from his daily two hours of turning a handle in the tomato-canning factory, will deliberately revert to a more primitive way of life and solace his creative instincts with a bit of fretwork, pottery-glazing, or hand-loom-weaving. And why is this picture an absurdity—as it is, of course? Because of a principle that is not always recognized, though always acted upon: that so long as the machine is *there*, one is under an obligation to use it. No one draws water from the well when he can turn on the tap.

One sees a good illustration of this in the matter of travel. Everyone who has travelled by primitive methods in an undeveloped country knows that the difference between that kind of travel and modern travel in trains, cars, etc., is the difference between life and death. The nomad who walks or rides, with his baggage stowed on a camel or an ox-cart, may suffer every kind of discomfort, but at least he is living while he is travelling; whereas for the passenger in an express train or a luxury liner his journey is an interregnum, a kind of temporary death. And yet so long as the railways exist, one has got to travel by train—or by car or airplane. Here am I, forty miles from London. When I want to go up to London why do I not pack my luggage on to a mule and set out on foot, making a two days of it? Because, with the Green Line buses whizzing past me every ten minutes, such a journey *would* be intolerably irksome. In order that one may enjoy primitive methods of travel, it is necessary that no other method should be available. No human being ever wants to do anything in a more cumbrous way than is necessary. Hence the absurdity of that picture of Utopians saving their souls with fretwork. In a world where everything could be done by machinery, everything would be done by machinery. Deliberately to revert to primitive methods to use archaic tools, to put silly little difficulties in your own way, would be a piece of dilettantism, of pretty-pretty arty and craftiness. It would be like solemnly sitting down to eat your dinner with stone implements. Revert to handwork in a machine age, and you are back in Ye Olde Tea Shoppe or the Tudor villa with the sham beams tacked to the wall.

The tendency of mechanical progress, then, is to frustrate the human need for effort and creation. It makes unnecessary and even impossible the activities of the eye and the hand. The apostle of 'progress' will sometimes declare that this does not matter, but you can usually drive him into a corner by pointing out the horrible lengths to which the process can be carried. Why, for instance, use your hands at all—why use them even for blowing your nose or sharpening a pencil? Surely you could fix some kind of steel and rubber contraption to your shoulders and let your arms wither into stumps of skin and bone? And so with every organ and every faculty. There is really no reason why a human being should do more than eat, drink, sleep, breathe, and pro-create; *everything* else could be done for him by machinery. Therefore the logical end of mechanical progress is to reduce the human being to something resembling a brain in a bottle. That is the goal towards which we are already moving, though, of course, we have no intention of getting there; just as a man who drinks a bottle of whisky a day does not actually intend to get cirrhosis of the liver.

The implied objective of 'progress' is—not *exactly*, perhaps, the brain in the bottle, but at any rate some frightful subhuman depth of softness and helplessness. And the unfortunate thing is that at present the word 'progress' and the word 'Socialism' are linked inseparably in almost everyone's mind. The kind of person who hates machinery also takes it for granted to hate Socialism; the Socialist is always in favour of mechanization, rationalization, modernization—or at least thinks that he ought to be in favour of them. Quite recently, for instance, a prominent ILP'er con-fessed to me with a sort of wistful shame—as though it were something faintly improper—that he was 'fond of horses'. Horses, you see, belong to the vanished agri-cultural past, and all sentiment for the past carries with it a vague smell of heresy. I do not believe that this need necessarily be so, but undoubtedly it is so. And in itself it is quite enough to explain the alienation of decent minds from Socialism.

A generation ago every intelligent person was in some sense a revolutionary; nowadays it would be nearer the mark to say that every intelligent person is a reac-tionary. In this connection it is worth comparing H. G. Wells's *The Sleeper Awakes* with Aldous Huxley's *Brave New World*, written thirty years later. Each is a pes-simistic Utopia, a vision of a sort of prig's paradise in which all the dreams of the 'pro-gressive' person come true. Considered merely as a piece of imaginative construction *The Sleeper Awakes* is, I think, much superior, but it suffers from vast contradictions because of the fact that Wells, as the arch-priest of 'progress', cannot write with any conviction *against* 'progress'. He draws a picture of a glittering, strangely sinister world in which the privileged classes live a life of shallow gutless hedonism, and the workers, reduced to a state of utter slavery and sub-human ignorance, toil like troglodytes in caverns underground. As soon as one examines this idea—it is further developed in a splendid short story in *Stories of Space and Time*—one sees its incon-sistency. For in the immensely mechanized world that Wells is imagining, why should the workers have to work harder than at present? Obviously the tendency of the machine is to eliminate work, not to increase it. In the machine-world the workers might be enslaved, ill-treated, and even under-fed, but they certainly would not be condemned to ceaseless manual toil; because in that case what would be the function of the machine? You can have machines doing all the work or human beings doing all the work, but you can't have both. Those armies of underground workers, with their blue uniforms and their debased, half-human language, are only put in 'to make your flesh creep'. Wells wants to suggest that 'progress' might take a wrong turning; but the only evil he cares to imagine is inequality—one class grabbing all the wealth and power and oppressing the others, apparently out of pure spite. Give it quite a small twist, he seems to suggest, overthrow the privileged class—change over from world-capitalism to Socialism, in fact—and all will be well. The machine-civilization is to continue, but its products are to be shared out equally. The thought he dare not face is that the machine itself may be the enemy. So in his more characteristic Utopias (*The Dream, Men Like Gods,* etc.), he returns to optimism and to a vision of humanity, 'liberated' by the machine, as a race of enlightened sunbathers whose sole topic of conversation is their own superiority to their ancestors.

Brave New World belongs to a later time and to a generation which has seen through the swindle of 'progress'. It contains its own contradictions (the most important of them is pointed out in Mr John Strachey's *The Coming Struggle for Power*), but it is at least a memorable assault on the more fat-bellied type of perfectionism. Allowing for the exaggerations of caricature, it probably expresses what a majority of thinking people feel about machine-civilization.

The sensitive person's hostility to the machine is in one sense unrealistic, because of the obvious fact that the machine has come to stay. But as an attitude of mind there is a great deal to be said for it. The machine has got to be accepted, but it is probably better to accept it rather as one accepts a drug—that is, grudgingly and suspiciously. Like a drug, the machine is useful, dangerous, and habit-forming. The oftener one surrenders to it, the tighter its grip becomes. You have only to look about you at this moment to realize with what sinister speed the machine is getting us into its power.

To begin with, there is the frightful debauchery of taste that has already been effected by a century of mechanization. This is almost too obvious and too generally admitted to need pointing out. But as a single instance, take taste in its narrowest sense—the taste for decent food. In the highly mechanized countries, thanks to tinned food, cold storage, synthetic flavoring matters, etc., the palate is almost a dead organ. As you can see by looking at any greengrocer's shop, what the majority of English people mean by an apple is a lump of highly-colored cotton wool from America or Australia; they will devour these things, apparently with pleasure, and let the English apples rot under the trees. It is the shiny, standardized, machine-made look of the American apple that appeals to them; the superior taste of the English apple is something they simply do not notice. Or look at the factory-made, foil-wrapped cheese and 'blended' butter in any grocer's; look at the hideous rows of tins which usurp more and more of the space in any food-shop, even a dairy; look at a sixpenny Swiss roll or a twopenny ice-cream; look at the filthy chemical by-product that people will pour down their throats under the name of beer. Wherever you look you will see some slick machine-made article triumphing over the old-fashioned article that still tastes of something other than sawdust. And what applies to food applies also to furniture, houses, clothes, books, amusements, and everything else that makes up our environment. There are now millions of people, and they are increasing every year, to whom the blaring of a radio is not only a more acceptable but a more *normal* background to their thoughts than the lowing of cattle or the song of birds. The mechanization of the world could never proceed very far while taste, even the taste-buds of the tongue, remained uncorrupted, because in that case most of the products of the machine would be simply unwanted. In a healthy world there would be no demand for tinned foods, aspirins, gramophones, gaspipe chairs, machine guns, daily newspapers, telephones, motor-cars, etc., etc.; and on the other hand there would be a constant demand for the things the machine cannot produce.

But meanwhile the machine is here, and its corrupting effects are almost irresistible. One inveighs against it, but one goes on using it. Even a bare-arse savage, given the chance, will learn the vices of civilization within a few months. Mechanization leads to the decay of taste, the decay of taste leads to the demand for machine-made articles and hence to more mechanization, and so a vicious circle is established.

But in addition to this there is a tendency for the mechanization of the world to proceed as it were automatically, whether we want it or not. This is due to the fact that in modern Western man the faculty of mechanical invention has been fed and stimulated till it has reached almost the status of an instinct. People invent new machines and improve existing ones almost unconsciously, rather as a somnambulist will go on working in his sleep. In the past, when it was taken for granted that life on this planet is harsh or at any rate laborious, it seemed the natural fate to go on using the clumsy implements of your forefathers, and only a few eccentric persons, centuries apart, proposed innovations; hence throughout enormous ages such things as the ox-cart, the plough, the sickle, etc., remained radically unchanged. It is on record that screws have been in use since remote antiquity and yet that it was not till the middle of the nineteenth century that anyone thought of making screws with points on them, for several thousand years they remained flat-ended and holes had to be drilled for them before they could be inserted. In our own epoch such a thing would be unthinkable. For almost every modern Western man has his inventive faculty to some extent developed; the Western man invents machines as naturally as the Polynesian islander swims. Give a Western man a job of work and he immediately begins devising a machine that would do it for him; give him a machine and he thinks of ways of improving it.

I understand this tendency well enough, for in an ineffectual sort of way I have that type of mind myself. I have not either the patience or the mechanical skill to devise any machine that would work, but I am perpetually seeing, as it were, the ghosts of possible machines that might save me the trouble of using my brain or muscles. A person with a more definite mechanical turn would probably construct some of them and put them into operation. But under our present economic system, whether he constructed them—or rather, whether anyone else had the benefit of them—would depend upon whether they were commercially valuable. The Socialists are right, therefore, when they claim that the rate of mechanical progress will be much more rapid once Socialism is established. Given a mechanical civilization the process of invention and improvement will always continue, but the tendency of capitalism is to slow it down, because under capitalism any invention which does not promise fairly immediate profits is neglected; some, indeed, which threaten to reduce profits are suppressed almost as ruthlessly as the flexible glass mentioned by Petronius. Establish Socialism—remove the profit principle—and the inventor will have a free hand. The mechanization of the world, already rapid enough, would be or at any rate could be enormously accelerated.

And this prospect is a slightly sinister one, because it is obvious even now that the process of mechanization is out of control. It is happening merely because humanity has got the habit. A chemist perfects a new method of synthesizing rubber, or a mechanic devises a new pattern of gudgeon-pin. Why? Not for any clearly understood purpose, but simply from the impulse to invent and improve, which has now become instinctive. Put a pacifist to work in a bomb-factory and in two months he will be devising a new type of bomb. Hence the appearance of such diabolical things as poison gases, which are not expected even by their inventors to be beneficial to humanity. Our attitude towards such things as poison gases *ought* to be the attitude of the king of Brobdingnag towards gunpowder; but because we live in a mechanical and scientific age we are infected with the notion that, whatever else happens, 'progress' must continue and knowledge must never be suppressed. Verbally, no doubt, we would agree that machinery is made for man and not man for machinery; in practice any attempt to check the development of the machine appears to us an attack on knowledge and therefore a kind of blasphemy.

And even if the whole of humanity suddenly revolted against the machine and decided to escape to a simpler way of life, the escape would still be immensely difficult. It would not do, as in Butler's *Erewhon*, to smash every machine invented after a certain date; we should also have to smash the habit of mind that would, almost involuntarily, devise fresh machines as soon as the old ones were smashed. And in all of us there is at least a tinge of that habit of mind. In every country in the world the large army of scientists and technicians, with the rest of us panting at their heels, are marching along the road of 'progress' with the blind persistence of a column of ants. Comparatively few people want it to happen, plenty of people actively want it *not* to happen, and yet it is happening. The process of mechanization has itself become a machine, a huge glittering vehicle whirling us we are not certain where, but probably towards the padded Wells-world and the brain in the bottle.

This, then, is the case against the machine. Whether it is a sound or unsound case hardly matters. The point is that these or very similar arguments would be echoed by every person who is hostile to machine-civilization. And unfortunately, because of that nexus of thought, 'Socialism-progress-machinery-Russia-tractor-hygiene-machinery-progress', which exists in almost everyone's mind, it is usually the *same* person who is hostile to Socialism. The kind of person who hates central heating and gaspipe chairs is also the kind of person who, when you mention Socialism, murmurs something about 'beehive state' and moves away with a pained expression. So far as my observation goes, very few Socialists grasp why this is so, or even that it *is* so. Get the more vocal type of Socialist into a corner, repeat to him the substance of what I have said in this chapter, and see what kind of answer you get. As a matter of fact you will get several answers; I am so familiar with them that I know them almost by heart.

In the first place he will tell you that it is impossible to 'go back' (or to 'put back the hand of progress' — as though the hand of progress hadn't been pretty violently put back several times in human history!), and will then accuse you of being a medievalist and begin to descant upon the horrors of the Middle Ages, leprosy, the Inquisition, etc. As a matter of fact, most attacks upon the Middle Ages and the past generally by apologists of modernity are beside the point, because their essential trick is to project a modern man, with his squeamishness and his high standards of comfort, into an age when such things were unheard of. But notice that in any case this is not an answer. For a dislike of the mechanized future does not imply the smallest reverence for any period of the past. D. H. Lawrence, wiser than the medievalist, chose to idealize the Etruscans about whom we know conveniently little. But there is no need to idealize even the Etruscans or the Pelasgians, or the Aztecs, or the Sumerians, or any other vanished and romantic people. When one pictures a desirable civilization, one pictures it merely as an objective; there is no need to pretend that it has ever existed in space and time. Press this point home, explain that you wish to aim at making life simpler and harder instead of softer and more complex, and the Socialist will usually assume that you want to revert to a 'state of nature' — meaning some stinking Paleolithic cave: as though there were nothing between a flint scraper and the steel mills of Sheffield, or between a skin coracle and the *Queen Mary*.

Finally, however, you will get an answer which is rather more to the point and which runs roughly as follows: 'Yes, what you are saying is all very well in its way. No doubt it would be very noble to harden ourselves and do without aspirins and central heating and so forth. But the point is, you see, that nobody seriously wants it. It would mean going back to an agricultural way of life, which means beastly hard work and isn't at all the same thing as playing at gardening. I don't want hard work, you don't want hard work — nobody wants it who knows what it means. You only talk as you do because you've never done a day's work in your life,' etc., etc.

Now this in a sense is true. It amounts to saying, 'We're soft — for God's sake let's stay soft!' which at least is realistic. As I have pointed out already, the machine has got us in its grip and to escape will be immensely difficult. Nevertheless this answer is really an evasion, because it fails to make dear what we mean when we say that we 'want' this or that. I am a degenerate modern semi-intellectual who would die if I did not get my early morning cup of tea and my *New Statesman* every Friday. Clearly I do not, in a sense, 'want' to return to a simpler, harder, probably agricultural way of life. In the same sense I don't 'want' to cut down my drinking, to pay my debts, to take enough exercise, to be faithful to my wife, etc., etc. But in another and more permanent sense I do want these things, and perhaps in the same sense I want a civilization in which 'progress' is not definable as making the world safe for little fat men. [...]

Chapter 13

And finally, is there anything one can do about it? [...]

The distaste for 'progress' and machine-civilization which is so common among sensitive people is only defensible as an attitude of mind. It is not valid as a reason for rejecting Socialism, because it presupposes an alternative which does not exist. When you say, 'I object to mechanization and standardization—therefore I object to Socialism', you are saying in effect, 'I am free to do without the machine if I choose', which is nonsense. We are all dependent upon the machine, and if the machines stopped working most of us would die. You may hate the machine-civilization, probably you are right to hate it, but for the present there can be no question of accepting or rejecting it. The machine-civilization *is here*, and it can only be criticized from the inside, because all of us are inside it. It is only romantic fools who flatter themselves that they have escaped, like the literary gent in his Tudor cottage with bathroom h. and c., and the he-man who goes off to live a 'primitive' life in the jungle with a Mannlicher rifle and four wagon-loads of tinned food.

And almost certainly the machine-civilization will continue to triumph. There is no reason to think that it will destroy itself or stop functioning of its own accord. For some time past it has been fashionable to say that war is presently going to 'wreck civilization' altogether; but, though the next full-sized war will certainly be horrible enough to make all previous ones seem a joke, it is immensely unlikely that it will put a stop to mechanical progress. It is true that a very vulnerable country like England, and perhaps the whole of western Europe, could be reduced to chaos by a few thousand well-placed bombs, but no war is at present thinkable which could wipe out industrialization in all countries simultaneously. We may take it that the return to a simpler, free, less mechanized way of life, however desirable it may be, is not going to happen. This is not fatalism, it is merely acceptance of facts. It is meaningless to oppose Socialism on the ground that you object to the beehive State, for the beehive State *is here*. The choice is not, as yet, between a human and an inhuman world. It is simply between Socialism and Fascism, which at its very best is Socialism with the virtues left out.

The job of the thinking person, therefore, is not to reject Socialism but to make up his mind to humanize it. Once Socialism is in a way to being established, those who can see through the swindle of "progress" will probably find themselves resisting. In fact, it is their special function to do so. In the machine-world, they have got to be a sort of permanent opposition...

THE QUESTION CONCERNING TECHNOLOGY

MARTIN HEIDEGGER

Heidegger (1889-1976) was an influential German philosopher. His work covered metaphysics, existentialism, and phenomenology, but his key focus was ontology, or the nature of 'being'. His most famous book was his first, *Being and Time* (1927). All of Heidegger's writings are notoriously difficult—some would say, nearly incomprehensible. The following essay, originally published in 1949 and modified over the subsequent 5 years, attempts to analyze the nature of technology and its effect on humanity.

Section 1

In what follows we shall be *questioning* concerning technology. Questioning builds a way. We would be advised, therefore, above all to pay heed to the way, and not to fix our attention on isolated sentences and topics. The way is a way of thinking. All ways of thinking, more or less perceptibly, lead through language in a manner that is extraordinary. We shall be questioning concerning *technology*, and in so doing we should like to prepare a 'free' relationship to it. The relationship will be free if it opens our human existence to the essence of technology. When we can respond to this essence, we shall be able to experience the technological within its own bounds.

Technology is not equivalent to the 'essence' of technology. When we are seeking the essence of "tree," we have to become aware that 'That' which pervades every tree, as tree, is not *itself* a tree that can be encountered among all the other trees.

Likewise, the essence of technology is by no means anything technological. Thus we shall never experience our relationship to the essence of technology so long as we merely conceive and push forward the technological, put up with it, *or evade it. Everywhere we remain unfree and chained to technology, whether we passionately affirm or deny it*. But we are delivered over to it in the worst possible way when we regard it as something neutral; for this conception of it, to which today we particularly like to do homage, makes us utterly blind to the essence of technology.

According to ancient doctrine, the essence of a thing is considered to be *what* the thing is. We ask the question concerning technology when we ask what it is. Everyone knows the two statements that answer this question. One says: "Technology is a means to an end." The second says: "Technology is a human activity." The two definitions of technology belong together. For to posit ends and procure and utilize the means to them is a human activity. The manufacture and utilization of equipment, tools, and machines, the manufactured and used things themselves, and the needs and ends that they serve, all belong to what technology is. The whole complex of these contrivances is technology. Technology itself is a contrivance, or, in Latin, an *instrumentum*.

The current conception of technology, according to which it is [both] a means and a human activity, can therefore be called the instrumental and anthropological definition of technology.

Who would ever deny that it is correct? It is in obvious conformity with what we are envisioning when we talk about technology. The instrumental definition of technology is indeed so uncannily correct that it even holds for modern technology, of which, in other respects, we maintain with some justification that it is, in contrast to the older handwork technology, *something completely different* and therefore new. Even the power plant with its turbines and generators is a man-made means to an end established by man. Even the jet aircraft and the high-frequency apparatus are means to ends. A radar station is of course less simple than a weather vane. To be sure, the construction of a high-frequency apparatus requires the interlocking of various processes of technical-industrial production. And certainly a sawmill in a secluded valley of the Black Forest is a primitive means compared with the hydroelectric plant in the Rhine River.

But this much remains correct: Modern technology too is a means to an end. That is why the instrumental conception of technology conditions every attempt to bring man into the right relation to technology. Everything depends on our manipulating technology in the proper manner as a means. We will, as we say, "get" technology "spiritually in hand." We will master it. The will to mastery becomes all the more urgent the more technology threatens to slip from human control.

But suppose now that technology was no mere means, how would it stand with the will to master it? Yet we said, did we not, that the instrumental definition of technology is correct? To be sure. The correct always fixes upon something pertinent in whatever is under consideration. However, in order to be correct, this fixing by no means needs to uncover the thing in question in its essence. Only at the point where such an uncovering happens does the true come to pass. For that reason the merely correct is not yet the true. Only the true brings us into a free relationship with that which concerns us from out of its essence. Accordingly, the correct instrumental definition of technology still does not show us technology's essence. In order that we may arrive at this, or at least come close to it, we must seek the true by way of the correct. We must ask: What is the instrumental itself? Within what do such things as means and end belong? A means is that whereby something is effected and thus attained. Whatever has an effect as its consequence is called a 'cause'. But not only that by means of which something else is effected is a cause. The end in keeping with which the kind of means to be used is determined is also considered a cause. Wherever ends are pursued and means are employed, wherever instrumentality reigns, there reigns causality. ...

[Take an example of causality: A silversmith creates a silver chalice, as for a religious ceremony. According to Aristotle, there are four distinct 'causes' involved with this act of creation: (1) the *material* cause, i.e. the material in the cup; (2) the *formal* cause, i.e. the chalice-shape or form; (3) the *efficient* cause, i.e. the maker, the silver-smith; and (4) the *final* cause, i.e. the end purpose of the thing which caused it to be needed, such as a specific religious ceremony. All four causes are present in every event.]

According to this example, the four causes are responsible for the silver chalice's lying ready before us as a sacrificial vessel. Lying before and lying ready characterize the 'presencing' of something that 'presences'. The four ways of being responsible bring something into appearance. They let it come forth into presencing. They set it free to that place and so start it on its way, namely, into its complete arrival. The principal characteristic of being 'responsible' is this starting something on its way into arrival. It is in the sense of such a starting something on its way into arrival that being responsible is an 'occasioning' or an inducing to go forward. On the basis of a look at what the Greeks experienced in being responsible, we now give this verb "to occasion" a more inclusive meaning, so that it now is the name for the essence of causality, thought as the Greeks thought it.

But in what, then, does the playing in unison of the "four ways of occasioning" play? They let what is not yet present arrive into presencing. Accordingly, they are unifiedly ruled over by a **'bringing'** that brings what presences into appearance. Plato tells us what this bringing is in a sentence from the <u>Symposium</u> (205b) : *he gar toi ek tou me onton eis to on ionti hotoioun aitia pasa esti poiesis*: "Every occasion for whatever passes over and goes forward into presencing from that which is not presencing is *poiesis*, is bringing-forth." [Alt. translation: "Everything that is responsible for creating something out of nothing is a kind of poetry (*poiesis*)."]

It is of utmost importance that we think of bringing-forth in its full scope and at the same time in the sense in which the Greeks thought it. Not only handcraft manufacture, not only artistic and poetical bringing into appearance and concrete imagery, is a bringing-forth, *poiesis*. *Physis* [nature] also, the arising of something from *out of itself*, is a bringing forth, *poiesis*. *Physis* is indeed *poiesis* in the highest sense. For what presences by means of *physis* has the bursting open belonging to bringing-forth, e.g., the bursting of a blossom into bloom, in itself. In contrast, what is brought forth by the artisan or the artist, e.g., the silver chalice, has the bursting open belonging to bringing-forth *not* in itself, but *in another*, in the craftsman or artist.

The modes of occasioning, the four causes, are at play, then, within bringing-forth. Through bringing-forth, the growing things of nature as well as whatever is completed through the crafts and the arts come at any given time to their appearance.

But how does bringing-forth happen, be it in nature or in handwork and art? What is the bringing-forth in which the fourfold way of occasioning plays? Occasioning has to do with the presencing of that which at any given time comes to appearance in bringing-forth. Bringing-forth brings hither out of concealment forth into disclosure (unconcealment). Bringing-forth comes to pass only insofar as something concealed becomes disclosed. This coming occurs within what we call **'revealing'**. The Greeks have the word *aletheia* for revealing. The Romans translate this with *veritas*. We say "truth" and usually understand it as the correctness of an idea.

Section 2

But where have we strayed to? We are questioning concerning technology, and we have arrived now at *aletheia*, at revealing. What has the essence of technology to do with revealing? The answer: *everything*. For every bringing-forth is grounded in revealing. Bringing-forth, indeed, gathers within itself the four modes of occasioning—causality—and rules them throughout. Within its domain belong end and means, belongs instrumentality. Instrumentality is considered to be the fundamental characteristic of technology. If we inquire, step by step, into what technology, represented as means, actually is, then we shall arrive at 'revealing'. The possibility of all productive manufacturing lies in revealing.

Technology is therefore no mere means. *Technology is a way of revealing.* If we give heed to this, then another whole realm for the essence of technology will open itself up to us. It is the realm of revealing, i.e. of truth.

This prospect strikes us as strange. Indeed, it should do so, should do so as persistently as possible and with so much urgency that we will finally take seriously the simple question of what the name "technology" means. The word stems from the Greek: *Technikon* means that which belongs to *techne*. We must observe two things with respect to the meaning of this word. One is that *techne* is the name not only for the activities and skills of the craftsman, but also for the arts of the mind and the fine arts. *Techne* belongs to bringing-forth, to *poiesis*; it is something poietic.

The other point that we should observe with regard to *techne* is even more important. From earliest times until Plato the word *techne* is linked with the word *episteme*. Both words are names for 'knowing' in the widest sense. They mean to be entirely at home in something, to understand and be expert in it. Such knowing provides an opening up. As an opening up it is a revealing. *Techne* is a mode of *aletheuein*. It reveals whatever does not bring itself forth and does not yet lie here before us, whatever can look and turn out now one way and now another. Whoever builds a house or a ship or forges a sacrificial chalice reveals what is to be brought forth, according to the perspectives of the four modes of occasioning. This revealing gathers together in advance the aspect and the matter of ship or house, with a view to the finished thing envisioned as completed, and from this gathering determines the manner of its construction. Thus what is decisive in *techne* does not lie at all in making and manipulating nor in the using of means, but rather in the aforementioned revealing. It is as revealing, and not as manufacturing, that *techne* is a bringing-forth.

Thus the clue to what the word *techne* means and to how the Greeks defined it leads us into the same context that opened itself to us when we pursued the question of what instrumentality as such in truth might be.

Technology is a mode of revealing. Technology comes 'to presence' in the realm where revealing and disclosure take place, where *aletheia*, truth, happens.

In opposition to this definition of the essential domain of technology, one can object that it indeed holds for Greek thought and that at best it might apply to the techniques of the handicraftsman, but that it simply does not fit modern machine-powered technology. And it is precisely such modern technology that is the disturbing thing, that moves us to ask the question concerning technology per se. It is said that modern technology is something incomparably different from all earlier technologies because it is based on *modern physics* as an exact science. Meanwhile we have come to understand more clearly that the *reverse* holds true as well: Modern physics, as experimental, is dependent upon technical apparatus and upon progress in the building of apparatus. The establishing of this mutual relationship between technology and physics is correct. But it remains a merely historiographical establishing of facts and says nothing about that in which this mutual relationship is grounded. The decisive question still remains: Of what *essence* is modern technology that it happens to think of putting exact science to use?

What is *modern* technology? *It too is a revealing.* Only when we allow our attention to rest on this fundamental characteristic does that which is new in modern technology show itself to us.

And yet the revealing that holds sway throughout modern technology does *not* unfold into a bringing-forth in the sense of *poiesis*. The revealing that rules in modern technology is a *challenge*, which puts to nature the unreasonable demand that it supply *energy* that can be extracted and stored as such. But does this not hold true for the old windmill as well? No. Its sails do indeed turn in the wind; they are left entirely to the wind's blowing. But the windmill does not unlock energy from the air currents in order to store it.

In contrast, a tract of land is challenged into the putting out of coal and ore. The earth now reveals itself as a coal-mining district, the soil as a mineral deposit. The field that the peasant formerly cultivated and set in order appears differently than it did when to set in order still meant to take care of and to maintain. *The work of the peasant does not challenge the soil of the field.* In the sowing of the grain it places the seed in the keeping of the forces of growth and watches over its increase. But meanwhile even the [modern] cultivation of the field has come under the grip of another kind of setting-in-order, which *sets upon nature*. It sets upon it in the sense of challenging it. Agriculture is now the mechanized food industry. Air is now set upon to yield nitrogen, the earth to yield ore, ore to yield uranium, for example; uranium is set upon to yield atomic energy, which can be released either for destruction or for peaceful use.

This setting-upon...is always itself directed from the beginning toward furthering something else, i.e., toward driving on to the maximum yield at the minimum expense. The coal that has been hauled out in some mining district has not been supplied in order that it may simply be present somewhere or other. It is stockpiled; that is, it is on call, ready to deliver the sun's warmth that is stored in it. The sun's warmth is challenged forth for heat, which in turn is ordered to deliver steam whose pressure turns the wheels that keep a factory running.

The hydroelectric plant is set into the current of the Rhine. It sets the Rhine to supplying its hydraulic pressure, which then sets the turbines turning. This turning sets those machines in motion whose thrust sets going the electric current for which the long-distance power station and its network of cables are set up to dispatch electricity. In the context of the interlocking processes pertaining to the orderly disposition of electrical energy, even the Rhine itself appears as something at our command. The hydroelectric plant is not built into the Rhine River as was the old wooden bridge that joined bank with bank for hundreds of years. Rather the river is dammed up into the power plant. What the river is now, namely, a water power supplier, derives from out of the essence of the power station. In order that we may even remotely consider the monstrousness that reigns here, let us ponder for a moment the contrast that speaks out of the two titles, "The Rhine" as dammed up into the *power* works, and "The Rhine" as uttered out of the *art* work, in Holderlin's hymn by that name. But, it will be replied, the Rhine is still a river in the landscape, is it not? Perhaps. But how? In no other way than as an object on call for inspection by a tour group ordered there by the vacation industry.

The revealing that rules throughout modern technology has the character of a 'setting-upon', in the sense of a challenging-forth. That challenging happens in that the energy concealed in nature is unlocked, what is unlocked is transformed, what is transformed is stored up, what is stored up is, in turn, distributed, and what is distributed is switched about ever anew. Unlocking, transforming, storing, distributing, and switching about are ways of revealing. But the revealing never simply comes to an end. Neither does it run off into the indeterminate. The revealing reveals to itself its own manifoldly interlocking paths, through regulating their course. This regulating itself is, for its part, everywhere secured. Regulating and securing even become the chief characteristics of the challenging revealing.

What kind of disclosure is it, then, that is peculiar to that which comes to stand forth through this setting-upon that challenges? Everywhere everything is **ordered** to "stand by," to be immediately at hand, indeed to stand there just so that it may be on call for a further ordering. Whatever is ordered about in this way has its own standing. We call it the 'standing-reserve'. The word expresses here something more, and something more essential, than mere "stock." The name "standing-reserve" assumes the rank of an inclusive category. It designates nothing less than the way in which everything presences that is wrought upon by the challenging revealing. Whatever stands by in the sense of standing-reserve no longer stands over against us as object.

Yet an airliner that stands on the runway is surely an object. Certainly. We can represent the machine so. But then it conceals itself as to what and how it is. Revealed, it stands on the taxi strip only as standing-reserve, inasmuch as it is ordered to ensure the possibility of transportation. For this it must be in its whole structure and in every one of its constituent parts, on call for duty, i.e., ready for takeoff.

The fact that now, wherever we try to point to modern technology as the challenging revealing, the words "setting-upon," "ordering," "standing-reserve," obtrude and accumulate in a dry, monotonous, and therefore oppressive way, has its basis in what is now coming to utterance.

Who accomplishes the challenging setting-upon through which what we call the real is revealed as standing-reserve? Obviously, *man*. To what extent is man capable of such a revealing? Man can indeed conceive, fashion, and carry through this or that in one way or another. But man *does not have control over* disclosure itself, in which at any given time the real shows itself or withdraws. The fact that the real has been showing itself in the light of Ideas ever since the time of Plato, Plato did not bring about. The thinker only responded to what addressed itself to him.

Only to the extent that man for his part is already challenged to exploit the energies of nature can this ordering revealing happen. If man is challenged, ordered, to do this, then does not *man himself* belong, even more originally than nature, within the standing-reserve? The current talk about 'human resources', about the supply of patients for a clinic, gives evidence of this. The forester who, in the wood, measures the felled timber and to all appearances walks the same forest path in the same way as did his grandfather is today *commanded by profit-making* in the lumber industry, whether he knows it or not. He is made subordinate to the orderability of cellulose, which for its part is challenged forth by the need for paper, which is then delivered to newspapers and illustrated magazines. The latter, in their turn, set public opinion to swallowing what is printed, so that a set configuration of opinion becomes available on demand. Yet precisely because man is challenged more originally than are the energies of nature, i.e., into the process of ordering, he *never* is transformed into *mere* standing-reserve. Since man drives technology forward, he takes part in ordering as a way of revealing. But the disclosure itself, within which ordering unfolds, is never a human handiwork, any more than is the realm through which man is already passing every time he as a subject relates to an object.

Section 3

Where and how does this revealing happen if it is no mere handiwork of man? We need not look far. We need only apprehend in an unbiased way 'That' which has already claimed man and has done so, so decisively that he can only be man at any given time as the one so claimed. Wherever man opens his eyes and ears, unlocks his heart, and gives himself over to meditating and striving, shaping and working, entreating and thanking, he finds himself everywhere already brought into the disclosed. The disclosure of the disclosed has already come to pass whenever it calls man forth into the modes of 'revealing' allotted to him. ...

Modern technology as an ordering revealing is, then, no merely human doing. Therefore we must take that challenging—the one that sets upon man to order the real as standing-reserve—in accordance with the way in which it shows itself. That challenging gathers *man* into ordering. This gathering concentrates man, upon ordering the real, as standing-reserve. ...

We now name that challenging claim which gathers man there to order the self-revealing as standing-reserve: "*Gestell*", or, ***Enframing***.

According to ordinary usage, the word *Gestell* [frame] means some kind of apparatus, e.g., a bookrack. *Gestell* is also the name for a skeleton. And the employment of the word *Gestell* that is now required of us seems equally eerie, not to speak of the arbitrariness with which words of a mature language are thus misused. Can anything be more strange? Surely not. Yet this strangeness is an old usage of thinking. And indeed thinkers accord with this usage precisely at the point where it is a matter of thinking that which is highest. We, late born, are no longer in a position to appreciate the significance of Plato's daring to use the word *eidos* [Idea] for that which in everything and in each particular thing endures as present. For *eidos*, in the common speech, meant the outward aspect that a visible thing offers to the physical eye. Plato exacts of this word, however, something utterly extraordinary: that it name what precisely is not and never will be perceivable with physical eyes. But even this is by no means the full extent of what is extraordinary here. For *idea* names not only the nonsensuous aspect of what is physically visible. Aspect (*idea*) names and is, also, that which constitutes the essence in the audible, the tasteable, the tactile, in everything that is in any way accessible.

Compared with the demands that Plato makes on language and thought in this and other instances, the use of the word *Gestell* as the name for the *essence of modern technology*, which we now venture here, seems almost harmless. Even so, the usage now required remains something exacting and is open to misinterpretation.

Enframing means the gathering together of that setting-upon which *sets upon man*, i.e., challenges him forth, to reveal the real, in the mode of ordering, as standing-reserve. Enframing means that way of revealing which holds sway in the essence of modern technology and which is *itself* nothing technological. On the other hand, all those things that are so familiar to us and are standard parts of an assembly, such as rods, pistons, and chassis, belong to the technological. The assembly itself, however, together with the aforementioned stockparts, falls within the sphere of technological activity; and this activity always merely responds to the challenge of Enframing, but it never comprises Enframing itself or brings it about.

The word *stellen* [to set upon] in the name *Gestell* [Enframing] not only means challenging. At the same time it should preserve the suggestion of another *Stellen* from which it stems, namely, that 'producing' and 'presenting' which, in the sense of *poiesis*, lets what presences come forth into disclosure. This producing that brings forth—e.g., the erecting of a statue in the temple precinct—and the challenging ordering now under consideration are indeed fundamentally different, and yet they remain related in their essence. Both are ways of revealing, of *aletheia*. In Enframing, that disclosure comes to pass in conformity with which the work of modern technology reveals the real as standing-reserve. This work is therefore *neither* only a human activity *nor* a mere means within such activity. The merely instrumental, merely anthropological definition of technology is therefore in principle *untenable*. And it cannot be rounded out by being referred back to some metaphysical or religious explanation that undergirds it.

It remains true, nonetheless, that man in the technological age is, in a particularly striking way, challenged forth into revealing. That 'revealing' concerns *nature, above all, as the chief storehouse of the standing energy reserve*. Accordingly, man's ordering attitude and behavior display themselves first in the rise of modern physics as an exact science. Modern science's way of 'representing' *pursues and entraps nature as a calculable coherence of forces*. Modern physics is not experimental physics because it applies apparatus to the questioning of nature. Rather the reverse is true. Because physics, indeed already as pure theory, sets nature up to exhibit itself as a coherence of forces calculable in advance, it therefore orders its experiments precisely for the purpose of asking whether and how nature reports itself when set up in this way.

But after all, mathematical physics arose almost two centuries before [modern] technology. How, then, could it have already been set upon by modern technology and placed in its service? The facts testify to the contrary. Surely technology got under way only when it could be supported by exact physical science. Reckoned chronologically, this is correct. Thought historically, it does not hit upon the truth.

The modern physical theory of nature prepares the way first not simply for technology but for the essence of modern technology. For already in physics the challenging gathering-together into ordering revealing holds sway. But in it, that gathering does not yet come expressly to appearance. Modern physics is the herald of Enframing, a herald whose origin is still unknown. The essence of modern technology has for a long time been *concealing itself*, even where power machinery has been invented, where electrical technology is in full swing, and where atomic technology is well under way.

All coming to presence, not only modern technology, keeps itself everywhere concealed to the last. Nevertheless, it remains, with respect to its holding sway, that which precedes all: the earliest. The Greek thinkers already knew of this when they said: That which is earlier with regard to the arising that holds sway becomes manifest to us men only later. That which is primally *early* shows itself only *ultimately* to men. Therefore, in the realm of thinking, a painstaking effort to think through still more primally what was primally thought is not the absurd wish to revive what is past, but rather the sober readiness to be astounded before the coming of what is early.

Chronologically speaking, modern physical science begins in the seventeenth century. In contrast, machine-power technology develops only in the second half of the eighteenth century. But modern technology, which for chronological reckoning is the later, is, from the point of view of the essence holding sway within it, the historically earlier.

If modern physics must resign itself ever increasingly to the fact that its realm of representation remains inscrutable and incapable of being visualized, this resignation is not dictated by any committee of researchers. It is challenged forth by the rule of Enframing, which *demands that nature be orderable as standing-reserve*. Hence physics, in all its retreating from the representation turned only toward objects that has alone been standard till recently, will never be able to renounce this one thing: that nature reports itself in some way or other that is identifiable through calculation and that it remains orderable as a system of information. This system is determined, then, out of a causality that has changed once again. Causality now displays neither the character of the occasioning that brings forth nor the nature of the *causa efficiens*, let alone that of the *causa formalis*. It seems as though causality is shrinking into a reporting—a reporting challenged forth—of standing-reserves that must be guaranteed either simultaneously or in sequence.

Because the essence of modern technology lies in Enframing, modern technology must employ exact physical science. Through its so doing, the deceptive *illusion* arises that modern technology is applied physical science. This illusion can maintain itself only so long as neither the essential origin of modern science nor indeed the essence of modern technology is adequately found out through questioning.

Section 4

We are questioning concerning technology in order to bring to light our relationship to its essence. The *essence* of modern technology shows itself in what we call *Enframing*. But simply to point to this is still in no way to answer the question concerning technology, if to answer means to respond, in the sense of correspond, to the essence of what is being asked about.

Where do we find ourselves brought to, if now we think one step further regarding what Enframing itself actually is? It is nothing technological, nothing on the order of a machine. It is the way in which the real reveals itself as standing-reserve.

Again we ask: Does this revealing happen somewhere beyond all human doing? No. But neither does it happen exclusively *in* man, or decisively *through* man.

Enframing is the gathering together that belongs to that setting-upon which sets upon man and puts him in position to reveal the real, in the mode of ordering, as standing-reserve. As the one who is challenged forth in this way, *man stands within the essential realm of Enframing*. He can never take up a relationship to it only subsequently. Thus the question as to how we are to arrive at a relationship to the essence of technology, asked in this way, *always* comes too late. But *never* too late comes the question as to whether we actually experience ourselves as the ones whose activities everywhere, public and private, are challenged forth by Enframing. Above all, never too late comes the question as to *whether* and *how* we actually admit ourselves into that wherein Enframing itself comes to presence.

The essence of modern technology starts man upon the way of that revealing through which the real everywhere, more or less distinctly, becomes standing-reserve. "To start upon a way" means "to send" in our ordinary language. We shall call that sending-that-gathers which first starts man upon a way of revealing, *'destining'*. It is from out of this destining that the essence of all history is determined. History is neither simply the object of written chronicle nor simply the fulfillment of human activity. That activity first becomes history as something destined. And it is only the destining into objectifying representation that makes the historical accessible as an object for historiography, i.e., for a science, and on this basis makes possible the current equating of the historical with that which is chronicled.

Enframing, as a challenging-forth into ordering, 'sends' into a way of revealing. Enframing is an enacting of destining, as is *every* way of revealing. Bringing-forth, *poiesis*, is also a destining in this sense.

Always the disclosure of 'that which is' goes upon a way of revealing. *Always* the 'destining of revealing' *holds complete sway over man*. But that destining is *never a fate that compels*. For man becomes truly free only insofar as he belongs to the realm of destining and so becomes one who listens and hears, and not one who is simply constrained to obey. ...

The essence of modern technology lies in Enframing. Enframing belongs within the destining of revealing. These sentences express something different from the talk that we hear more frequently, to the effect that technology is the fate of our age, where "fate" means the inevitableness of an unalterable course.

But when we consider the essence of technology, then we experience Enframing as a 'destining of revealing'. In this way we are already sojourning within the open space of destining, a destining that in no way confines us to a stultified compulsion to push on blindly with technology or, what comes to the same thing, to rebel helplessly against it and curse it as the work of the devil. Quite to the contrary, when we once open ourselves expressly to the *essence* of technology, we find ourselves unexpectedly taken into a freeing claim.

Man is endangered from out of destining. The destining of revealing is as such, in every one of its modes, and therefore necessarily, *danger*.

In whatever way the destining of revealing may hold sway, the disclosure in which everything that is shows itself at any given time harbors the danger that man may recoil at the disclosed and may misinterpret it. Thus where everything that presences exhibits itself in the light of a cause-effect coherence, even God can, for representational thinking, lose all that is exalted and holy, the mysteriousness of his distance. In the light of causality, God can sink to the level of a cause, of *causa efficiens*. He then becomes, even in theology, the god of the philosophers, namely, of those who define the disclosed and the concealed in terms of the causality of making, without ever considering the essential origin of this causality.

In a similar way the disclosure, in accordance with which nature presents itself as a calculable complex of the effects of forces, can indeed permit correct determinations; but precisely through these successes the danger can remain that in the midst of all that is correct, the true will withdraw.

The destining of revealing (Enframing) is in itself not just any danger, but danger as such.

Yet when destining reigns in the mode of Enframing, it is the *supreme danger*. This danger attests itself to us in two ways. As soon as what is disclosed no longer concerns man even as object, but does so, rather, *exclusively* as standing-reserve, and man in the midst of objectlessness is nothing but the orderer of the standing-reserve, then he comes to the very brink of a precipitous fall; that is, he comes to the point where *he himself will have to be taken as standing-reserve*. Meanwhile man, precisely as the one so threatened, *exalts himself* to the posture of lord of the earth. In this way the impression comes to prevail that everything man encounters exists only insofar as it is his construct. This illusion gives rise in turn to one final delusion: It seems as though man everywhere and always encounters only himself. In truth, however, precisely *nowhere* does man today any longer encounter himself, i.e., his essence. Man stands so decisively in attendance on the challenging-forth of Enframing that he does not apprehend Enframing as a claim, that he fails to see himself as the one spoken to, and hence also fails in every way to hear in what respect he exists, from out of his essence, in the realm of an exhortation or address, and thus *can never* encounter only himself.

But Enframing does not simply endanger man in his relationship to himself and to everything that is. As a destining, it banishes man into that kind of revealing which is an 'ordering'. Where this ordering holds sway, it drives out every other possibility of revealing. Above all, Enframing conceals that [benign] revealing which, in the sense of *poiesis*, lets what presences come forth into appearance. As compared with that other revealing, the "setting-upon that challenges forth" thrusts man into a relation to that which is, that is at once antithetical and rigorously ordered. Where Enframing holds sway, regulating and securing of the standing-reserve mark all revealing. They no longer even let their own fundamental characteristic appear, namely, this revealing as such.

Thus the challenging Enframing not only conceals a former way of revealing, 'bringing-forth', but it conceals *revealing itself*, and with it, 'That' wherein disclosure, i.e., truth, comes to pass.

Enframing blocks the shining-forth and holding-sway of truth. The destining that sends into ordering is consequently the extreme danger. What is dangerous is not technology. There is no demonry of technology, but rather there is the mystery of its essence. The essence of technology, as a destining of revealing, is the danger. The transformed meaning of the word "Enframing" will perhaps become somewhat more familiar to us now if we think Enframing in the sense of destining and danger.

The threat to man does not come in the first instance from the potentially lethal machines and apparatus of technology. The actual threat has *already* affected man in his essence. The rule of Enframing threatens man with the possibility that it could be denied to him to enter into a more original revealing and hence to experience the call of a more primal truth.

Section 5

Thus, where Enframing reigns, there is *danger* in the highest sense.

> *But where danger is, grows*
> *The saving power also.*

Let us think carefully about these words of Holderlin. What does it mean "to save"? Usually we think that it means only to seize hold of a thing threatened by ruin, in order to secure it in its former continuance. But the verb "to save" says more. "To save" is to fetch something home into its essence, in order to bring the essence for the first time into its genuine appearing. If the essence of technology, Enframing, is the extreme danger, and if there is truth in Holderlin's words, then the rule of Enframing cannot exhaust itself solely in blocking all appearing of truth. Rather, precisely the essence of technology must harbor *in itself* the growth of the saving power. But in that case, might not an adequate look into what Enframing is as a destining of revealing bring into appearance the saving power in its arising?

In what respect does the saving power grow there also where the danger is? Where something grows, there it takes root, from thence it thrives. Both happen concealedly and quietly and in their own time. But according to the words of the poet we have no right whatsoever to expect that there, where the danger is, we should be able to lay hold of the saving power immediately and without preparation. Therefore we must consider now, in advance, in what respect the saving power does most profoundly take root and thence thrive even in that wherein the extreme danger lies. In order to consider this, it is necessary, as a last step upon our way, to look with yet clearer eyes into the danger. Accordingly, we must once more question concerning technology. For we have said that in technology's essence thrives the saving power.

But how shall we behold the saving power in the essence of technology, so long as we do not consider in what sense of "essence" it is that Enframing is actually the essence of technology?

Thus far we have understood "essence" in its current meaning. In the academic language of philosophy, "essence" means *what* something is; in Latin, *quid. Quidditas*, 'whatness', provides the answer to the question concerning essence. For example, what pertains to all kinds of trees—oaks, beeches, birches, firs—is the same "treeness." Under this inclusive genus—the "universal"—fall all real and possible trees. Is then the essence of technology, Enframing, the common genus for everything technological? If that were the case then the steam turbine, the radio transmitter, and the cyclotron would each be an Enframing. But the word "Enframing" does not mean here a tool or any kind of apparatus. Still less does it mean the general concept of such resources. The machines and apparatus are no more cases and kinds of Enframing than are the man at the switchboard and the engineer in the drafting room. Each of these in its own way indeed belongs as stockpart, available resource, or executer, within Enframing; but Enframing is never the essence of technology in the sense of a genus. Enframing is a way of revealing, having the character of destining, namely, the way that challenges forth. The revealing that brings forth (*poiesis*) is also a way that has the character of destining.

The challenging revealing [i.e. Enframing] has its origin as a destining in bringing-forth. But at the same time Enframing, in a way characteristic of a destining, blocks *poiesis*.

Thus Enframing, as a destining of revealing, is indeed the essence of technology, but never in the sense of genus and *essentia*. If we pay heed to this, something astounding strikes us: It is *technology itself* that makes the demand on us to think in another way what is usually understood by "essence." But in what way?

If we speak of the "essence of a house" and the "essence of a state," we do not mean a generic type; rather we mean the ways in which house and state hold sway, administer themselves, develop and decay... Socrates and Plato already think the essence of something as 'what essences', what comes to presence, in the sense of what endures. But they think what endures as what remains permanently. And they find what endures permanently in what, as that which remains, tenaciously persists throughout all that happens. That which remains they discover, in turn, in the aspect, for example, the Idea "house."

The Idea "house" displays what anything is that is fashioned as a house. Particular, real, and possible houses, in contrast, are changing and transitory derivatives of the Idea and thus belong to what does not endure.

But it can never in any way be established that enduring is based solely on what Plato thinks as *idea* and Aristotle thinks as *to ti on einai* ("that which any particular thing has always been"), or what metaphysics in its most varied interpretations thinks as *essentia*.

All essencing endures. But is enduring only *permanent* enduring? Does the essence of technology endure in the sense of the *permanent* enduring of an Idea that hovers over everything technological, thus making it seem that by technology we mean some mythological abstraction? The way in which technology 'essences' lets itself be seen only from out of that permanent enduring in which Enframing comes to pass as a destining of revealing. ... And if we now ponder more carefully than we did before what it is that actually endures and perhaps alone endures, we may venture to say: Only what is 'granted' endures. That which endures primally out of the earliest beginning is what *grants*.

As the essencing of technology, Enframing is that which endures. Does Enframing hold sway at all in the sense of granting? No doubt the question seems a horrendous blunder. For according to everything that has been said, Enframing is, rather, a destining that gathers together into the revealing that challenges forth.

Challenging is anything *but* a granting. So it seems, so long as we do not notice that the challenging-forth into the ordering of the real as standing-reserve still remains a destining that starts man upon a way of revealing. As this destining, the coming to presence of technology gives man entry into That which, of himself, he can neither invent nor in any way make.

But if this destining, Enframing, is the extreme danger, not only for man's coming to presence, but for all revealing as such, should this destining still be called a granting? Yes, most emphatically, if in this destining the *saving power* is said to grow. Every destining of revealing comes to pass from out of a granting, and is as such a granting. For it is granting that first conveys to man that share in revealing which the coming-to-pass of revealing needs. As the one so needed and used, man is given to belong to the coming-to-pass of truth. The granting that sends into revealing is as such the saving power. For the saving power lets man see and enter into the highest dignity of his essence. This dignity lies in keeping watch over the disclosure — and with it, from the first, the concealment — of all coming to presence on this earth. It is precisely in Enframing, which threatens to sweep man away into ordering as the supposed single way of revealing, and so thrusts man into the danger of the surrender of his free essence — it is precisely in this extreme danger that the innermost indestructible belongingness of man within granting may come to light, provided that we, for our part, begin to pay heed to the coming to presence of technology.

Thus the coming to presence of technology harbors in itself what we least suspect, the possible arising of the saving power.

Everything, then, depends upon this: that we ponder this arising and that, recollecting, we *watch over it.* How can this happen? Above all through our catching sight of what comes to presence in technology, instead of merely staring at the technological. So long as we represent technology as an instrument, we remain held fast in the will to master it. We press on past the essence of technology.

When, however, we ask how the instrumental comes to presence as a kind of causality, then we experience this coming to presence as the destining of a revealing.

When we consider, finally, that the coming to presence of the essence of technology comes to pass in the granting that needs and uses man so that he may share in revealing, then the following becomes clear:

The essence of technology is in a lofty sense *ambiguous.* Such ambiguity points to the mystery of all revealing, i.e., of truth.

On the one hand, Enframing challenges forth into the frenziedness of ordering that blocks every view into the coming-to-pass of revealing and so radically endangers the relation to the essence of truth.

On the other hand, Enframing comes to pass for its part in the granting that lets man endure — as yet unexperienced, but perhaps more experienced in the future — that he may be the *one who is needed and used for the safekeeping of the coming to presence of truth.* Thus does the arising of the saving power appear.

The irresistibility of ordering and the restraint of the saving power draw past each other like the paths of two stars in the course of the heavens. But precisely this, their passing by, is the hidden side of their nearness.

When we look into the ambiguous essence of technology, we behold the constellation, the stellar course of the mystery.

The question concerning technology is the question concerning the constellation in which revealing and concealing, in which the coming to presence of truth, comes to pass.

But what help is it to us to look into the constellation of truth? We look into the danger and see the growth of the saving power. Through this we are not yet saved. But we are thereupon summoned to hope in the growing light of the saving power. How can this happen? Here and now and in little things, that we may foster the saving power in its increase. This includes holding always before our eyes the extreme danger.

The coming to presence of technology threatens 'revealing', threatens it with the possibility that *all* revealing will be consumed in ordering, and that everything will present itself only in the disclosedness of standing-reserve. Human activity can never directly counter this danger. Human achievement alone can never banish it. But human reflection can ponder the fact that all saving power must be of a higher essence than what is endangered, though at the same time kindred to it.

Section 6

But might there not perhaps be a *more primally granted* revealing that could bring the saving power into its first shining forth in the midst of the danger, a revealing that in the technological age rather conceals than shows itself?

There was a time when it was not technology alone that bore the name *techne*. Once that revealing that brings forth truth into the splendor of radiant appearing also was called *techne*.

Once there was a time when the bringing-forth of the true into the *beautiful* was called *techne*. And the *poiesis* of the fine arts also was called *techne*.

In Greece, at the outset of the destining of the West, the arts soared to the supreme height of the revealing granted them. They brought the presence of the gods, brought the dialogue of divine and human destinings, to radiance. And art was simply called *techne*. It was a single, manifold revealing.

The arts were not derived from the artistic. Art works were not enjoyed aesthetically. Art was not a sector of cultural activity.

What, then, was art—perhaps only for that brief but magnificent time? Why did art bear the modest name *techne*? Because it was a revealing that brought forth, and therefore belonged within *poiesis*. It was finally that revealing which holds complete sway in all the fine arts, in poetry, and in everything poetical that obtained *poiesis* as its proper name.

The same poet from whom we heard the words

> *But where danger is, grows*
> *The saving power also.*

says to us:

> *... poetically dwells man upon this earth.*

The poetical brings the true into the splendor of what Plato in the <u>Phaedrus</u> [250e] calls *ekphanestaton*, 'that which shines forth most purely'. The poetical thoroughly pervades every art, every revealing of coming to presence into the beautiful.

Could it be that the fine arts are called to poetic revealing? Could it be that revealing lays claim to the arts most primally, so that they for their part may expressly foster the growth of the saving power, may awaken and found anew our look into that which grants and our trust in it?

Whether art may be granted this highest possibility of its essence in the midst of the extreme danger, no one can tell. Yet we can be astounded. Before what? Before this other possibility: that *the frenziedness of technology may entrench itself everywhere* to such an extent that someday, throughout everything technological, *the essence of technology may come to presence in the coming-to-pass of truth.*

Because the essence of technology is nothing technological, essential reflection upon technology and decisive confrontation with it must happen in a realm that is, on the one hand, akin to the essence of technology and, on the other, fundamentally different from it.

Such a realm is art. But certainly only if reflection on art, for its part, does not shut its eyes to the constellation of truth after which we are questioning.

Thus questioning, we bear witness to the crisis that in our sheer preoccupation with technology we do not yet experience the coming to presence of technology, that in our sheer aesthetic-mindedness we no longer guard and preserve the coming to presence of art. Yet the more questioningly we ponder the essence of technology, the more mysterious the essence of art becomes.

The closer we come to the danger, the more brightly do the ways into the saving power begin to shine and the more questioning we become. For questioning is the piety of thought.

ONE-DIMENSIONAL MAN

HERBERT MARCUSE

Marcuse (1898-1979) was one of Heidegger's most famous students. He was a member of the Frankfurt School of philosophy, focusing on social and political theory rather than abstract philosophical concepts. The following book was published in 1964.

Introduction to the First Edition

The Paralysis of Criticism: Society without Opposition

Does not the threat of an atomic catastrophe which could wipe out the human race also serve to protect the very forces which perpetuate this danger? The efforts to prevent such a catastrophe overshadow the search for its potential causes in contemporary industrial society. These causes remain unidentified, unexposed, unattacked by the public because they recede before the all too obvious threat from without—to the West from the East, to the East from the West. Equally obvious is the need for being prepared, for living on the brink, for facing the challenge. We submit to the peaceful production of the means of destruction, to the perfection of waste, to being educated for a defense which deforms the defenders and that which they defend.

If we attempt to relate the causes of the danger to the way in which society is organized and organizes its members, we are immediately confronted with the fact that advanced industrial society becomes richer, bigger, and better as it perpetuates the danger. The defense structure makes life easier for a greater number of people and extends man's mastery of nature. Under these circumstances, our mass media have little difficulty in selling particular interests as those of all sensible men. The political needs of society become individual needs and aspirations, their satisfaction promotes business and the commonweal, and the whole appears to be the very embodiment of Reason.

And yet *this society is irrational as a whole*. Its productivity is destructive of the free development of human needs and faculties, its peace maintained by the constant threat of war, its growth dependent on the repression of the real possibilities for pacifying the struggle for existence—individual, national, and international. This repression, so different from that which characterized the preceding, less developed stages of our society, operates today not from a position of natural and technical immaturity but rather from a position of strength. The capabilities (intellectual and material) of

CONFRONTING TECHNOLOGY

contemporary society are immeasurably greater than ever before — which means that the scope of society's domination over the individual is immeasurably greater than ever before. Our society distinguishes itself by conquering the centrifugal social forces with Technology rather than Terror, on the dual basis of an *overwhelming efficiency* and an *increasing standard of living*.

To investigate the roots of these developments and examine their historical alternatives is part of the aim of a critical theory of contemporary society, a theory which analyzes society in the light of its used and unused or abused capabilities for improving the human condition. But what are the standards for such a critique?

Certainly value judgments play a part. The established way of organizing society is measured against other possible ways, ways which are held to offer better chances for alleviating man's struggle for existence; a specific historical practice is measured against its own historical alternatives. From the beginning, any critical theory of society is thus confronted with the problem of historical objectivity, a problem which arises at the two points where the analysis implies value judgments:

1. The judgment that human life is worth living, or rather can be and ought to be made worth living. This judgment underlies all intellectual effort; it is the *a priori* of social theory, and its rejection (which is perfectly logical) rejects theory itself;

2. The judgment that, in a given society, specific possibilities exist for the amelioration of human life and specific ways and means of realizing these possibilities. Critical analysis has to demonstrate the objective validity of these judgments, and the demonstration has to proceed on empirical grounds. The established society has available an ascertainable quantity and quality of intellectual and material resources. How can these resources be used for the optimal development and satisfaction of individual needs and faculties with a minimum of toil and misery? Social theory is historical theory, and history is the realm of chance in the realm of necessity. Therefore, among the various possible and actual modes of organizing and utilizing the available resources, which ones offer the greatest chance of an optimal development? [...]

But here, advanced industrial society confronts the critique with a situation which seems to deprive it of its very basis. Technical progress, extended to a whole system of domination and coordination, creates forms of life (and of power) which appear to reconcile the forces opposing the system, and to *defeat or refute all protest* in the name of the historical prospects of freedom from toil and domination. Contemporary society seems to be capable of *containing* [i.e. restricting] social change — qualitative change which would establish essentially *different* institutions, a *new direction* of the productive process, *new modes* of human existence. This containment of social change is perhaps the most singular achievement of advanced industrial society; the general acceptance of the national purpose, bipartisan policy, the decline of pluralism, the collusion of business and labor within the strong state testify to the integration of opposites which is the result as well as the prerequisite of this achievement. [...]

In the face of apparently contradictory facts, the critical analysis continues to insist that *the need for qualitative change is as pressing as ever before*. Needed by whom? The answer continues to be the same: by the society as a whole, for every one of its members. The union of growing productivity and growing destruction; the brinkmanship of annihilation; the surrender of thought, hope, and fear to the decisions of the powers that be; the preservation of misery in the face of unprecedented wealth constitute the most impartial indictment—even if they are not the *raison d'etre* of this society but only its by-product: its *sweeping rationality*, which propels efficiency and growth, *is itself irrational*.

The fact that the vast majority of the population accepts, and is made to accept, this society does not render it less irrational and less reprehensible. The distinction between true and false consciousness, real and immediate interest still is meaningful. But this distinction itself must be validated. Men must come to see it and to find their way from false to true consciousness, from their immediate to their real interest. They can do so only if they live in need of changing their way of life, of *denying the positive* [benefits of society], of *refusing*. It is precisely this need which the established society manages to repress to the degree to which it is capable of "delivering the goods" on an increasingly large scale, and using the scientific conquest of nature for the scientific conquest of man. [...]

This ambiguous situation involves a still more fundamental ambiguity. *One-Dimensional Man* will vacillate throughout between two contradictory hypotheses : (1) that advanced industrial society is capable of *containing* [restricting] qualitative change for the foreseeable future; (2) that forces and tendencies exist which may break this containment, and explode the society. I do not think that a clear answer can be given. Both tendencies are there, side by side—and even the one in the other. The first tendency is dominant, and whatever preconditions for a reversal may exist are being used to prevent it. Perhaps an accident may alter the situation, but unless the recognition of what is being done and what is being prevented subverts the consciousness and the behavior of man, not even a catastrophe will bring about the change.

The analysis is focused on advanced industrial society, in which the technical apparatus of production and distribution (with an increasing sector of automation) functions, not as the sum-total of mere instruments which can be isolated from their social and political effects, but rather as a system which determines *a priori* the product of the apparatus as well as the operations of servicing and extending it. In this society, the productive apparatus tends to become totalitarian to the extent to which it determines not only the socially needed occupations, skills, and attitudes, but also individual needs and aspirations. It thus obliterates the opposition between the private and public existence, between individual and social needs. *Technology serves to institute new, more effective, and more pleasant forms of social control and social cohesion*. The totalitarian tendency of these controls seems to assert itself in still another sense—by spreading to the less developed and even to the pre-industrial areas of the world, and by creating similarities in the development of capitalism and communism.

In the face of the totalitarian features of this society, *the traditional notion of the "neutrality" of technology can no longer be maintained*. Technology as such cannot be isolated from the use to which it is put; the technological society is a system of domination which operates already in the concept and construction of techniques.

The way in which a society organizes the life of its members involves an initial choice between historical alternatives which are determined by the inherited level of the material and intellectual culture. The choice itself results from the play of the dominant interests. It anticipates specific modes of transforming and utilizing man and nature and rejects other modes. It is one "project" of realization among others. But once the project has become operative in the basic institutions and relations, it tends to become exclusive and to determine the development of the society as a whole. As a *technological* universe, advanced industrial society is a *political* universe, the latest stage in the realization of a specific historical project—namely, the *experience, transformation, and organization of nature as the mere stuff of domination.*

As the project unfolds, it shapes the entire universe of discourse and action, intellectual and material culture. In the medium of technology, culture, politics, and the economy merge into an omnipresent system which swallows up or repulses all alternatives. The productivity and growth potential of this system stabilize the society and contain technical progress within the framework of domination. *Technological rationality has become political rationality.*

Chapter 1 – The New Forms of Control

A comfortable, smooth, reasonable, democratic un-freedom prevails in advanced industrial civilization, a token of technical progress. Indeed, what could be more rational than the *suppression of individuality* in the mechanization of socially necessary but painful performances; the concentration of individual enterprises in more effective, more productive corporations; the regulation of free competition among unequally equipped economic subjects; the curtailment of prerogatives and national sovereignties which impede the international organization of resources. That this technological order also involves a political and intellectual coordination may be a regrettable and yet promising development.

The rights and liberties which were such vital factors in the origins and earlier stages of industrial society yield to a higher stage of this society: they are losing their traditional rationale and content. Freedom of thought, speech, and conscience were essentially *critical* ideas, designed to replace an obsolescent material and intellectual culture by a more productive and rational one. Once institutionalized, these rights and liberties shared the fate of the society of which they had become an integral part. [...]

Today political power asserts itself through its power over the machine process and over the technical organization of the apparatus. The government of advanced and advancing industrial societies can maintain and secure itself only when it succeeds in mobilizing, organizing, and exploiting the technical, scientific, and mechanical productivity available to industrial civilization. And this productivity mobilizes *society as a whole*, above and beyond any particular individual or group interests. The brute fact that the machine's physical power surpasses that of the individual, and of any particular group of individuals, *makes the machine the most effective political instrument in any society whose basic organization is that of the machine process.* But the political trend may be reversed; essentially the power of the machine is only the stored-up and projected power of man.

Contemporary industrial civilization demonstrates that it has reached the stage at which "the free society" can no longer be adequately defined in the traditional terms of economic, political, and intellectual liberties, not because these liberties have become insignificant, but because they are too significant to be confined within the traditional forms. New modes of realization are needed, corresponding to the new capabilities of society.

Such new modes can be indicated only in *negative* terms because they would amount to the negation of the prevailing modes. Thus economic freedom would mean freedom *from* the economy—from being controlled by economic forces and relationships; freedom from the daily struggle for existence, from 'earning a living'. Political freedom would mean liberation of the individuals *from* politics over which they have no effective control. Similarly, intellectual freedom would mean the restoration of individual thought now absorbed by mass communication and indoctrination, abolition of "public opinion" together with its makers. The unrealistic sound of these propositions is indicative, not of their utopian character, but of the strength of the forces which prevent their realization. The most effective and enduring form of warfare against liberation is the *implanting of material and intellectual needs* that perpetuate obsolete forms of the struggle for existence. [...]

In the last analysis, the question of what are true and false needs must be answered by the individuals themselves, but only in the last analysis; that is, if and when they are free to give their own answer. As long as they are kept incapable of being *autonomous*, as long as they are indoctrinated and manipulated, their answer to this question cannot be taken as their own. By the same token, however, no tribunal can justly arrogate to itself the right to decide which needs should be developed and satisfied. Any such tribunal is reprehensible, although our revulsion does not do away with the question: how can the people who have been the object of effective and productive domination by themselves create the conditions of freedom?

The more rational, productive, technical, and total the repressive administration of society becomes, the more unimaginable the means and ways by which the administered individuals might break their servitude and seize their own liberation. To be sure, to impose Reason upon an entire society is a paradoxical and scandalous idea—although one might dispute the righteousness of a society which ridicules this idea while making its own population into objects of total control. All liberation depends on the consciousness [i.e. awareness] of servitude, and the emergence of this consciousness is always hampered by the predominance of needs and satisfactions which, to a great extent, have become the individual's own. The process always replaces one system of preconditioning by another; the optimal goal is the replacement of false needs by true ones, the abandonment of repressive satisfaction.

The distinguishing feature of advanced industrial society is its effective suffocation of those needs which demand liberation—liberation also from that which is tolerable and rewarding and comfortable—while it sustains and absolves the destructive power and repressive function of the affluent society. Here, the social controls exact the overwhelming need for the production and consumption of waste; the need for stupefying work where it is no longer a real necessity; the need for modes of relaxation which soothe and prolong this stupefication; the need for maintaining such deceptive liberties as free competition at administered prices, a free press which censors itself, free choice between brands and gadgets.

Under the rule of a repressive whole, *'liberty' can be made into a powerful instrument of domination*. The range of choice open to the individual is *not* the decisive factor in determining the degree of human freedom, but *what* can be chosen and what *is* chosen by the individual. The criterion for free choice can never be an absolute one, but neither is it entirely relative. Free election of masters does not abolish the masters or the slaves. Free choice among a wide variety of goods and services does not signify freedom if these goods and services sustain social controls over a life of toil and fear—that is, if they sustain alienation. And the spontaneous reproduction of superimposed needs by the individual does not establish autonomy; it only testifies to the efficacy of the controls.

We are again confronted with one of the most vexing aspects of advanced industrial civilization: the *rational character of its irrationality*. Its productivity and efficiency, its capacity to increase and spread comforts, to turn waste into need, and destruction into construction, the extent to which this civilization transforms the object world into an extension of man's mind and body makes the very notion of alienation questionable. The people recognize themselves in their commodities; they find their soul in their automobile, hi-fi set, split-level home, kitchen equipment. The very mechanism which ties the individual to his society has changed, and social control is anchored in the new needs which it has produced.

The prevailing forms of social control are technological in a new sense. To be sure, the technical structure and efficacy of the productive and destructive apparatus has been a major instrumentality for subjecting the population to the established social division of labor throughout the modern period. Moreover, such integration has always been accompanied by more obvious forms of compulsion: loss of livelihood, the administration of justice, the police, the armed forces. It still is. But in the contemporary period, the technological controls appear to be the very embodiment of Reason for the benefit of all social groups and interests to such an extent that all contradiction seems irrational and all counteraction impossible.

No wonder then that, in the most advanced areas of this civilization, the social controls have been subconsciously absorbed to the point where even individual protest is affected at its roots. The intellectual and emotional refusal "to go along" appears neurotic and impotent. This is the socio-psychological aspect of the political event that marks the contemporary period: the passing of the historical forces which, at the preceding stage of industrial society, seemed to represent the possibility of new forms of existence. [...]

I have just suggested that the concept of alienation seems to become questionable when the individuals identify themselves with the existence which is imposed upon them and have in it their own development and satisfaction. This identification is not illusion but reality. However, the reality constitutes a more progressive stage of alienation. The latter has become entirely objective; the subject which is alienated is swallowed up by its alienated existence. There is only one dimension, and it is everywhere and in all forms. The achievements of progress defy ideological indictment as well as justification; before their tribunal, the "false consciousness" of their rationality becomes the true consciousness.

This absorption of ideology into reality does not, however, signify the "end of ideology." On the contrary, in a specific sense advanced industrial culture is *more* ideological than its predecessor, inasmuch as today the ideology is in the process of production itself. In a provocative form, this proposition reveals the *political aspects* of the prevailing technological rationality. The productive apparatus and the goods and services which it produces "sell" or impose the social system as a whole. The means of mass transportation and communication, the commodities of lodging, food, and clothing, the irresistible output of the entertainment and information industry carry with them prescribed attitudes and habits, certain intellectual and emotional reactions which bind the consumers more or less pleasantly to the producers and, through the latter, to the whole. The products indoctrinate and manipulate; they promote a false consciousness which is immune against its falsehood. And as these beneficial products become available to more individuals in more social classes, the indoctrination they carry ceases to be publicity; it becomes a way of life. It is a 'good' way of life—much better than before—and as a good way of life, it militates against qualitative change. Thus emerges a pattern of *one-dimensional thought and behavior* in which ideas, aspirations, and objectives that, by their content, transcend the established universe of discourse and action are either repelled or reduced to terms of this universe. They are redefined by the rationality of the given system and of its quantitative extension.

Outside the academic establishment, the "far-reaching change in all our habits of thought" is more serious. It serves to coordinate ideas and goals with those exacted by the prevailing system, to enclose them in the system, and to repel those which are irreconcilable with the system. The reign of such a one-dimensional reality does *not* mean that materialism rules, and that the spiritual, metaphysical, and bohemian occupations are petering out. On the contrary, there is a great deal of "Worship together this week," "Why not try God," Zen, existentialism, and beat ways of life, etc. But such modes of protest and transcendence are *no longer contradictory to the status quo* and *no longer negative.* They are rather the ceremonial part of practical behaviorism, its harmless negation, and are *quickly digested by the status quo* as part of its healthy diet.

"Progress" is not a neutral term; it moves toward specific ends, and these ends are defined by the possibilities of ameliorating the human condition. Advanced industrial society is approaching the stage where continued progress would demand the radical subversion of the prevailing direction and organization of progress. This stage would be reached when material production (including the necessary services) becomes automated to the extent that all vital needs can be satisfied while necessary labor time is reduced to marginal time. From this point on, technical progress would transcend the realm of necessity, where it served as the instrument of domination and exploitation which thereby limited its rationality; technology would become subject to the free play of faculties in the struggle for the pacification of nature and of society. [...]

The most advanced areas of industrial society exhibit throughout these two features: (1) a trend toward *consummation* of technological rationality, and (2) intensive efforts to *contain* this trend within the established institutions. Here is the internal contradiction of this civilization: *the irrational element in its rationality*. It is the token of its achievements. The industrial society which makes technology and science its own, is organized for the ever-more-effective domination of man and nature, for the ever-more-effective utilization of its resources. It becomes irrational when the success of these efforts opens new dimensions of human realization. Organization for peace is different from organization for war; the institutions which served the struggle for existence cannot serve the pacification of existence. Life as an end is qualitatively different from life as a means. [...]

When this point is reached, domination—in the guise of affluence and liberty—extends to all spheres of private and public existence, integrates all authentic opposition, absorbs all alternatives. Technological rationality reveals its political character as it becomes the great vehicle of better domination, creating a truly totalitarian universe in which society and nature, mind and body are kept in a state of permanent mobilization for the defense of this [technological] universe.

FORBIDDEN GAMES

THEODORE ROSZAK

Roszak (1933-2011) was a professor of history at California State University. Rising to prominence within the counterculture movement of the 1960s, his writings have provided numerous critiques of modern society. Roszak has written a number of influential books, including *Where the Wasteland Ends* (1972), *Person/Planet* (1979), *The Cult of Information* (1986), and *The Voice of the Earth* (1992). The following essay is one of his earlier writings, dating from 1966.

Those of us who find ourselves distressed or even horrified at the shape that the technological society is forcing upon our lives find ourselves again and again brought up short by the familiar cliché that technology—in both its mechanical and its organizational aspects—is, after all, a neutral force that can be wielded for man's well-being as well as for his harm. It is a cliché that is bound to leave us unsatisfied as a final verdict on the nature of technological development. For suppose one followed its suggestion and surrendered analysis in favor of a simple-minded act of double-entry tabulation (balancing off "good things" against "bad things"). Surely, sooner or later, any reasonably curious mind would begin to wonder why the technological account finishes with a liability column of such significant size. Or, indeed, those with a utopian bent to their curiosity might wonder why there is a liability column at all. Why haven't all of man's inventiveness and organizational skill worked out for the best, as an unmixed blessing?

Unless we insist upon an examination of the human perversity that has again and again cheated us of the full promise of technology, we will have to settle down to the grotesque complacency of the prominent American physicist who found a strange consolation in the fate that would befall mankind in the radioactive environment following World War III. The fall-out would, he concluded, only shorten man's life span by about the same amount that modern medicine had lengthened it in the last sixty years. Science giveth and science taketh away: blessed be the name of science.

In such an attitude may lie the key to technology's indiscriminate scattering of goods and evils. What is it that offends us so much in this easy casting-up of life and death accounts but the technician's implicit denial that technology in all its manifold works has anything to do, *essentially*, with promoting human welfare? Sometimes it does, and sometimes it doesn't: technology produces penicillin *and* the H-bomb ...

and lets the chips fall where they may. It is the same attitude that is reflected in the consoling conclusion of another technician that a thermonuclear disaster, while producing untold devastation to human life and culture, might at the same time lead the way to a renaissance of the arts by clearing away the dead hand of the cultural past.

What the technicians seem to be saying (and they seem very content to be saying it) is that technology is some manner of chance force that has as its main object the unfolding of its own inherent capacities. Its benefits and its destructive features are mere byproducts. But, meanwhile, what goes without question is the assertion that the technology *must* be elaborated; it *must* go on "progressing." The technicians leave it to others to put their achievements to good use...or to pick up the pieces: a task that becomes increasingly more difficult as technology rapidly proliferates effects beyond anyone's anticipation or full understanding.

But if the technician's exclusive and often compulsive pursuit of progress is taken to have no inherent responsibility for the social results of that progress, then how are we to understand what the technician is about? What is the psychology of his project?

From the humanistic critic's viewpoint, the technician's lack of a clear, moral purpose may make technology seem a monstrosity, a chaos, a blind, erratic force throwing up unaccountable effects without reason or purpose. But what such a characterization overlooks is the fact that the technological project, for all the harm it does, has a very definite structure and discipline to it, an order that ought to be familiar to us all. The scientists, economists, engineers, entrepreneurs, managers, systems analysts, and bureaucrats who contribute to technological excellence are behaving far from blindly and erratically; their lives and careers are not lacking in purpose. They are seeking to do whatever they lay their hands to faster, more economically, and on a larger scale with less friction and with better control over more and more variable factors. And for all of these usually competitive objectives they can produce definite quantitative measures that are often ingeniously refined and marvelously discriminating. Moreover, each technician is pursuing a career within an established and articulated hierarchy which has a well developed scale of rewards and honors and some critical public to enforce legal or professional rules and standards. These are, of course, arbitrary—in the sense that the question "why" cannot be sensibly applied to them any more than it can be applied to, say, the standards of dress or the rules of courtship in our society.

We have a name for the sort of human activity that absorbs people in the orderly pursuit of arbitrary—usually competitive—goals according to arbitrary rules. We call it a "game." *Why* must an economy grow, *why* must profit be maximized, *why* must every bureaucracy expand and concentrate control, *why* must scientific truth and organizational efficiency and industrial productivity be ceaselessly elaborated? As soon as we grasp the fact that the expansion of technology is, for those who participate in the project, a game, these questions have the same logical status as the question: "Why is first base 90 feet from home plate?" or "Why is climbing a difficult mountain better than climbing an easy mountain?" And, of course, there is no answer to such questions, other than to reassert the rule.

The elaboration of the technological society is, then, a game played by its own rules and for its own well-established goals. To be sure, because the players manipulate people and resources, their game is bound to have effects "beyond" itself—in the "real" world outside its perimeters. But whether those effects are good or harmful is as irrelevant to the game as what happens to a baseball once it has been hit outside the ball park: perhaps a father catches it for his son and makes the boy happy; perhaps it hits a bystander in the head and kills him. "Wonderful," we may say, or "too bad"... but the game goes on.

One objects, of course: an activity embracing so many people and resources should not be merely a game. Nonetheless, technological progress *is* a game, and those who participate in making the technological society work, and in perfecting its techniques, possess the psychology of people playing games. It is the structure and discipline of game-playing that bound their understanding of themselves and their work. The important questions, then, are: How did technological progress become a game? And how can that game be stopped...or, at least, its rules altered?

None of this will sound new to those familiar with the work of the mathematicians Von Neumann and Morgenstern and of the school of game theorists that follows from them. Since the end of World War II their work has made clear that both war and business—the two greatest technological projects—have all the characteristics of games. But the purpose of the game theorists has emphatically *not* been to develop a radical critique of military and economic activity. They do not, for example, ask the question whether such activities ought properly to be games, or how their rules ought to be "moralized." In game theory there is no such category as "forbidden games." Rather, game theory has only reinforced the game structure and practice of war and business by elaborating more elegant strategic principles for their pursuit. Its effect has been to estrange business and war further from their flesh-and-blood implications by imposing a rarefied and intriguingly playful set of mathematical abstractions upon the real world in which men work and live and die.

What we want to know here is not how to play, more cunningly and self-consciously, games that should perhaps not be played at all, but rather how technological progress ever became a game in the first place and so lost its contact with the reality it molds...and too often torments.

TOOLS FOR CONVIVIALITY

IVAN ILLICH

Illich (1926-2002) was, like Roszak, a counterculture intellectual and philosopher. Born in Austria, he became a Catholic priest, working in New York, Mexico, and Puerto Rico. He left the Church and turned toward radical social criticism, and his penetrating analysis of large-scale institutions earned him wide acclaim. His notable books include *Deschooling Society* (1971), *Energy and Equity* (1974), *Medical Nemesis* (1975), and *Toward a History of Needs* (1978). The present book, *Tools for Conviviality*, was published in 1973.

Introduction

During the next several years I intend to work on an epilogue to the industrial age. I want to trace the changes in language, myth, ritual, and law which took place in the current epoch of packaging and of schooling. I want to describe the fading monopoly of the industrial mode of production and the vanishing of the industrially generated professions this mode of production serves.

Above all I want to show that *two-thirds of mankind still can avoid passing through the industrial age*, by choosing right now a postindustrial balance in their mode of production which the hyperindustrial nations will be forced to adopt as an alternative to chaos. To prepare for this task I submit this essay for critical comment.

For several years at CIDOC in Cuernavaca, Mexico we have conducted critical research on the monopoly of the industrial mode of production and have tried to define conceptually alternative modes that would fit a postindustrial age. During the late sixties this research centered on educational devices. By 1970 we had found that:

(1.) Universal education through compulsory schooling is not possible.

(2.) Alternative devices for the production and marketing of mass education are technically more feasible and ethically less tolerable than compulsory graded schools. Such new educational arrangements are now on the verge of replacing traditional school systems in rich and in poor countries. They are potentially more effective in the conditioning of jobholders and consumers in an industrial economy. They are therefore more attractive for the management of present societies, more seductive for the people, and insidiously destructive of fundamental values.

(3.) A society committed to high levels of shared learning and critical personal intercourse must set pedagogical limits on industrial growth.

I have published the results of this research in a previous volume of World Perspectives, entitled *Deschooling Society*. I clarified some of the points left ill-defined in that book by writing an article published in the *Saturday Review* of April 19, 1971.

Our analysis of schooling has led us to recognize the mass production of education as a paradigm for other industrial enterprises, each producing a service commodity, each organized as a public utility, and each defining its output as a basic necessity. At first our attention was drawn to the compulsory insurance of professional health care, and to systems of public transport, which tend to become compulsory once traffic rolls above a certain speed. We found that the industrialization of any service agency leads to destructive side effects analogous to the unwanted secondary results well known from the overproduction of goods. We had to face a set of limits to growth in the service sector of any society as inescapable as the limits inherent in the industrial production of artifacts. We concluded that a set of limits to industrial growth is well formulated only if these limits apply both to goods and to services which are produced in an industrial mode. So we set out to clarify these limits.

I here submit the concept of a multidimensional balance of human life which can serve as a framework for evaluating man's relation to his tools. In each of several dimensions of this balance it is possible to identify a *natural scale. When an enterprise grows beyond a certain point on this scale, it first frustrates the end for which it was originally designed, and then rapidly becomes a threat to society itself.* These scales must be identified, and the parameters of human endeavors within which human life remains viable must be explored.

Society can be destroyed when further growth of mass production renders the milieu hostile, when it extinguishes the free use of the natural abilities of society's members, when it isolates people from each other and locks them into a man-made shell, when it undermines the texture of community by promoting extreme social polarization and splintering specialization, or when cancerous acceleration enforces social change at a rate that rules out legal, cultural, and political precedents as formal guidelines to present behavior. Corporate endeavors which thus threaten society cannot be tolerated. At this point it becomes irrelevant whether an enterprise is nominally owned by individuals, corporations, or the state, because no form of management can make such fundamental destruction serve a social purpose.

Our present ideologies are useful to clarify the contradictions which appear in a society which relies on the capitalist control of industrial production; they do not, however, provide the necessary framework for analyzing the crisis in the industrial mode of production itself. I hope that one day a general theory of industrialization will be stated with precision, that it will be formulated in terms compelling enough to withstand the test of criticism. Its concepts ought to provide a common language for people in opposing parties who need to engage in the assessment of social programs or technologies, and who want to restrain the power of man's tools when they tend to overwhelm man and his goals. Such a theory should help people *invert* [i.e. overthrow] the present structure of major institutions. I hope that this essay will enhance the formulation of such a theory.

It is now difficult to imagine a modern society in which industrial growth is balanced and kept in check by several complementary, distinct, and equally scientific modes of production. Our vision of the possible and the feasible is so restricted by industrial expectations that any alternative to more mass production sounds like a return to past oppression or like a Utopian design for noble savages. In fact, however, the vision of new possibilities requires only the recognition that scientific discoveries can be used in at least two opposite ways. The first leads to specialization of functions, institutionalization of values and centralization of power and turns people into the accessories of bureaucracies or machines. The second enlarges the range of each person's competence, control, an initiative, limited only by other individuals' claims to an equal range of power and freedom.

To formulate a theory about a future society both very modern and not dominated by industry, it will be necessary to recognize *natural scales and limits*. We must come to admit that only within limits can machines take the place of slaves; beyond these limits they lead to a new kind of serfdom. Only within limits can education fit people into a man-made environment: beyond these limits lies the universal schoolhouse, hospital ward, or prison. Only within limits ought politics to be concerned with the distribution of maximum industrial outputs, rather than with equal inputs of either energy or information. Once these limits are recognized, it becomes possible to articulate the triadic relationship between *persons*, *tools*, and a *new collectivity. Such a society, in which modern technologies serve politically interrelated individuals rather than managers, I will call "convivial."*

After many doubts, and against the advice of friends whom I respect, I have chosen "convivial" as a technical term to designate a *modern society of responsibly limited tools*. In part this choice was conditioned by the desire to continue a discourse which had started with its Spanish cognate. I am aware that in English "convivial" now seeks the company of 'tipsy jollyness', which is distinct from that indicated by the Old English Dictionary and opposite to the austere meaning of modern "*eutrapelia*," (or graceful playfulness) which I intend. By applying the term "convivial" to tools rather than to people, I hope to forestall confusion.

"Austerity," which says something about people, has also been degraded and has acquired a bitter taste, while for Aristotle or Aquinas it marked the foundation of friendship. For Aquinas "austerity" is a complementary part of a more embracing virtue, which he calls friendship or joyfulness. It is the fruit of an apprehension that things or tools could destroy rather than enhance *eutrapelia* in personal relations.

Chapter 2 — Convivial Reconstruction

The symptoms of accelerated crisis are widely recognized. Multiple attempts have been made to explain them. I believe that this crisis is rooted in a major twofold experiment which has failed, and I claim that the resolution of the crisis begins with a recognition of the failure. For a hundred years we have tried to make machines work for men and to school men for life in their service. Now it turns out that machines do not "work" and that people cannot be schooled for a life at the service of machines.

The hypothesis on which the experiment was built must now be discarded. The hypothesis was that machines can replace slaves. The evidence shows that, used for this purpose, machines enslave men. Neither a dictatorial proletariat nor a leisure mass can escape the dominion of constantly expanding industrial tools.

The crisis can be solved only if we learn to invert the present deep structure of tools; if we give people tools that guarantee their right to work with high, independent efficiency, thus simultaneously eliminating the need for either slaves or masters and enhancing each person's range of freedom. People need new tools to work with rather than tools that "work" for them. They need technology to make the most of the energy and imagination each has, rather than more well-programmed energy slaves.

I believe that society must be *reconstructed* to enlarge the contribution of autonomous individuals and primary groups to the total effectiveness of a new system of production designed to satisfy the human needs which it also determines. In fact, the institutions of industrial society do just the opposite. As the power of machines increases, the role of persons more and more decreases to that of mere consumers.

In an age of scientific technology, the convivial structure of tools is a necessity for survival in full justice which is both distributive and participatory. This is so because science has opened new energy sources. Competition for inputs must lead to destruction, while their central control in the hands of a Leviathan would sacrifice equal control over inputs to the semblance of an equal distribution of outputs. Rationally designed convivial tools have become the basis for participatory justice.

But this does not mean that the transition from our present to a convivial mode of production can be accomplished without serious threats to the survival of many people. At present the relationship between people and their tools is suicidally distorted. The survival of Pakistanis depends on Canadian grain, and the survival of New Yorkers on world-wide exploitation of natural resources. The birth pangs of a convivial world society will inevitably be violently painful for hungry Indians and for helpless New Yorkers. I will later argue that the transition from the present mode of production, which is overwhelmingly industrial, toward conviviality may start suddenly. But for the sake of the survival of many people it will be desirable that the transition does not happen all at once. I argue that survival in justice is possible only at the cost of those sacrifices implicit in the adoption of a convivial mode of production and the universal renunciation of unlimited progeny, affluence, and power on the part of both individuals and groups. This price cannot be extorted by some despotic Leviathan, nor elicited by social engineering. People will rediscover the value of joyful sobriety and liberating austerity only if they relearn to depend on each other rather than on energy slaves. The price for a convivial society will be paid only as the result of a political process which reflects and promotes the society-wide inversion of present industrial consciousness. This political process will find its concrete expression not in some taboo, but in a series of temporary agreements on one or the other concrete limitation of means, constantly adjusted under the pressure of conflicting insights and interests.

In this volume I want to offer a methodology by which to recognize means which have turned into ends. My subject is tools and not intentions. [...]

In the past, convivial life for some inevitably demanded the servitude of others. Labor efficiency was low before the steel ax, the pump, the bicycle, and the nylon fishing line. Between the High Middle Ages and the Enlightenment, the alchemic dream misled many otherwise authentic Western humanists. The illusion prevailed that the machine was a laboratory-made homunculus [i.e. robot], and that it could do our labor instead of slaves. It is now time to correct this mistake and shake off the illusion that men are born to be slaveholders and that the only thing wrong in the past was that not all men could be equally so. By reducing our expectations of machines, however, we must guard against falling into the equally damaging rejection of all machines as if they were works of the devil.

A convivial society should be designed to allow all its members the most autonomous action by means of tools least controlled by others. People feel joy, as opposed to mere pleasure, to the extent that their activities are creative; while the growth of tools beyond a certain point increases regimentation, dependence, exploitation, and impotence. I use the term "tool" broadly enough to include not only simple hardware such as drills, pots, syringes, brooms, building elements, or motors, and not just large machines like cars or power stations; I also include among tools productive institutions such as factories that produce tangible commodities like corn flakes or electric current, and productive systems for intangible commodities such as those which produce "education," "health," "knowledge," or "decisions." I use this term because it allows me to subsume into one category all rationally designed devices, be they artifacts or rules, codes or operators, and to distinguish all these planned and engineered instrumentalities from other things such as basic food or implements, which in a given culture are not deemed to be subject to rationalization. School curricula or marriage laws are no less purposely shaped social devices than road networks.

Tools are intrinsic to social relationships. An individual relates himself in action to his society through the use of tools that he actively masters, or by which he is passively acted upon. To the degree that he masters his tools, he can invest the world with his meaning; to the degree that he is mastered by his tools, the shape of the tool determines his own self-image. Convivial tools are those which give each person who uses them the greatest opportunity to enrich the environment with the fruits of his or her vision. Industrial tools deny this possibility to those who use them and they allow their designers to determine the meaning and expectations of others. Most tools today cannot be used in a convivial fashion.

Hand tools are those which adapt man's metabolic energy to a specific task. They can be multipurpose, like some primitive hammers or good modern pocket knives, or again they can be highly specific in design such as spindles, looms, or pedal-driven sewing machines, and dentists' drills. They can also be complex such as a transportation system built to get the most in mobility out of human energy—for instance, a bicycle system composed of a series of man-powered vehicles, such as pushcarts and three-wheel rickshaws, with a corresponding road system equipped with repair stations and perhaps even covered roadways. Hand tools are mere transducers of the energy generated by man's extremities and fed by the intake of air and of nourishment.

Power tools are moved, at least partially, by energy converted outside the human body. Some of them act as amplifiers of human energy: the oxen pull the plow, but man works with the oxen—the result is obtained by pooling the powers of beast and man. Power saws and motor pulleys are used in the same fashion. On the other hand, the energy used to steer a jet plane has ceased to be a significant fraction of its power output. The pilot is reduced to a mere operator guided by data which a computer digests for him. The machine needs him for lack of a better computer; or he is in the cockpit because the social control of unions over airplanes imposes his presence.

Tools foster conviviality to the extent to which they can be easily used, by any-body, as often or as seldom as desired, for the accomplishment of a purpose chosen by the user. The use of such tools by one person does not restrain another from using them equally. They do not require previous certification of the user. Their existence does not impose any obligation to use them. They allow the user to express his meaning in action.

Every aspect of industrial societies has become part of a larval [i.e. hidden] system for escalating production and increasing the demand necessary to justify the total social cost. For this reason, criticism of bad management, official dishonesty, insufficient research, or technological lag distracts public attention from the one issue that counts: *careful analysis of the basic structure of tools as means.* It is equally distracting to suggest that the present frustration is primarily due to the private ownership of the means of production, and that the public ownership of these same factories under the tutelage of a planning board could protect the interest of the majority and lead society to an equally shared abundance. As long as Ford Motor Company can be condemned simply because it makes Ford rich, the illusion is bolstered that the same factory could make the public rich. As long as people believe that the public can profit from cars, they will not condemn Ford for making cars. The issue at hand is not the juridical ownership of tools, but rather the discovery of the characteristic of some tools which make it impossible for anybody to "own" them. The concept of 'ownership' cannot be applied to a tool that cannot be controlled.

The issue at hand, therefore, is what tools can be controlled in the public interest. Only secondarily does the question arise whether private control of a potentially useful tool is in the public interest.

Certain tools are destructive no matter who owns them, whether it be the Mafia, stockholders, a foreign company, the state, or even a workers' commune. Networks of multilane highways, long-range, wide-bandwidth transmitters, strip mines, or compulsory school systems are such tools. Destructive tools must inevitably increase regimentation, dependence, exploitation, or impotence, and rob not only the rich but also the poor of conviviality, which is the primary treasure in many so-called "underdeveloped" areas.

Chapter 3 – The Multiple Balance

Radical Monopoly

When overefficient tools are applied to facilitate man's relations with the physical environment, they can destroy the balance between man and nature. Overefficient tools corrupt the environment. But tools can also be made overefficient in quite a different way. They can upset the relationship between what people need to do by themselves and what they need to obtain ready-made. In this second dimension overefficient production results in radical monopoly.

By radical monopoly I mean a kind of dominance by one product that goes far beyond what the concept of monopoly usually implies. Generally we mean by "monopoly" the exclusive control by one corporation over the means of producing (or selling) a commodity or service. Coca-Cola can create a monopoly over the soft-drink market in Nicaragua by being the only maker of soft drinks which advertises with modern means. Nestle might impose its brand of cocoa by controlling the raw material, some car maker by restricting imports of other makes, a television channel by licensing. Monopolies of this kind have been recognized for a century as dangerous by-products of industrial expansion, and legal devices have been developed in a largely futile attempt to control them. Monopolies of this kind restrict the choices open to the consumer. They might even compel him to buy one product on the market, but they seldom simultaneously abridge his liberties in other domains. A thirsty man might desire a cold, gaseous, and sweet drink and find himself restricted to the choice of just one brand. He still remains free to quench his thirst with beer or water. Only if and when his thirst is translated without meaningful alternatives into the need for a Coke would the monopoly become radical. By "radical monopoly" I mean the dominance of one *type* of product rather than the dominance of one brand. I speak about radical monopoly when one industrial production process exercises an exclusive control over the satisfaction of a pressing need, and excludes non-industrial activities from competition.

Cars can thus monopolize traffic. They can shape a city into their image—practically ruling out locomotion on foot or by bicycle in Los Angeles. They can eliminate river traffic in Thailand. That motor traffic curtails the right to walk, not that more people drive Chevies than Fords, constitutes radical monopoly. What cars do to people by virtue of this radical monopoly is quite distinct from and independent of what they do by burning gasoline that could be transformed into food in a crowded world. It is also distinct from automotive manslaughter. Of course cars burn gasoline that could be used to make food. Of course they are dangerous and costly. But the radical monopoly cars establish is destructive in a special way. *Cars create distance.* Speedy vehicles of all kinds render space scarce. They drive wedges of highways into populated areas, and then extort tolls on the bridge over the remoteness between people that was manufactured for their sake. This monopoly over land turns space into car fodder. It destroys the environment for feet and bicycles. Even if planes and buses could run as nonpolluting, non-depleting public services, their inhuman velocities would degrade man's innate mobility and force him to spend more time for the sake of travel.

Schools tried to extend a radical monopoly on learning by redefining it as education. As long as people accepted the teacher's definition of reality, those who learned outside school were officially stamped "uneducated." Modern medicine deprives the ailing of care not prescribed by doctors. Radical monopoly exists where a major tool rules out natural competence. Radical monopoly imposes compulsory consumption and thereby restricts personal autonomy. It constitutes a special kind of social control because it is enforced by means of the imposed consumption of a standard product that only large institutions can provide.

The control of *undertakers* over burial shows how radical monopoly functions and how it differs from other forms of culturally defined behavior. A generation ago, in Mexico, only the opening of the grave and the blessing of the dead body were performed by professionals: the gravedigger and the priest. A death in the family created various demands, all of which could be taken care of within the family. The wake, the funeral, and the dinner served to compose quarrels, to vent grief, and to remind each participant of the fatality of death and the value of life. Most of these were of a ritual nature and carefully prescribed—different from region to region. Recently, funeral homes were established in the major cities. At first undertakers had difficulty finding clients because even in large cities people still knew how to bury their dead. During the sixties the funeral homes obtained control over new cemeteries and began offering package deals, including the casket, church service, and embalming. Now legislation is being passed to make the mortician's ministrations compulsory. Once he gets hold of the body, the funeral director will have established a radical monopoly over burial, as medicine is at the point of establishing one over dying.

The current debate over *health-care delivery* in the United States clearly illustrates the entrenchment of a radical monopoly. Each political party in the debate makes sick-care a burning public issue and thereby relegates health care to an area about which politics has nothing important to say. Each party promises more funds to doctors, hospitals, and drugstores. Such promises are not in the interest of the majority. They only serve to increase the power of a minority of professionals to prescribe the tools men are to use in maintaining health, healing sickness, and repressing death. More funds will strengthen the hold of the health industry over public resources and heighten its prestige and arbitrary power. Such power in the hands of a minority will produce only an increase in suffering and a decrease in personal self-reliance. More money will be invested in tools that only postpone unavoidable death and in services that abridge even further the civil rights of those who want to heal each other. More money spent under the control of the health profession means that more people are operationally conditioned into playing the role of the sick, a role they are not allowed to interpret for themselves. Once they accept this role, their most trivial needs can be satisfied only through commodities that are scarce by professional definition.

People have a native capacity for healing, consoling, moving, learning, building their houses, and burying their dead. Each of these capacities meets a need. The means for the satisfaction of these needs are abundant so long as they depend primarily on what people can do for themselves, with only marginal dependence on commodities. These activities have use-value without having been given exchange-value. Their exercise at the service of man is not considered labor.

These basic satisfactions become scarce when the social environment is transformed in such a manner that basic needs can no longer be met by abundant competence. The establishment of radical monopoly happens when people give up their native ability to do what they can do for themselves and for each other, in exchange for something "better" that can be done for them only by a major tool. Radical monopoly reflects the industrial institutionalization of values. It substitutes the standard package for the personal response. It introduces new classes of scarcity and a new device to classify people according to the level of their consumption. This redefinition raises the unit cost of valuable service, differentially rations privilege, restricts access to resources, and makes people dependent. Above all, by depriving people of the ability to satisfy personal needs in a personal manner, radical monopoly creates radical scarcity of personal—as opposed to institutional—service.

Against this radical monopoly *people need protection*. They need this protection whether consumption is imposed by the private interests of undertakers, by the government for the sake of hygiene, or by the self-destructive collusion between the mortician and the survivors, who want to do the best thing for their dear departed. They need this protection even if the majority is now sold on the professional's services. Unless the need for protection from radical monopoly is recognized, its multiple implementation can break the tolerance of man for enforced inactivity and passivity.

It is not always easy to determine what constitutes compulsory consumption. The monopoly held by schools is not established primarily by a law that threatens punishment to parent or child for truancy. Such laws exist, but school is established by other tactics: by discrimination against the unschooled, by centralizing learning tools under the control of teachers, by restricting public funds earmarked for baby-sitting to salaries for graduates from normal schools. Protection against laws that impose education, vaccination, or life prolongation is important, but it is not sufficient. Procedures must be used that permit any party who feels threatened by compulsory consumption to claim protection, whatever form the imposition takes. Like intolerable pollution, intolerable monopoly cannot be defined in advance. The threat can be anticipated, but the definition of its precise nature can result only from people's participation in deciding what may not be produced.

Protection against this general monopoly is as difficult as protection against pollution. People will face a danger that threatens their own self-interest but not one that threatens society as a whole. Many more people are against 'cars' than are against driving them. They are against cars because they pollute and because they monopolize traffic. They drive cars because they consider the pollution created by one car insignificant, and because they do not feel personally deprived of freedom when they drive. It is also difficult to be protected against monopoly when a society is already littered with roads, schools, or hospitals, when independent action has been paralyzed for so long that the ability for it seems to have atrophied, and when simple alternatives seem beyond the reach of the imagination. Monopoly is hard to get rid of when it has frozen not only the shape of the physical world but also the range of behavior and of imagination. Radical monopoly is generally discovered only when it is too late.

Commercial monopoly is broken at the cost of the few who profit from it. Usually, these few manage to evade controls. The cost of radical monopoly is already borne by the public and will be broken only if the public realizes that it would be better off paying the costs of ending the monopoly than by continuing to pay for its maintenance. But the price will not be paid unless the public learns to value the potential of a convivial society over the illusion of progress. It will not be paid voluntarily by those who confuse conviviality with intolerable poverty.

Some of the symptoms of radical monopoly are reaching public awareness, above all the degree to which frustration grows faster than output in even the most highly developed countries and under whatever political regime. Policies aimed to ease this frustration may easily distract attention from the general nature of the monopoly at its roots, however. The more these reforms succeed in correcting superficial abuses, the better they serve to bolster the monopoly I am trying to describe.

The first palliative is *consumer protection*. Consumers cannot do without cars. They buy different makes. They discover that most cars are unsafe at any speed. So they organize to get safer, better, and more durable cars and to get more as well as wider and safer roads. Yet when consumers gain more confidence in cars, *the victory only increases society's dependence* on high-powered vehicles — public or private — and frustrates even more those who have to, or would prefer to, walk.

While the organized self-protection of the addict-consumer immediately raises the quality of the dope and the power of the peddler, it also may lead ultimately to limits on growth. Cars may finally become too expensive to purchase and medicines too expensive to test. By exacerbating the contradictions inherent in this institutionalization of values, majorities can more easily become aware of them. Discerning consumers who are discriminatory in their purchasing habits may finally discover that they can do better by doing things for themselves.

The second palliative proposed to cure growing frustration with growing output is *planning*. The illusion is common that planners with socialist ideals might somehow create a socialist society in which industrial workers constitute a majority. The proponents of this idea overlook the fact that anti-convivial and manipulative tools can fit into a socialist society in only a very limited measure. Once transportation, education, or medicine is offered by a government free of cost, its use can be enforced by moral guardians. The underconsumer can be blamed for sabotage of the national effort. In a market economy, someone who wants to cure his flu by staying in bed will be penalized only through loss of income. In a society that appeals to the "people" to meet centrally determined production goals, resistance to the consumption of medicine becomes an act of public immorality. Protection against radical monopoly depends on a political consensus opposed to growth. Such a consensus is diametrically opposed to the issues now raised by political oppositions, since these converge in the demand to *increase* growth and to provide more and better things for more completely disabled people.

People must learn to live within bounds. This cannot be taught. Survival depends on people learning fast what they cannot do. They must learn to abstain from unlimited progeny, consumption, and use. It is impossible to educate people for voluntary poverty or to manipulate them into self-control. It is impossible to teach joyful renunciation in a world totally structured for higher output and the illusion of 'declining costs'.

When ends become subservient to the tools chosen for their sake, the user first feels frustration and finally either abstains from their use or goes mad. Compulsory maddening behavior in Hades was considered the ultimate punishment reserved for blasphemy. Sisyphus was forced to keep rolling a stone uphill, only to see it roll back down. When maddening behavior becomes the standard of a society, people learn to compete for the right to engage in it. Envy blinds people and makes them compete for addiction.

Wherever the maximum velocity of any one type of commuter vehicle grows beyond a certain mph, the travel time and the cost of transportation for the median commuter is increased. If the maximum velocity at any one point of a commuter system goes beyond a certain mph, most people are obliged to spend more time in traffic jams, or waiting for connections, or recovering from accidents. They will also have to spend more time paying for the transportation system they are compelled to use.

The *critical velocity* depends to a certain extent on a variety of factors: geography, culture, market controls, level of technology, and money flow. With so many variables affecting a quantity, it would seem that its value could fluctuate over a very wide range. Just the contrary is true. Once it is understood that we refer to any vehicular velocity in the transportation of people within a community, we find that the range within which the critical velocity can vary is very narrow. It is, in fact, so narrow and so low that it seems improbable and not worth the time of most traffic engineers to worry about.

Commuter transportation leads to negative returns when it admits, anywhere in the system, speeds much above those reached on a bicycle. Once the barrier of bicycle velocity is broken at any point in the system, the total per capita monthly time spent at the service of the travel industry increases.

Counterfoil research is concerned first with an analysis of increasing marginal disutility and the menace of growth. It is then concerned with the discovery of general systems of institutional structure which optimize convivial production. This kind of research meets psychological resistance. *Growth has become addictive.* Like heroin addiction, the habit distorts basic value judgments. *Addicts of any kind are willing to pay increasing amounts for declining satisfactions.* They have become tolerant to escalating marginal disutility. They are blind to deeper frustration because they are absorbed in playing for always mounting stakes.

Chapter 5 – Political Inversion

If within the very near future man cannot set limits to the interference of his tools with the environment, and practice effective birth control, the next generations will experience the gruesome apocalypse predicted by many ecologists. Faced with these impending disasters, society can stand in wait of survival within limits set and enforced by bureaucratic dictatorship. *Or* it can engage in a political process by the use of legal and political procedures. Ideologically biased interpretations of the past have made the recognition of political process increasingly difficult. Liberty has been interpreted as a right to power tools, a right claimed without reasonable limitation by individuals and private associations in capitalist countries and by the state in socialist societies. Recovery becomes feasible only if the fundamental structure of Western societies is clearly recognized and reclaimed. Analogous efforts to recover entirely different formal structures will become necessary when former political or cultural colonies shake off the Western mode of production.

The bureaucratic management of human survival is unacceptable on both ethical and political grounds. It would also be as futile as former attempts at mass therapy. This does not, of course, mean that a majority might not at first submit to it. People could be so frightened by the increasing evidence of growing population and dwindling resources that they would voluntarily put their destiny into the hands of Big Brothers. Technocratic caretakers could be mandated to set limits on growth in every dimension, and to set them just at the point beyond which further production would mean utter destruction. Such a *kakotopia* could maintain the industrial age at the highest endurable level of output.

Man would live in a plastic bubble that would protect his survival and make it increasingly worthless. Since man's tolerance would become the most serious limitation to growth, the alchemist's endeavor would be renewed in the attempt to produce a monstrous type of man, fit to live among reason's dreams. A major function of engineering would become the psychogenetic tooling of man himself as a condition for further growth. People would be confined from birth to death in a world-wide schoolhouse, treated in a world-wide hospital, surrounded by television screens, and the manmade environment would be distinguishable in name only from a world-wide prison.

The alternative to managerial fascism is a political process by which people decide how much of any scarce resource is the most any member of society can claim; a process in which they agree to keep limits relatively stationary over a long time, and by which they set a premium on the constant search for new ways to have an ever larger percentage of the population join in doing ever more with ever less. Such a political choice of a *frugal society* remains a pious dream unless it can be shown that it is not only necessary but also possible: (1) to define concrete procedures by which more people are enlightened about the nature of our present crisis and will come to understand that limits are necessary and a convivial life style desirable; (2) to bring the largest number of people into now-suppressed organizations which claim their right to a frugal life style and keep them satisfied and therefore committed to convivial life; and (3) to discover and revalue the political or legal tools that are accepted within a society and learn how to use them to establish and protect convivial life where it emerges. Such procedures may sound idealistic at the present moment. This is not proof that they cannot become effective as the present crisis deepens.

1. *Myths and Majorities*

The ultimate obstacle to the restructuring of society is not the lack of information about which limits are needed, nor the lack of people who would accept them if they became inevitable, but the power of political myths.

Almost everyone in rich societies is a destructive consumer. Almost everyone is, in some way, engaged in aggression against the milieu. Destructive consumers constitute a numerical majority. Myth transforms them into a political one. Numerical majorities come to form a mythical voting bloc on a nonexistent issue; "they" are invoked as the unbeatable guardians of vested interest in growth. This mythical majority paralyzes political action. At closer inspection, "they" are a number of reasonable individuals. One is an ecologist who takes a jet plane to a conference on protecting the environment from further pollution. Another is an economist who knows that growing efficiency renders work increasingly scarce; he tries to create new sources of employment. Neither of them has the same interests as the slum-dweller in Detroit who purchases his color TV on time. The three belong no more to a voting bloc that will defend growth than clerks, repairmen, and salesmen are somehow politically homogenized because each fears for his job, needs a car, and wants medicine for his children.

There can be no such thing as a majority opposed to an issue that has not arisen. A majority agitating for limits to growth is as ludicrous a concept as one demanding growth at all cost. Majorities are not created by shared ideologies. They develop out of enlightened self-interest. The most that even the best of ideologies can do is interpret this interest. The stance each man or woman takes when a social problem becomes an overwhelming threat depends on two factors: the first is how a smoldering conflict erupts into a political issue demanding attention and partisan action; the second is the existence of new elites which can provide an interpretative framework for new — and hitherto unexpected — alignments of interest.

2. *From Breakdown to Chaos*

I can only conjecture on how the breakdown of industrial society will ultimately become a critical issue. But I can make rather firm statements about the qualifications for providing guidance within the coming crisis. I believe that *growth will grind to a halt*. The *total collapse* of the industrial monopoly on production will be the result of synergy in the failure of the multiple systems that fed its expansion. This expansion is maintained by the illusion that careful systems engineering can stabilize and harmonize present growth, while in fact it pushes all institutions simultaneously toward their critical watershed. *Almost overnight* people will lose confidence not only in the major institutions but also in the miracle prescriptions of the would-be crisis managers. The ability of present institutions to define values such as education, health, welfare, transportation, or news will suddenly be extinguished because it will be recognized as an illusion.

This crisis may be triggered by an unforeseen event, as the Great Depression was touched off by the Wall Street Crash. Some fortuitous coincidence will render publicly obvious the structural contradictions between stated purposes and effective results in our major institutions. People will suddenly find obvious what is now evident to only a few: *that the organization of the entire economy toward the "better" life has become the major enemy of the* good *life*. Like other widely shared insights, this one will have the potential of turning public imagination inside out. Large institutions can quite suddenly lose their respectability, their legitimacy, and their reputation for serving the public good. It happened to the Roman Church in the Reformation, to Royalty in the Revolution. *The unthinkable became obvious overnight*: that people could and would behead their rulers.

Sudden change is of a different order than feedback or evolution. Observe the whirlpools below a waterfall. For many seasons the eddies stay in the same place no matter whether the water is high or low. Then, suddenly, one more stone falls into the basin, the entire array changes, and the old can never be reconstructed. People who invoke the specter of a hopelessly growth-oriented majority seem incapable of envisaging political behavior in a crash. Business ceases to be as usual when the populace loses confidence in industrial productivity, and not just in paper currency.

It is still possible to face the breakdown of each of our various systems in a separate perspective. No remedy seems to work, but we can still find resources to support every remedy proposed. Governments think they can deal with the breakdown of utilities, the disruption of the educational system, intolerable transportation, the chaos of the judicial process, the violent disaffection of the young. Each is dealt with as a separate phenomenon, each is explained by a different report, each calls for a new tax and a new program. Squabbles about alternative remedies give credibility to both: free schools vs. public schools double the demand for education; satellite cities vs. monorails for commuters make the growth of cities seem inexorable; higher professional standards in medicine vs. more paramedical professions further aggrandize the health professions. Since each of the proposed remedies appeals to some, the usual solution is an attempt to try both. The result is a further effort to make the pie grow, and to forget that it is pie in the sky.

The Coolidge approach to the warnings of the Depression is now applied to the signs of a much more radical crisis. General systems analysis is trusted to relate the institutional breakdowns to each other, which only leads to more planning, centralization, and bureaucracy in order to achieve control over population, affluence, and inefficient industry. Unemployment in the manufacturing sector is supposed to be compensated for by growth in the output of decisions, controls, and therapies. Fascination with industry and mechanical production still blinds people to the possibility of a postindustrial society in which several distinct modes of production would complement each other. Trying to bring about an era which is both hyperindustrial and ecologically feasible, they accelerate the breakdown of several other nonphysical and equally fundamental dimensions of the balance of life.

It would be a mere exercise in geomancy to predict which series of events will play the role of the Wall Street Crash as catalyst of the first crisis of, not just in, industrial society. But it would be folly not to expect in the very near future an event whose effects will jam the growth of tools. When this happens, the noise that accompanies the crash will distract attention from seeing it in proper perspective.

We still have a chance to understand the causes of the coming crisis, and to prepare for it. If we are to anticipate its effects, we must investigate how sudden change can bring about the emergence into power of previously submerged social groups. It is not calamity as such that creates these groups; it is much less calamity that brings about their emergence; but calamity weakens the prevailing powers which have excluded the submerged from participation in the social process. It is the power of surprise that weakens control, that shakes up the established controllers, and brings to the top those people who have not lost their bearings.

3. *Insight into Crisis*

Forces tending to limit production are already at work within society. Public, counterfoil research can significantly help these individuals become more cohesive and self-conscious in their indictment of growth they consider destructive. We can anticipate that their voices will acquire new resonance when the crisis of overproductive society becomes acute. They form no constituency, but they are spokesmen for a majority of which everyone is a potential member. The more unexpectedly the crisis comes, the more suddenly their velleities [i.e. inclinations] can turn into a program. But the ability to direct events at that moment depends on how well these minorities grasp the profound nature of the crisis, and know how to state it in effective language: *to declare what they want, what they can do,* and *what they do not need.* The critical use of ordinary language is the first pivot in a political inversion. A second pivot is needed.

Further growth *must lead to a multiple catastrophe.* That people would accept multiple limits to growth without catastrophe seems highly improbable. The inevitable catastrophic event could be either a *crisis in civilization* or its *end*: end by annihilation or end in B. F. Skinner's world-wide concentration camp run by a T. E. Frazier. The foreseeable catastrophe will be a true crisis—that is, the occasion for a choice—only if at the moment it strikes the necessary social demands can be effectively expressed. They must be represented by people who can demonstrate that the breakdown of the current industrial illusion is for them a condition for choosing an effective and convivial mode of production. The preparation of such groups is the key task of new politics at the present moment.

I have already argued that these groups must be prepared to provide a logically coherent analysis of the catastrophic event and to communicate it in ordinary language. I have argued that they must be prepared to propose the necessity for a *bounded society* in practical terms that have general appeal. Sacrifice must be shown as the inevitable price for different groups of people to get what they want—or at least to be liberated from what has become intolerable. But beyond using words to describe the limits as both necessary and appealing, the leadership of these groups must be prepared to use a social tool that is fit to ordain what is good enough for all. It must be a tool which, like language, is respected by all; a tool which, like language, does not lose its power because of the purpose to which it has been put in recent history; a tool which, like language, possesses a fundamental structure that misuse cannot totally corrupt.

4. *Sudden Change*

When I speak about emerging interest groups and their preparation, I am not speaking of action groups, or of a church, or of new kinds of experts. I am above all *not* speaking about one political party which could assume power at a moment of crisis. Management of the crisis would make catastrophe irreversible. A well-knit, well-trained party can establish its power at the moment of a crisis in which the choice to be made is one within an over-all system. Such was the Great Depression. What was at issue was control over the tools of production. Such were the events which brought the Marxists to power in Eastern Europe. But the crisis I have described as imminent is not a crisis *within* industrial society, but a crisis *of the industrial mode of production itself.* The crisis I have described confronts people with a choice between convivial tools and being crushed by machines. The only response to this crisis is a full recognition of its depth and an acceptance of inevitable self-limitations. The more varied the perspectives from which this insight is shared by interest groups and the more disparate the interests that may be protected only by a reduction of power within society, the greater the probability that the inevitable will be recognized as such.

The proponents of a bounded society have no need to put together some kind of majority. A voting majority in a democracy is not motivated by the explicit commitment of all its members to some specific ideology or to some particular value. A voting majority in favor of a specific institutional limitation would have to be composed of very disparate elements: those seriously aggrieved by some aspect of overproduction, those who do not profit from it, and those who may have objections to the overall organization of society — but not directly to the specific limit being set. How this functions in times of normal politics can be well illustrated by the example of *school*. Some people are childless and resent the school tax. Others feel they are taxed more heavily and served less well than their peers in another district. Others object to tax support of schools since they want to send their children to parochial schools. Others object to compulsory schooling as such: some because it does harm to the young and others because it fosters discrimination. All these people could form a voting majority, but not a party or a sect. Under present circumstances they might succeed in cutting school down to size, but thereby they would merely assure its more legitimate survival. A majority vote to limit one major institution tends to be conservative when business is as usual.

But a majority can have the contrary effect in a crisis which affects society on a deeper level. The joint arrival of several institutions at their critical watershed is the beginning of such a crisis. The crash that will follow must make it clear that *industrial society as such* — and not just its separate institutions — *has outgrown the range of its effectiveness.*

The nation-state has become so powerful that it cannot perform its stated functions. Just as [Vietnamese] General Vo Nguyen Giap could use the U.S. military machine to win his war, so the multinational corporations and professions can now use the law and the two-party system to establish their empire. But while democracy in the United States can survive a victory by Giap, it cannot survive one by ITT [International Telephone and Telegraph] and its like. As a total crisis approaches, it becomes more obvious that the nation-state has grown into the holding corporation for a multiplicity of self-serving tools, and the political party into an instrument to organize stockholders for the occasional election of boards and presidents. In this situation, parties support each voter's right to claim higher levels of individual consumption and to enforce thereby higher levels of industrial consumption. People can claim cars, but the appropriation of society's over-all resources by a transportation system which determines that cars are useful is left to the decision of experts. Such parties support a state whose only purpose is the support of an increasing GNP, and they are obviously useless at the moment of a general crash.

Reconstruction for poor countries means adopting a set of negative design criteria within which their tools are kept, in order to advance directly into a postindustrial era of conviviality. The limits to choose are of the same order as those which hyperindustrialized countries will have to adopt for the sake of survival and at the cost of their vested interest. Such social reconstruction cannot be supported by a high-powered army, both because the maintenance of such an army would foil reconstruction and because no such army would be powerful enough. *Defense of conviviality is possible only if undertaken by the people with tools they control.* Imperialist mercenaries can poison or maim but never conquer a people who have chosen to set boundaries to their tools for the sake of conviviality.

PART V: ECOLOGICAL CRITIQUES AND NEO-PRIMITIVISM

THE PROBLEM OF TECHNOLOGY

GREEN ANARCHY JOURNAL

Green Anarchy (www.greenanarchy.org) is among the more prominent venues for the neo-primitivism movement, and the attack on
modern industrial society. It has been in publication since 2001. The
following short editorial statement on technology was printed in 2004.

All green anarchists question technology at some level. While there are those who still
suggest the notion of "green" or "appropriate" technology and search for rationales to
cling to forms of domestication, most reject technology completely.

Technology is more than wires, silicon, plastic, and steel. It is a complex system
involving division of labor, resource extraction, and exploitation for the benefit of
those who implement its process. The interface with, and result of, technology is
always an alienated, mediated, and distorted reality. Despite the claims of postmodern
apologists and other technophiles, technology is not neutral. The values and goals of
those who produce and control technology are always embedded within it.

Technology is distinct from simple tools in many regards. A simple tool is a temporary usage of an element within our immediate surroundings used for a specific
task. Tools do not involve complex systems which alienate the user from the act.
Implicit in technology is this separation, creating an unhealthy and mediated experience which leads to various forms of authority.

Domination increases every time a new "time-saving" technology is created, as it
necessitates the construction of more technology to support, fuel, maintain and repair
the original technology. This technological system seems to have an existence independent from the humans who created it. Discarded by-products of the technological
society are polluting both our physical and our psychological environments. Lives are
stolen in service of the Machine and the toxic effluent of the technological system's
fuels—both are choking us.

Technology is now replicating itself, with something resembling a sinister sen-
tience. Technological society is a planetary infection, propelled forward by its own
momentum, rapidly ordering a new kind of environment: one designed for mechanical
efficiency and technological expansionism alone.

The technological system methodically destroys, eliminates, or subordinates the
natural world, constructing a world fit only for machines. The ideal for which the
technological system strives is the mechanization of everything it encounters.

ECOSOPHY, TECHNOLOGY, AND LIFESTYLE

ARNE NAESS

Naess (1912-2009) is a well-known Norwegian philosopher, and a founder of the environmental school of philosophy called *deep ecology*. This was one of the first systematic philosophies of nature — an 'ecosophy'. Deep ecology holds that things in nature possess *intrinsic value*, i.e. value in their own right, apart from any possible human use. It also argues that humans are intimately connected to their natural environment, and cannot consider themselves as distinct or isolated beings. The 8-Point Platform of deep ecology is: (1) Human and non-human life have intrinsic value. (2) Diversity of life is a value. (3) Reduce diversity only for "vital human needs." (4) Flourishing nature requires reduction in human population. (5) Human interference is excessive and worsening. (6) Social, political, and economic policies must change. (7) 'Life quality' over 'standard of living'. (8) Knowledge implies a duty to *act*. The following passage is from Naess' 1989 book.

Technology and Lifestyle

The technological developments in modern industrial societies have resulted in continuous pressures towards a kind of lifestyle repugnant not only to supporters of the deep ecology movement but to those in most alternative movements. Some of the reasons for such a confrontation are fairly obvious: modern industrial technology is a centralizing factor, it tends towards bigness, it decreases the area within which one can say 'self-made is well-made', it attaches us to big markets, and forces us to seek an ever-increasing income. The administrative technologies are adapted to the physical technologies and encourage more and more impersonal relations.

Those who resist such modern developments have technological symbols in common: the bicycle, home-baked bread, the recycling of goods. In what follows I shall only mention some principles but otherwise refer to the growing body of important literature covering parts of this enormous and complex field of inquiry: technology, lifestyle, economy, politics. The deep ecology movement confronts issues in this realm daily.

Energy consciousness means consciousness of using limited resources, delight in being able to satisfy needs for energy, concern about waste, and concern about the poor and underprivileged for whom energy requirements are a major threat. Where we, who are not poor, live in close and direct relation to nature, and where we are active in providing energy from natural resources, energy consciousness adds to the feeling and experience of richness of the Earth.

In modern industrial life, hot water is tapped in large quantities without the joy of being fabulously rich and without the joy of sometimes enjoying extravagance. This holds even among those who work for water conservation, and are fully, even if rather abstractly, aware of the crisis due to thoughtless misuse of a limited resource.

In Nordic countries energy consciousness was developed even from childhood through life in cabins as part of classical *friluftsliv* ('free air life'). When returning from the cabin to live with 'ordinary' ways of using energy, the lack of joy of richness and the unbelievable waste have always had a strong impact. Clearly the cabin tradition is one of the ecosophically ['ecologically philosophical'] most potent sources of permanent alertness towards the destructive misbehaviors of modern life. It is scarcely an overstatement that the private consumption of energy in Norway could be reduced by 80% without affecting the satisfaction of needs, and with an increase of *joyful* energy consciousness. To be realistic, the change must be seen as one that takes many years, even generations, and there is at the moment no strong trend in favor of life quality comprising energy joys.

Within an ecologically interested minority in many industrial countries the use of wood for heating has been rapidly increasing. Especially if the wood has been collected personally, it favors joyful energy consciousness. In this situation as in many others, a certain amount of knowledge is required in order to avoid an unecological result: undue pollution of the atmosphere. Again, an active interest is required: one has to reflect about the proper use of ventilation.

The above ecosophical critique of 'average' industrial lifestyle applies with heavier emphasis to the average lifestyle of the economic elites. The fashionable lifestyle we can learn about under the heading 'Living' in *Time* magazine might more appropriately have the heading 'Dying' in so far as the universalisation and implementation of the norms imply a catastrophic decrease in living conditions of most kinds of living beings.

(a) **The non-existence of purely technical advance**

When a so-called 'purely technical' improvement is discovered, it is falsely assumed that the individual and society must regulate themselves accordingly: technique, in part, determines its own development. It is treated as if it were *autonomous*. Certain subordinate areas of technical development can be favored and others hindered through political means, but when a 'breakthrough' takes place, we are expected to conform, and adjust society appropriately as soon as possible.

It 'ought to be relatively simple' to solve, for instance, the social problems of automation 'through reschooling and planned retraining within the framework of the extensive public welfare organs'. Certain pressure groups are said to attempt to stop or delay this 'natural development'. The use of the word 'natural' is typical of an interpretation of society as subject to laws of *man-made* nature to which mankind must submit. When a technical 'advance' is made in a leading industrial country, is it *natural* that the thousands of cultures and sub-cultures on this globe ultimately adapt themselves to one group's 'progress'?

Within Marxist literature, the assumption is sometimes made that technical development of the means of production essentially determines all other development. The mode of production can come out of step with the means of production: the 'contradiction' must and will be resolved by reworking the *mode* of production (a broad Marxist term which encompasses social relationships), not the *techniques*.

Even in a traditional society with technical tasks it is 'unnatural' (in many senses of the word) to stop the search for technical improvements. It is against our active nature, our personal and cultural unfolding. However, the *evaluation* of a technical change in such a society is relational: it is relative to social and cultural goals. If a technician points to a specific machine part and says: 'there, now you can see the purely technical advance!' this can only be interpreted as a highly condensed lecture. To prove that progress has been made, the technician will naturally not limit the substantiation to the anatomy of the machine part. He or she will point out saved labor time and other social consequences.

Improvement of technique implies improvement within the framework of a cultural pattern. That which threatens this framework should not be interpreted as improvement, and should thus be rejected. In the industrial societies, these social consequences are not given enough consideration. *There is no such thing as purely technical progress.*

Those who maintain that technological development must run its course whether we like it or not are mistaken both historically and empirically. Why didn't the advanced technical inventiveness of old China change the social structure, for example? A society is capable of rejecting a more 'advanced' or 'higher' technique on account of its social and other consequences. The Chinese rejected banking and certain agricultural tools for this very reason. A lack of critical evaluation of technique is the harbinger of a society's dissolution. A technique has to be *culturally* tested.

Technique in the industrial countries is guided by narrow economic considerations by a small elite of the population. Technical development is driven in widely dissimilar directions in response to prices for different raw materials and energy, and the cost and make-up of the labor force. Our helplessness in questions of technical 'development' is a myth—a very useful myth for those introducing expensive new technology. Technology is chosen, but not by consideration of society as a whole.

One speaks of *laws* for technical development which are independent of other factors. Weighty objections to such a view have been advanced in recent years. In today's capitalist countries (including Russia, with its state capitalism), large profit margins in agriculture are intimately associated with a technology which entails excessive demands on the environment, ruining the soil in the long run.

Technical development is a fragment of total development, and it partakes in an intimate interaction with a host of factors. Social anthropology and related areas of study supply instructive examples of how ideological, and particularly religious, attitudes influence the directions taken by technical change. The subject is neglected in our technical schools, but the breakthrough of ecosophical thought implies a renaissance for the idea of technique submitted to the ideals of a world-view. The idea can be restated as *technique submitted to evaluation in normative systems.*

If a technique is said to express an improvement or a technical advance many tests are relevant. Here are some questions that must be raised:

(1) Is it conducive or dangerous to health?
(2) How meaningful, capable of variations, conducive to the self-determination and inventiveness of the worker?
(3) Does it strengthen cooperation and harmonious togetherness with other workers?
(4) Which other techniques does the technique require in order to be effective as part of greater units of technology? What is the quality of these techniques?
(5) Which raw materials are indispensable? Are they locally or regionally available? How easy is the access to them? Which tools are indispensable? How are they obtained?
(6) How much energy does the technique require? What is the amount of waste? What kind of energy?
(7) Does the technique pollute directly or indirectly? How much and what kind?
(8) How much capital is required? How big must the undertaking be? How vulnerable in times of crisis?
(9) How much administration is required? How much dependent upon hierarchical arrangements?
(10) Does it promote equality or class differences at the place of work or more generally?

Langdon Winner opens the first chapter, 'Autonomy and mastery,' of his book *Autonomous Technology* (1977) with a quotation from Paul Valery: "So the whole question comes down to this: can the human mind master what the human mind has made?" An excellent opening, but it may be added that the general trend of modern technological developments has perhaps not been masterminded by anybody, by any group or any constellation of humans. *It may have developed largely 'by itself.'*

(b) 'The environmental crisis can be technically resolved . . .'

A widespread assumption in influential circles of the industrial countries is that overcoming the environmental crisis is a technical problem: it does not presuppose changes in consciousness or economic system. This assumption is one of the pillars of the 'shallow' ecological movement.

Opposition to further economic growth in the industrial states is unnecessary, it is said, and continued growth is often simply taken for granted. Technical development will reduce pollution to tolerable levels and prevent serious resource depletion. Present forests may be dying, but we can find or create new kinds of trees that thrive on acid rain, or we can find ways to live entirely without trees.

Our governments are incessantly asked to provide good, liberal conditions for centralized, highly technical industry which obeys the 'laws' of the world market and the political pattern of the dominating Eastern and Western industrial lands. The 'concise, factual, professional manner' is on a level *isolated from a discussion of values*.

Those who believe in the possibility of a technical solution often refrain from discussing a radical transformation to *soft* technology. There is little demand on the market, so why bother? The market suggests a preference for hard technology: tremendous new energy sources, a more extreme 'efficiency program' based upon centralization, or technical solutions to population growth.

W. Modell, MD, New York, has indicated to a group of pharmaceutical manufacturers that, by studying organisms which live in the poisonous atmosphere of volcanoes or the near boiling waters of a geyser, we can find substances which could render future conditions on a devastated Earth livable for mankind. The animals which now live in sewage may supply us with knowledge so that we too could survive in sewage-like conditions. Dr Modell concludes with the hope that none of these possibilities will need to be realized. The approach is characteristic of the one-sided technical approach to our crisis, but I am glad to say that Dr Modell is not entirely serious about his solutions.

The essential ingredients for a technocracy are present when the individual and the organizations in which the individual functions become more occupied with means than with ends, and more occupied with subordinate ends (buildings) than fundamental ones (homes). The more the ability to dwell upon intrinsic value diminishes, the faster consciousness turns from immediate experience to planning for the coming times. Although the intrinsic values are ostensibly still the central themes, the procurement of effective means is the principal occupation. The undesirable consequences of this become more and more aggravated as the individual consumer has less and less to do with production. The techniques are 'improved' constantly, requiring great sacrifices of time and energy. Unnoticed, the time spent upon goals withers away. The headlong rush after means takes over: the improvements are illusory.

A crucial objective of the coming years is, therefore, *decentralization* and differentiation as a means to *increased local autonomy* and, ultimately, as a means to unfolding the rich potentialities of the human person.

The great representative of intermediate technology, E. F. Schumacher, spoke of 'production of the masses' as opposed to 'mass production'. The expression 'local production' is also appropriate, as 'the masses' is often associated with many people in a homogeneous milieu. There are masses of small communities, but the techniques will vary greatly if the message of ecosophy is taken seriously. In the same light, 'advanced technology' should be seen as *technology which advances the basic goals of each culture*, not anything more complicated or difficult for its own sake.

Schumacher emphasizes that production of the masses mobilizes the inestimable resources which ordinary human beings possess: brains and skilled hands. And the means of production of the masses assist them with first-class tools. The technology of mass production is in itself violent, ecologically harmful, ultimately self-destructive in its consumption of non-renewable resources and stupefying for the human person (*Small is Beautiful*, 1973).

(c) Soft technology and ecosophy

'To tread lightly on Earth' is a powerful slogan in the deep ecological movement, and slogans such as 'soft technology' are obvious corollaries. Which technologies satisfy maximally both the requirements of reduced interference with nature and satisfaction of human vital needs? Clearly the requirements cannot both be maximally satisfied without getting into conflicts. It is a major concern to find a kind of equilibrium, and the proposals are dependent on geographical and social diversity of life conditions.

A widening circle of technically proficient people are devoting their attention to the discovery of ecologically satisfactory techniques. The increase in interest evidenced by those who direct research in industry and governmental machinery is proceeding more slowly, and grants are minuscule compared with the amounts received by projects indifferent or blatantly irresponsible to ecosophy.

There are many useful works outlining the distinctive qualities of soft technology. What is often missing in such overviews is a discussion of the *transition path* between our present society and one which would make full use of soft and appropriate technologies. Johan Galtung (1978) outlines a way to utilize both *alpha* structures (big, centralized, hierarchical), and *beta* structures ('small is beautiful') as instruments for a composite change to a way in which the former will be phased out gradually, as structures move slowly from the vertical to the horizontal. He asks for a mix of technologies, thus a realistic and immediate alternative (see Table 4.1).

Item	Alpha (large)	Beta (small)
Food	Build down trade in food, drop cash crop practices; build down agribusiness	Try to restore the old system that the food is grown within the horizon-local autarchy; also local preservation and storage; collectivise ground that can be used for food
Clothing	Build down international textile business	Try to restore patterns of local handicraft: symbiosis with food production
Shelter	Build down housing business; transfer more work to homes to help dissolve centre/periphery distinction	Try to restore local building patterns with local materials; collectivise ground that can be used for housing
Medical care	Rural clinics, control of drugs	Positive health care: participation, less separation between healthy and ill
Transportation / Communication	Less centralised, two-way patterns, collective means of transport	Try to restore patterns of walking, talking, bicycling, more car-free areas, cable TV, local media
Energy	Better distribution of centres for large-scale energy production	Solar/wind/wave/biogas networks
Defence	Democratized armies, better distribution of commanding positions	Local defence patterns, nonviolent groups
Comprehension	Maximum transparency through citizen participation and reporting	Small-size units comprehensible by anybody

Table 4.1 (from Galtung)

In spite of the comprehensive nature of the list, the confrontation between standardization and diversity could be further highlighted. Decentralization, and emphasis upon local resources, climate, and other characteristics would result in variations of a technique within the same ecosophically sane technology. The same applies to the products of the techniques. Diminishing standardization and increasing diversity follow.

The demand for expert aid to carry out planned transitions to softer technology is greater than the supply in Great Britain and elsewhere. Work procedures are being reworked. Volvo's experiments with smaller factories, improved external milieus, more all-round tasks and more responsible decision making on the job are well-known.

But the dark outlook for an early transformation to soft technology in Europe may be especially associated with three restraining political factors: the fear for reduced industrial-economic profitability, the fear for reduced material standard of living, and the fear of unemployment. The last factor would appear to be paradoxical, as there seems to be universal agreement that a transition to soft technology would increase the demand for labor, and improve the opportunities for workers. The counterargument reveals a vulgar empiricism: it is said that historically the development of non-soft technology over the last fifty years has occurred simultaneously with a decrease in unemployment. But where is the substance of such a connection?

In technical circles, it is often said that a radical transition to soft technology is politically unrealistic and unnecessarily drastic. But the importance of the many *small* changes in the environment caused by hard technology is presently underestimated. For example: even if we totally avoid large oil catastrophes at sea, the many completely 'normal' small leaks can result in a multitude of tiny detrimental effects upon organisms which will ultimately be catastrophic for living conditions. The minor spills and leaks are calculated to release between five and ten million tons of oil into the seas each year. If the consumption of oil in the Third World increases to the European level within thirty years, living conditions will degenerate ten or more times faster than the present rate.

(d) The invasion of hard technology in the Third World

Faced with the dominance of hard technology, some are proud of the fact that it is necessary to ask if it can be universalized. Can all countries follow in our footsteps? Will people in poorer countries and the generations of the future have a chance to live in our (seemingly) magnificent way?

If not, should we not subscribe to the following norm: 'Choose a level of standard of living such that you realistically may desire that all fellow humans reach the same level if they want'? With the rate of destruction of woods and degrading quality and quantity of good soils, and the prospect of human population at least reaching 8 billion, there is no universalisability present, no planet available for that. The average level in rich industrial countries is unjustifiable and irrational considering its very uncertain relation to level of life *quality*.

One central question in the Third World is: how much industrial techniques can we import from the leading industrial nations without being obliged to open the doors to undesirable characteristics of their social structures? Do we have to develop a weapon industry like that of the industrial countries to prevent our domination by them? Is it necessary to develop a western technocracy in order to survive as a self-determining nation?

For many years, the answers were overwhelmingly optimistic. The leaders of these countries would say: 'We can assimilate whatever we find technically useful, if we take care to retain our own ideology and our own value priorities. Our cultures will remain unharmed.' This could be termed the 'skim the cream' theory.

The military and administrative elite of the Third World has since 1945 to a great extent been educated in the industrial countries and has adopted our predominant ideology, including a distaste for local traditions and cultural diversity in general. The optimism in this case ultimately rests upon an evaluation about how little there was to lose if the ideologies of the industrial countries happened to be introduced together with the techniques: the integrated notion of technology, remember, includes both.

Today, they have made a near total about-face. If one adopts a technique from the leading industrial societies, e.g. a specific method for treatment of cancer, experience has shown that it *cannot be imported in isolation*—it presupposes much more importation. And this supportive import is not purely technical. New patterns of human association, and other subcultures of work are assumed. In short: *cultural invasion* and *increased dependence*. One's own culture is gradually eroded.

In Tibet, Sikkim, and Bhutan, among other until recently isolated places, the leaders throughout the ages have been aware of the 'domino theory'. Tibet is a dramatic example. Tibet managed to remain isolated for many years. When the leaders of Tibet felt themselves threatened by the new China, they sought contact with the industrial states, to increase their chances for a military defense of Tibetan society and culture. Too late! Their cultural sovereignty has been destroyed. But in Bhutan in the 1980s the government is considering technology and influence from the outside only with extreme caution. For example, any students who go abroad for higher education must, immediately upon their return, spend six months traveling through the countryside for a *re-education* on the actual conditions and values of the people of their own country.

The transfer of technologies from the industrial countries to the Third World has included dramatic, often tragic episodes. *The Careless Technology* (1972) illustrates the importance of regarding a culture as a whole, and supplies crystal-clear examples of actual results of thoughtless exportation of technology to the Third World. This thoughtlessness is implicit in the 1940s concept of 'underdeveloped countries'. One imagined that all cultures would and should develop technology in the same manner as the leading industrial countries.

A widely unnoticed, but, in the history of the world, meaningful, clash between spokespeople for hard and soft technology took place in India in the years following the Second World War. On one side stood a group of politicians with Nehru in the fore. They were inspired by the industrialization philosophy of the Soviet Union. On the other side was Gandhi. His social philosophy, *sarvodaya*, 'to the best for all', emphasized the importance of decentralized industrial life and extensive self-sufficiency in India's 500,000 villages. His greatest goal was the elimination of direct material and spiritual destitution. His propaganda for weaving looms is particularly well-known, but he also supported other artisan crafts. *Centralization and urbanization were, for him, evils*. The emphasis on large industry and all technology which deepened the division between a technical elite and workers stripped of their culture would lead to a proletarization of the cities, and increase in violence, and opposition between the Hindus and Muslims.

The contest revolved around the extent to which free India's politics would be based on the red or green dimensions. Both Nehru and Gandhi were aware of the implications in the choice of technology. After independence, the opposition blocks led to compromises between the red and the blue factions. It has been said that the two greatest catastrophes in India have been the *elimination of Buddhism* and the deaf ear turned towards the *green teachings of Gandhi*. This may be an exaggeration, but had priority been given to the technical development of the local community, India's material needs would in all likelihood have been met in the 1950s.

Ecosophy and technology: a summary

(1) Objects produced by labor of a technical nature are in intimate interaction, not only with the means and the mode of production, but with all essential aspects of cultural activity.

(2) Therefore technology is intimately related directly or indirectly to other social institutions, e.g. the sciences, the degree of centralized government, and beliefs about what is reasonable. *Change in technology implies change in culture.*

(3) The height of technical development is primarily judged by the leading industrial states in terms of how the techniques can be assimilated in the economies of these states. The more advanced Western science, e.g. quantum physics or electronics, a technique presupposes, the higher it is regarded. This untenable criterion of progressiveness is applied not only to our own technology but also to the technology of other cultures. This in turn leads to the general depreciation of the viability of foreign cultures.

(4) The ecosophical criteria for progressiveness in technology are relative to ultimate normative objectives. Therefore culture-neutral statements of the degree of advancement cannot be formulated.

(5) The ecosophical basis for an appraisal of technique is the satisfaction of vital needs in the diverse local communities.

(6) The objectives of the deep ecological movement do not imply any depreciation of technology or industry, but they imply general cultural control of developments.

(7) Technocracies—societies to an overwhelming degree determined by technique and technology—can arise as a consequence of extreme division of labor, and intimate merging of technologies of a higher order, combined with extremely specialized, centralized, and exclusive education of technologists. Although neither politicians, nor clergy, nor other groups with authority in the culture can test the explanations granted to the public, they can to some extent determine the political development. The extent of this influence is dependent upon many things: how much technical counter-

expertise can be mobilized, and how willing the mass media are to present these counter-reports in a generally understandable form.

(8) When a technique is replaced by another which requires more attention, education, and is otherwise more self-engaging and detached, the contact with the medium or milieu in which the technique acts is diminished. To the extent that this medium is *nature*, the engagement in nature is reduced in favor of engagement in the *technology*. The degree of inattentiveness or apathy *increases* and thus our awareness of the changes in nature caused by the technique decreases.

(9) The *degree of self-reliance* for individuals and local communities diminishes in proportion to the extent a technique or technology transcends the abilities and resources of the particular individuals or local communities. Passivity, helplessness, and dependence upon 'megasociety' and the world market increase.

FROM RELIGIOUS CONSCIOUSNESS TO TECHNOLOGICAL CONSCIOUSNESS

HENRYK SKOLIMOWSKI

Skolimowski (1930-) is a widely-published and influential philosopher, now retired from the University of Michigan-Ann Arbor. He
has written on science, technology, ethics, and spirituality, but is
best known as creator of a new, comprehensive school of thought
called *eco-philosophy*. The following article was published in 1989.

In this paper I will discuss the emergence of a new form of consciousness, namely
ecological consciousness. Though new in our times, this emerging consciousness is
steeped in the magma of old consciousnesses.

I will argue that consciousness is not a shapeless cloud that encompasses
humankind from time immemorial but rather that in various epochs of human history
it takes distinctive forms, modes or structures. This is also the conclusion of other
writers on the subject, such as Carl Gustav Jung, Mircea Eliade, Jean Gebser, Ken
Wilber, etc. who, in various books and publications, have attempted to reconstruct
human consciousness from prehistoric times to present times.

Let me underscore the point, namely that human consciousness, as it functions
within cultures, is *highly structured* or finely tuned. We do not have a consciousness
as such, but rather we live and function within *specific* modes of consciousness. These
modes are culture-bound on one level, and species-bound on another level.

I will not attempt to review the history of the forms of human consciousness in
this brief essay. I shall concentrate instead on the dialectics of consciousness which
has occurred during the last six centuries, and especially on the dialectics which has
occurred during the last few decades.

Some six hundred years ago, religious consciousness was dominant in Western
world. This consciousness was wedded to Christian cosmology; was in the jurisdiction of this cosmology. As we shall see later, every form of consciousness is connected
with, and a mirror image of, a cosmology [i.e. a worldview]—by which it is engen-

dered and which it articulates. Religious consciousness and Christian cosmology went hand in hand with each other, co-defined and supported each other.

Guided by religious consciousness, people were adjusted to the universe, which was overlooked and steered by God. Their daily affairs were regulated by the awareness of the omnipresence of God, and by the presence of his representatives—the clergy. The point to be emphasized is that the entire field of consciousness of the medieval person was shaped and pervaded by the images of God, by the idea of responsibility to God, and by the desire to be saved and redeemed in God's heaven. Then things began to change. After the messy, turbulent and effervescent Renaissance, a new epoch gathers momentum in the 17th century. The medieval Christian cosmology was questioned. In the process, religious consciousness was undermined and punctured in many places.

Secularism emerges as a new umbrella under which a new consciousness is being crystallized. We are witnessing the slow emergence of secular consciousness—with its marked opposition to the earlier religious consciousness. This secular consciousness was to be articulated in time in the form of *technological* consciousness. For the time being, (throughout the 17th century) this consciousness is groping for articulation, for a new distinctive shape.

The humanism of the Renaissance was the first step towards the new non-religious consciousness. Man began to be deemed the measure of all things. The arrogance of man was in great abundance during the Renaissance period. The geniuses of the Renaissance often possessed an ego as big as a mountain. Humanism meant the re-assertion of the individual, thus of the ego. When the ego is overgrown human arrogance emerges. The arrogance of the Renaissance was actually tempered by the wisdom of the ancients and by the religious constraints which were still very powerful in the 15th and 16th centuries.

The new crucial step in the rise of technological consciousness was the quantification of the cosmos. Only with the advent of the mechanistic worldview does the situation change dramatically. The combined efforts of Bacon, Galileo and Descartes (and scores of others, of course) resulted in a new cosmological matrix, a new worldview which conceives of the universe as a *clock-like mechanism*. From this point on, the developments are rapid and far-reaching and have staggering consequences.

The new cosmological matrix requires that all phenomena—to be recognized as valid—must be *physical* in nature or reducible to physical ones. The important relationships about these physical phenomena should be expressed in quantitative laws. Thus, the physical and the quantitative are enshrined. Over time this led to:

- The worship of objective knowledge;
- The increasing quantification of all phenomena;
- The narrowing of the focus of our vision and of our inquiry;
- The elimination of the sacred.

A further consequence is the growing process of *alienation*—a direct result of the process of atomization and quantification. As we split everything into separate atoms, larger wholes are disintegrated. We no longer have a sense of wholeness. But rather a sense of isolation, separation, detachment, in brief—alienation. The psychological alienation is the result of the conceptual alienation.

Another consequence of the mechanistic approach to the cosmos is the growing worship of physical power, indeed a sense of intoxication with power, and an obsession with it. A corruption of power, to signify the coercive physical power, is a consequence of mechanistic cosmology—which, with an extraordinary consistency eulogizes the physical, the quantitative, the manipulative, the controlling.

It is at this stage that mechanistic cosmology becomes absorbed by the entire culture, that is Western culture. It then comes to dominate the minds; leads to the emergence of a distinctive form of consciousness, which I call *technological consciousness*.

We must stress with due gravity that technology is *not* to be thought of as a chest of tools—indifferent in themselves; bad or good only according to our use. This is a rather atomistic and naive view of technology, which technology itself attempts to promulgate. At this stage of history, technology is so all-pervading that it is a form of consciousness. When we think technology we think "control and manipulation." *Technology is a vision of reality*—not the use of tools.

When we interact with the world via technology, we never think how to be benign and compassionate and loving, but always how to be efficient, controlling, assertive. This attitude of controlling and manipulating is now a part of the mental make-up of Western people.

Another intriguing or shall we say fascinating aspect of technology is that although technological consciousness is supposed to be secular through and through, it contains its own transcendental program, its own form of divinity, if you wish. It seeks the divinity of man here on earth—by liberating him from old yokes, giving him dignity and freedom; and by creating him/her in the image of the industrious god who can do everything for himself.

Now to seek fulfillment and realization on Earth, through our own effort, is an admirable project. The trouble begins when we amass too much power, with which we destroy natural habitats and by which we become so intoxicated that we forget our place on this planet. In the absence of higher values and some form of wisdom concerning human destiny, the amassing of power is a very dangerous thing, as it leads to unbridled arrogance and ultimately to hubris.

Let us now succinctly summarize the characteristics of technological consciousness. When we view its overall structure, its overall mode of operation, technological consciousness reveals itself to be:

(1) Objectivizing
(2) Atomizing
(3) Alienating
(4) Power dominating
(5) De-sacralizing
(6) Geared to the eschatology of consumption

This last point needs to be elaborated. What is the eschatology of consumption? 'Eschatology' is the subject which concerns itself with the ultimate ends and goals of human life. In the absence of far-reaching transcendental goals, while religious and spiritual values have collapsed, consumption has become an imperative of our life, an overall goal, a form of fulfillment, the focus of aspirations.

In its clumsy and indirect way, consumption has become a form of salvation, thus an eschatology. Consuming new toys, new cornflakes, new cars, new televisions, new makeups and new computers is not dangerous in itself. What is dangerous, and unhealthy to our psychic lives, is when this process of consumption becomes a kind of religious urge promising happiness, fulfillment, salvation. At this point technology has become a form of eschatology.

In brief, after we have emptied the universe of the sacred, of the spiritual, of intrinsic values; after we declared the physical, the objective, the coldly rational as our new deities; after the traditional cultural patterns have been disintegrated in the miasma of atomistic thinking what has emerged as our new eschatology is the drunken pursuit of power and of stupefying consumption.

What has also emerged is a new image of man—the Faustian man who celebrates his day by seeking gratification here and now. Faustian man maintains that one only lives once, therefore one lives dangerously, at whatever and whosesoever expense; even if it means the ruin of future generations and the destruction of ecological habitats. Faustian man is the human manifestation of the rapacious technology. Faustian man is a symbolic acknowledgement of the ascent of naked power and the simultaneous waning of human spirituality.

While analyzing the fallout of technological consciousness, we are not forgetting the other side of the coin, namely technology has been a noble dream that has turned out sour, that the mechanistic world view was once a ladder to freedom—from the constraint and oppressions of religious consciousness. Nor do we wish to question the obviously beneficial aspects of technology—comfort, the rise of the material standard of living, the sense of freedom of movement (if only illusory), the elimination of contagious diseases, an easy way of plumbing many of our problems.

Yet as we approach the 21st century, technological consciousness appears increasingly as a menace with its ruthless efficiency, its uncontrolled increase of power, and its lack of any compassionate accounting. Technological consciousness simply does not add up. It produces too many diseases of its own. And the price for comfort and other amenities is simply not worth paying. This is the realization to which we, as a society, have come to only gradually.

Technology is a story of success which has been so stupendous that it has become our nightmare. The pendulum has swung too far in the direction opposite to religious consciousness. We thus seek to re-establish a new balance, and in the process we attempt to balance our own lives.

Ecological consciousness, which is outlined in this essay, is a synthesis. Religious consciousness was the thesis. Technological consciousness was the anti-thesis. Ecological consciousness is the synthesis as it marks a return to the spiritual without submitting to religious orthodoxies and the religious dogma; and as it seeks social amelioration and justice for all without worshipping physical power and without celebrating the aggressive nature of the human person.

We are not claiming that ecological consciousness has arrived, is fully articulated, and comfortably dwells in us. Rather we are suggesting that in reaching out, we bring to fruition that which is dormant and which wants to be awakened and articulated. Our projection is part of our articulation, is an essential part of the creation of ecological consciousness. As we dream, so reality becomes.

2. The Rise of Ecological Consciousness

All modes of consciousness are rooted in history and have a historical character. They come at a certain point of history and disappear, or are profoundly modified, at another point of history. Thus consciousness is determined by history. But, in its turn, it determines history. In our times we have been witnessing the *greening of consciousness*. Simultaneous to this process of greening, another process is going on: the process of the greening of world religions, although the latter is less easily perceptible. Let us be aware that from the time of the Congress of Assisi (1986) where five major religions were represented, the ecological interpretation of world religions has acquired an important momentum.

The forerunner of ecological consciousness was the Ecology Movement, on the one hand, and various schools of humanistic psychology, on the other hand. In their respective ways they were against the temper of the mechanistic age. Both have emphasized holism and the irreducibility of large complex wholes to their underpinning components: ecological habits and human persons. Both these movements were a challenge thrown to the rationality of the mechanistic system. Both movements professed a new type of holistic rationality.

Moreover, in a certain sense both these movements possessed a religious flavor. They offered not only new intellectual vistas but a form of liberation. This liberation, although not always explicit, was meant to give us freedom from the deterministic and mechanistic shackles. Holism, which both movements emphasized, was the first step to liberation.

What are the chief characteristics of ecological consciousness? We shall enumerate six such characteristics and contrast them with the respective ones of technological consciousness. We shall not claim that these six characteristics completely define the scope and nature of ecological consciousness. We have to simplify. To understand is to simplify.

ECOLOGICAL CONSCIOUSNESS		TECHNOLOGICAL CONSCIOUSNESS
holistic	vs.	atomistic
qualitative	vs.	quantitative
spiritual	vs.	secular
reverential	vs.	objective
evolutionary	vs.	mechanistic
participatory	vs.	alienating

Let us now discuss these characteristics and see what it all means for our own lives and for our perception of the universe.

That a healthy and complete human person is a micro-universe which is *holistic* and *qualitative*—there can be no doubt. That a human being who seeks meaning beyond the triviality of consumption is on some kind of *spiritual* path—there can be no doubt either. The quest of meaning is a spiritual quest.

As for the attitude of *reverence*, any person who truly respects others, and truly appreciates the amazing alchemy of the universe, cannot be but reverential vis-à-vis the awesome spectacle of creation. Thus reverence is an aspect of seeing and understanding the universe—in depth and with a true appreciation.

To live in grace is to be in a continuous mode of reverential understanding. To live in grace is to think reverentially. To live in grace is to walk in beauty, as a song of American Indians proclaims.

Reverential understanding as well as the reverential attitude are not new creations of ecological consciousness. They have existed in traditional cultures and religions for a long time. We are only articulating them *de novo*.

The natural condition of the human person who is alive is to be enchanted by the world. Reverence is an acknowledgement of this enchantment.

A rational skeptic may respond at this point: "It is all very well to treat the human person in a holistic, reverential and spiritual manner—particularly if you are a humanistic softie, but how can we demonstrate, how can we *prove* that the physical universe is holistic, reverential, spiritual?"

We are rather fortunate that at this juncture of history, science at the cutting edge provides us with unambiguous evidence of the holistic nature of the universe. David Bohm's book *Wholeness and the Implicate Order* is a case in point. Bohm's views are rather well known so we shall not belabor them here. Suffice it to say that it can be consistently and legitimately maintained that the physical cosmos, in its stupendous evolution is one interconnected whole of which each part partakes in the glory of the whole, and somehow knows of the existence of all others. And Bohm is not the only physicist who assumes that by its very nature the universe is holistic, since we all came from the same fireball.

Equally significant, if not more striking, is the evidence emerging from astrophysicists who have been wondering why the structure of the universe is as it is. The balances in the overall structure of the universe are exquisitely tuned. It is now clear that if the universe had different density, dimensions and time-span, life could not have emerged. It has been calculated recently that if the composition of some of the basic elements in the universe had been altered by as little as 10%, the whole fabric of the cosmos would not be able to generate life. Thus this laboratory, called the cosmos, seems to have been uniquely set to generate life. Freeman Dyson writes: "As we look out into the universe and identify the many accidents of physics and astronomy that have worked together to our benefit, it almost seems as if the universe in some sense must have known that we were coming."

Thus the conclusion has slowly emerged... Why is the universe as it is? Because we are here. The universe is as it is because it meant to generate intelligent life. This insight led to the formulation of the Anthropic Principle (in 1981) which proclaims that the universe is home for man. Its composition, structure and dynamics were exquisitely balanced to enable life to emerge. As we search deeper and deeper into the underlying structure of the cosmic evolution, we are more and more convinced that "the coincidences" may not have been so coincidental.

The universe has brought life to celebrate itself. We are part of its glory. We neither wish to deny our special place, nor do we wish to be unduly arrogant about it. It has just happened that we are part of the flowering of the universe. To deny this special place to Homo sapiens in the name of the ideology of anti-anthropocentrism is a folly based on a new form of misanthropy; is, in fact, a form of inverted arrogance.

From the standpoint of ecological consciousness, the Anthropic Principle reaffirms the wholeness and unity of it all, and reassures us that we are legitimate dwellers of the cosmos, not some kind of cosmic freaks. In one sense, we are the justification of the nature of the universe: the tremendous cosmic changes have worked out to create life endowed with intelligence. On another level, the conception of the universe as home for man implies that we are its custodians, responsible for all there is, including our own fate.

3. From the Anthropic Principle to Reverence for Life

Our rational skeptic is now nodding his head and says: "Although we can somehow agree that the universe is holistic in nature, how can we *prove* that it is reverential and spiritual in nature? Where is the evidence for that?" We shall respond: the evidence is in our mind. To think beautifully about the universe is to think reverentially of it, is to behold it with reverence. The skeptic may object that this is not the kind of evidence he has expected. At this point we shall say that the skeptic has not grasped the subtlety of our argument. We are not saying that reverence is one of the *physical* aspects of the universe, to be found among other physical attributes. We have never committed ourselves to such a proposition. We are saying rather that reverence is an attribute of the mind. If we behold the universe reverentially, we dwell in a reverential universe. If we behold the universe mechanistically, we dwell in a mechanistic universe.

The mechanistic nature of the universe can be attested to by the existence of present mechanistic science and technology. How can the reverential nature of the universe be attested to—the skeptic may further press. We would like to say...by the existence of *reverential science* and *reverential technology*—although these two have not been much developed. It would not be true, however, to say that they do not exist.

Various systems of yoga represent reverential technology. And we should not raise an eyebrow or dismiss the idea because we are used to thinking about technology as physical tools engaging us with the physical world. Our interactions with the universe are numerous and subtle. *Any tool or technique which engages us with the universe is a form of technology.* Yoga systems are the techniques of the soul. Prayer is a form of technology—if it does something to you and facilitates your interactions with the universe—divine or otherwise.

Let us now further reflect on the nature of cosmology. Cosmology makes assumptions about the universe *in toto*. And then, acting on these assumptions, it finds in the universe what it has assumed to be the case. Such has been the story of most historically known cosmologies. Cosmologies *do not prove* the existence of this or that attribute which they assume about the cosmos. They proceed as if this attribute was inherent in the structure of the universe. And then they build large patterns of perception and knowledge which vindicate the assumed existence of a given attribute.

Nothing reveals itself in the cosmos unless we assume it to be the case. If we don't assume that the universe is physical in nature, we shall never be able to elicit this attribute from the universe. We have to start somewhere. Cosmology is the game of assumptions. These assumptions are not proven. They are made. And then acted upon.

This process of assumption-making or of creating cosmologies is a pre-scientific process, both in the historical and the epistemological senses. Science has little to say about this process because this process precedes science. Therefore this process is outside the jurisdiction of science. Western science comes into being only when a cosmology of a certain kind is assumed, namely *mechanistic* cosmology. Therefore, within the sphere of general cosmology, science cannot be an arbiter of the validity of other cosmologies because it is peculiarly partisan with regard to one cosmology.

Thus, we need not worry about the verdict of science while we involve ourselves in the creation of non-mechanistic cosmologies. If science attempts to interfere with our new cosmological designs, we can tell science to go to hell; or go back to mechanistic cosmology to which it belongs. For in trying to interfere with other cosmologies, mechanistic science oversteps its own domain and competence.

Returning to the point concerning the reverential nature of the universe. We need not be anti-scientific, or ignore the existence of science for that matter. We need to observe, however, that there is nothing in the structure and the language of science that forbids us to view the universe reverentially.

Cosmologies are a matter of will and of vision. If we develop the reverential attitude towards the universe, if we articulate these forms of thinking, perception and behavior which enable us to walk in beauty, we shall dwell in the reverential universe. The universe will be reverential because we would have made it so. Thus the universe is reverential if we have the capacities to interact with it reverentially. This is what ecological consciousness is about. It is about developing and articulating these capacities which enable us to dwell in a reverential universe, and ultimately to live in grace.

To the divine mind the cosmos is divine. To the crass mind the cosmos is crass. To the monkey's mind the cosmos is monkey-like. These propositions must be taken with all seriousness. For it is the mind which rules over the unruly cosmos. Whatever order we have found in the universe, it is one which the mind has invented. Whatever attributes we have found in the universe, these are the ones which the mind has conceived. The universe is neither big nor small; neither beautiful nor ugly. The glow of mind fills the void and makes its space divine.

It is only from the time when some minds tuned themselves into the sacramental or divine mode that the universe could be experienced as divine. When such superbly tuned minds appeared in India, they created the Upanishads. When such minds appeared in the ancient Hebrew world, they created the Bible. In ancient Greece these minds are exemplified by Pythagoras and Plato, who talked about godhead.

Now these minds, who had originally projected their divinity upon the cosmos, were so delighted with their creation that they decided to attribute this divinity to the cosmos itself. They made the cosmos itself divine. They claimed that the divinity is in the cosmos itself, especially after they invented Brahman, Jehovah and Godhead, whom they conceived as the Absolute Ground for Being from which everything springs.

What I am proposing, on the other hand, is the natural conception of divinity, or the *Noetic* conception of divinity, as the mind (*nous*) is the creator of all orders, including the divine or spiritual order. Mircea Eliade is exactly right when he says that the sacred is an element of the structure of consciousness, not a stage in the history of consciousness. Sacredness is an attribute of the mind, not an attribute of the cosmos. Only when we approach the universe with the reverential attitude and behold it by the mind that is sacred, do we find the universe sacred.

INDUSTRIAL SOCIETY AND ITS FUTURE

THEODORE KACZYNSKI

Kaczynski (1942-), also known as the Unabomber, was convicted of a series of mail-bombings in 1997; the bombs killed three people and injured 22. His arrest came about after the publication of a lengthy manifesto in the *Washington Post* in 1995, from which the following excerpts are taken. Kaczynski's brother recognized the writing style and turned him in to the FBI. Kaczynski's motive was based on a desire to bring down industrial, technological society, which, he argued, was on the brink of destroying humanity and the planet.

INTRODUCTION

1. The Industrial Revolution and its consequences have been a disaster for the human race. They have greatly increased the life-expectancy of those of us who live in "advanced" countries, but they have destabilized society, have made life unfulfilling, have subjected human beings to indignities, have led to widespread psychological suffering (in the Third World to physical suffering as well) and have inflicted severe damage on the natural world. The continued development of technology will worsen the situation. It will certainly subject human beings to greater indignities and inflict greater damage on the natural world, it will probably lead to greater social disruption and psychological suffering, and it may lead to increased physical suffering even in "advanced" countries.

2. The industrial-technological system may survive or it may break down. If it survives, it MAY eventually achieve a low level of physical and psychological suffering, but only after passing through a long and very painful period of adjustment and only at the cost of permanently reducing human beings and many other living organisms to engineered products and mere cogs in the social machine. Furthermore, if the system survives, the consequences will be inevitable: There is no way of reforming or modifying the system so as to prevent it from depriving people of dignity and autonomy.

3. If the system breaks down the consequences will still be very painful. But the bigger the system grows the more disastrous the results of its breakdown will be, so if it is to break down it had best break down sooner rather than later.

4. We therefore advocate a revolution against the industrial system. This revolution may or may not make use of violence; it may be sudden or it may be a relatively gradual process spanning a few decades. We can't predict any of that. But we do outline in a very general way the measures that those who hate the industrial system should take in order to prepare the way for a revolution against that form of society. This is not to be a POLITICAL revolution. Its object will be to overthrow not governments but the economic and technological basis of the present society.

5. In this article we give attention to only some of the negative developments that have grown out of the industrial-technological system. Other such developments we mention only briefly or ignore altogether. This does not mean that we regard these other developments as unimportant. For practical reasons we have to confine our discussion to areas that have received insufficient public attention or in which we have something new to say. For example, since there are well developed environmental and wilderness movements, we have written very little about environmental degradation or the destruction of wild nature, even though we consider these to be highly important.

THE POWER PROCESS

33. Human beings have a need (probably based in biology) for something that we will call the *power process*. This is closely related to the need for power (which is widely recognized) but is not quite the same thing. The power process has four elements. The three most clearcut of these we call goal, effort and attainment of goal. (Everyone needs to have goals whose attainment requires effort, and needs to succeed in attaining at least some of his goals.) The fourth element is more difficult to define and may not be necessary for everyone. We call it autonomy and will discuss it later (paragraphs 42-44).

34. Consider the hypothetical case of a man who can have anything he wants just by wishing for it. Such a man has power, but he will develop serious psychological problems. At first he will have a lot of fun, but by and by he will become acutely bored and demoralized. Eventually he may become clinically depressed. History shows that leisured aristocracies tend to become decadent. This is not true of fighting aristocracies that have to struggle to maintain their power. But leisured, secure aristocracies that have no need to exert themselves usually become bored, hedonistic and demoralized, even though they have power. This shows that power is not enough. One must have goals toward which to exercise one's power.

35. Everyone has goals; if nothing else, to obtain the physical necessities of life: food, water and whatever clothing and shelter are made necessary by the climate. But the leisured aristocrat obtains these things without effort. Hence his boredom and demoralization.

36. Nonattainment of important goals results in death if the goals are physical necessities, and in frustration if nonattainment of the goals is compatible with survival. Consistent failure to attain goals throughout life results in defeatism, low self-esteem or depression.

37. Thus, in order to avoid serious psychological problems, a human being needs goals whose attainment requires effort, and he must have a reasonable rate of success in attaining his goals.

AUTONOMY

42. Autonomy as a part of the power process may not be necessary for every individual. But most people need a greater or lesser degree of autonomy in working toward their goals. Their efforts must be undertaken on their own initiative and must be under their own direction and control. Yet most people do not have to exert this initiative, direction and control as single individuals. It is usually enough to act as a member of a SMALL group. Thus if half a dozen people discuss a goal among themselves and make a successful joint effort to attain that goal, their need for the power process will be served. But if they work under rigid orders handed down from above that leave them no room for autonomous decision and initiative, then their need for the power process will not be served. The same is true when decisions are made on a collective basis if the group making the collective decision is so large that the role of each individual is insignificant.

43. It is true that some individuals seem to have little need for autonomy. Either their drive for power is weak or they satisfy it by identifying themselves with some powerful organization to which they belong. And then there are unthinking, animal types who seem to be satisfied with a purely physical sense of power (the good combat soldier, who gets his sense of power by developing fighting skills that he is quite content to use in blind obedience to his superiors).

44. But for most people it is through the power process—having a goal, making an AUTONOMOUS effort and attaining the goal—that self-esteem, self-confidence and a sense of power are acquired. When one does not have adequate opportunity to go through the power process the consequences are (depending on the individual and on the way the power process is disrupted) boredom, demoralization, low self-esteem, inferiority feelings, defeatism, depression, anxiety, guilt, frustration, hostility, spouse or child abuse, insatiable hedonism, abnormal sexual behavior, sleep disorders, eating disorders, etc.

SOURCES OF SOCIAL PROBLEMS

45. Any of the foregoing symptoms can occur in any society, but in modern industrial society they are present on a massive scale. We aren't the first to mention that the world today seems to be going crazy. This sort of thing is not normal for human societies. There is good reason to believe that primitive man suffered from less stress and frustration and was better satisfied with his way of life than modern man is. It is true that not all was sweetness and light in primitive societies. Abuse of women was common among the Australian aborigines, transsexuality was fairly common among some of the American Indian tribes. But it does appear that GENERALLY SPEAKING the kinds

of problems that we have listed in the preceding paragraph were far less common among primitive peoples than they are in modern society.

46. We attribute the social and psychological problems of modern society to the fact that that society requires people to live under conditions radically different from those under which the human race evolved, and to behave in ways that conflict with the patterns of behavior that the human race developed while living under the earlier conditions. It is clear from what we have already written that we consider lack of opportunity to properly experience the power process as the most important of the abnormal conditions to which modern society subjects people. But it is not the only one. Before dealing with disruption of the power process as a source of social problems we will discuss some of the other sources.

47. Among the abnormal conditions present in modern industrial society are excessive density of population, isolation of man from nature, excessive rapidity of social change and the breakdown of natural small-scale communities such as the extended family, the village or the tribe.

48. It is well known that crowding increases stress and aggression. The degree of crowding that exists today and the isolation of man from nature are consequences of technological progress. All pre-industrial societies were predominantly rural. The Industrial Revolution vastly increased the size of cities and the proportion of the population that lives in them, and modern agricultural technology has made it possible for the Earth to support a far denser population than it ever did before. (Also, technology exacerbates the effects of crowding because it puts increased disruptive powers in people's hands. For example, a variety of noise-making devices: power mowers, radios, motorcycles, etc. If the use of these devices is unrestricted, people who want peace and quiet are frustrated by the noise. If their use is restricted, people who use the devices are frustrated by the regulations. But if these machines had never been invented there would have been no conflict and no frustration generated by them.)

49. For primitive societies the natural world (which usually changes only slowly) provided a stable framework and therefore a sense of security. In the modern world it is human society that dominates nature rather than the other way around, and modern society changes very rapidly owing to technological change. Thus there is no stable framework.

50. The conservatives are fools: They whine about the decay of traditional values, yet they enthusiastically support technological progress and economic growth. Apparently it never occurs to them that you can't make rapid, drastic changes in the technology and the economy of a society without causing rapid changes in all other aspects of the society as well, and that such rapid changes inevitably break down traditional values.

51. The breakdown of traditional values to some extent implies the breakdown of the bonds that hold together traditional small-scale social groups. The disintegration of small-scale social groups is also promoted by the fact that modern conditions often require or

tempt individuals to move to new locations, separating themselves from their communities. Beyond that, a technological society HAS TO weaken family ties and local communities if it is to function efficiently. In modern society an individual's loyalty must be first to the system and only secondarily to a small-scale community, because if the internal loyalties of small-scale communities were stronger than loyalty to the system, such communities would pursue their own advantage at the expense of the system.

52. Suppose that a public official or a corporation executive appoints his cousin, his friend or his coreligionist to a position rather than appointing the person best qualified for the job. He has permitted personal loyalty to supersede his loyalty to the system, and that is "nepotism" or "discrimination," both of which are terrible sins in modern society. Would-be industrial societies that have done a poor job of subordinating personal or local loyalties to loyalty to the system are usually very inefficient. (Look at Latin America.) Thus an advanced industrial society can tolerate only those small-scale communities that are emasculated, tamed and made into tools of the system.

53. Crowding, rapid change and the breakdown of communities have been widely recognized as sources of social problems. But we do not believe they are enough to account for the extent of the problems that are seen today.

54. A few pre-industrial cities were very large and crowded, yet their inhabitants do not seem to have suffered from psychological problems to the same extent as modern man. In America today there still are uncrowded rural areas, and we find there the same problems as in urban areas, though the problems tend to be less acute in the rural areas. Thus crowding does not seem to be the decisive factor.

55. On the growing edge of the American frontier during the 19th century, the mobility of the population probably broke down extended families and small-scale social groups to at least the same extent as these are broken down today. In fact, many nuclear families lived by choice in such isolation, having no neighbors within several miles, that they belonged to no community at all, yet they do not seem to have developed problems as a result.

56. Furthermore, change in American frontier society was very rapid and deep. A man might be born and raised in a log cabin, outside the reach of law and order and fed largely on wild meat; and by the time he arrived at old age he might be working at a regular job and living in an ordered community with effective law enforcement. This was a deeper change than that which typically occurs in the life of a modern individual, yet it does not seem to have led to psychological problems. In fact, 19th century American society had an optimistic and self-confident tone, quite unlike that of today's society.

57. The difference, we argue, is that modern man has the sense (largely justified) that change is IMPOSED on him, whereas the 19th century frontiersman had the sense (also largely justified) that he created change himself, by his own choice. Thus a pioneer settled on a piece of land of his own choosing and made it into a farm through his own effort. In those days an entire county might have only a couple of hundred

inhabitants and was a far more isolated and autonomous entity than a modern county is. Hence the pioneer farmer participated as a member of a relatively small group in the creation of a new, ordered community. One may well question whether the creation of this community was an improvement, but at any rate it satisfied the pioneer's need for the power process.

THE NATURE OF FREEDOM

93. We are going to argue that industrial-technological society cannot be reformed in such a way as to prevent it from progressively narrowing the sphere of human freedom. But, because "freedom" is a word that can be interpreted in many ways, we must first make clear what kind of freedom we are concerned with.

94. By "freedom" we mean the opportunity to go through the power process, with real goals not the artificial goals of surrogate activities, and without interference, manipulation or supervision from anyone, especially from any large organization. Freedom means being in control (either as an individual or as a member of a SMALL group) of the life-and-death issues of one's existence: food, clothing, shelter and defense against whatever threats there may be in one's environment. Freedom means having power; not the power to control other people but the power to control the circumstances of one's own life. One does not have freedom if anyone else (especially a large organization) has power over one, no matter how benevolently, tolerantly and permissively that power may be exercised. It is important not to confuse freedom with mere permissiveness.

95. It is said that we live in a free society because we have a certain number of constitutionally guaranteed rights. But these are not as important as they seem. The degree of personal freedom that exists in a society is determined more by the economic and technological structure of the society than by its laws or its form of government. Most of the Indian nations of New England were monarchies, and many of the cities of the Italian Renaissance were controlled by dictators. But in reading about these societies one gets the impression that they allowed far more personal freedom than our society does. In part this was because they lacked efficient mechanisms for enforcing the ruler's will: There were no modern, well-organized police forces, no rapid long-distance communications, no surveillance cameras, no dossiers of information about the lives of average citizens. Hence it was relatively easy to evade control.

96. As for our constitutional rights, consider for example that of freedom of the press. We certainly don't mean to knock that right; it is a very important tool for limiting concentration of political power and for keeping those who do have political power in line by publicly exposing any misbehavior on their part. But freedom of the press is of very little use to the average citizen as an individual. The mass media are mostly under the control of large organizations that are integrated into the system. Anyone who has a little money can have something printed, or can distribute it on the Internet

or in some such way, but what he has to say will be swamped by the vast volume of material put out by the media, hence it will have no practical effect. To make an impression on society with words is therefore almost impossible for most individuals and small groups. Take us (FC) for example. If we had never done anything violent and had submitted the present writings to a publisher, they probably would not have been accepted. If they had been accepted and published, they probably would not have attracted many readers, because it's more fun to watch the entertainment put out by the media than to read a sober essay. Even if these writings had had many readers, most of those readers would soon have forgotten what they had read as their minds were flooded by the mass of material to which the media expose them. In order to get our message before the public with some chance of making a lasting impression, we've had to kill people.

97. Constitutional rights are useful up to a point, but they do not serve to guarantee much more than what might be called the bourgeois conception of freedom. According to the bourgeois conception, a "free" man is essentially an element of a social machine and has only a certain set of prescribed and delimited freedoms; freedoms that are designed to serve the needs of the social machine more than those of the individual. Thus the bourgeois's "free" man has economic freedom because that promotes growth and progress; he has freedom of the press because public criticism restrains misbehavior by political leaders; he has a right to a fair trial because imprisonment at the whim of the powerful would be bad for the system. [...]

98. One more point to be made in this section: It should not be assumed that a person has enough freedom just because he SAYS he has enough. Freedom is restricted in part by psychological controls of which people are unconscious, and moreover many people's ideas of what constitutes freedom are governed more by social convention than by their real needs. For example, it's likely that many leftists of the oversocialized type would say that most people, including themselves, are socialized too little rather than too much, yet the oversocialized leftist pays a heavy psychological price for his high level of socialization.

INDUSTRIAL-TECHNOLOGICAL SOCIETY CANNOT BE REFORMED

111. The foregoing principles help to show how hopelessly difficult it would be to reform the industrial system in such a way as to prevent it from progressively narrowing our sphere of freedom. There has been a consistent tendency, going back at least to the Industrial Revolution, for technology to strengthen the system at a high cost in individual freedom and local autonomy. Hence any change designed to protect freedom from technology would be contrary to a fundamental trend in the development of our society. Consequently, such a change either would be a transitory one — soon swamped by the tide of history — or, if large enough to be permanent, would alter the nature of our whole society. Moreover, since society would be altered in a way that could not be predicted in advance there would be great risk. Changes large enough to

make a lasting difference in favor of freedom would not be initiated because it would be realized that they would gravely disrupt the system. So any attempts at reform would be too timid to be effective. Even if changes large enough to make a lasting difference were initiated, they would be retracted when their disruptive effects became apparent. Thus, permanent changes in favor of freedom could be brought about only by persons prepared to accept radical, dangerous and unpredictable alteration of the entire system. In other words by revolutionaries, not reformers.

112. People anxious to rescue freedom without sacrificing the supposed benefits of technology will suggest naive schemes for some new form of society that would reconcile freedom with technology. Apart from the fact that people who make such suggestions seldom propose any practical means by which the new form of society could be set up in the first place, even if the new form of society could be once established, it either would collapse or would give results very different from those expected.

113. So even on very general grounds it seems highly improbable that any way of changing society could be found that would reconcile freedom with modern technology. In the next few sections we will give more specific reasons for concluding that freedom and technological progress are incompatible.

RESTRICTION OF FREEDOM IS UNAVOIDABLE IN INDUSTRIAL SOCIETY

114. As explained in paragraphs 65-67, 70-73, modern man is strapped down by a network of rules and regulations, and his fate depends on the actions of persons remote from him whose decisions he cannot influence. This is not accidental or a result of the arbitrariness of arrogant bureaucrats. It is necessary and inevitable in any technologically advanced society. The system HAS TO regulate human behavior closely in order to function. At work, people have to do what they are told to do, when they are told to do it and in the way they are told to do it, otherwise production would be thrown into chaos. Bureaucracies HAVE TO be run according to rigid rules. To allow any substantial personal discretion to lower-level bureaucrats would disrupt the system, and lead to charges of unfairness due to differences in the way individual bureaucrats exercised their discretion. It is true that some restrictions on our freedom could be eliminated, but GENERALLY SPEAKING the regulation of our lives by large organizations is necessary for the functioning of industrial-technological society. The result is a sense of powerlessness on the part of the average person. It may be, however, that formal regulations will tend increasingly to be replaced by psychological tools that make us want to do what the system requires of us. (Propaganda, educational techniques, "mental health" programs, etc.)

115. The system HAS TO force people to behave in ways that are increasingly remote from the natural pattern of human behavior. For example, the system needs scientists, mathematicians and engineers. It can't function without them. So heavy pressure is put

on children to excel in these fields. It isn't natural for an adolescent human being to spend the bulk of his time sitting at a desk absorbed in study. A normal adolescent wants to spend his time in active contact with the real world. Among primitive peoples the things that children are trained to do tend to be in reasonable harmony with natural human impulses. Among the American Indians, for example, boys were trained in active outdoor pursuits—just the sort of thing that boys like. But in our society children are pushed into studying technical subjects, which most do grudgingly.

116. Because of the constant pressure that the system exerts to modify human behavior, there is a gradual increase in the number of people who cannot or will not adjust to society's requirements: welfare leeches, youth-gang members, cultists, anti-government rebels, radical environmentalist saboteurs, dropouts and resisters of various kinds.

117. In any technologically advanced society the individual's fate MUST depend on decisions that he personally cannot influence to any great extent. A technological society cannot be broken down into small, autonomous communities, because production depends on the cooperation of very large numbers of people. Such a society MUST be highly organized and decisions HAVE TO be made that affect very large numbers of people. When a decision affects, say, a million people, then each of the affected individuals has, on the average, only a one-millionth share in making the decision. What usually happens in practice is that decisions are made by public officials or corporation executives, or by technical specialists, but even when the public votes on a decision the number of voters ordinarily is too large for the vote of any one individual to be significant. Thus most individuals are unable to influence measurably the major decisions that affect their lives. There is no conceivable way to remedy this in a technologically advanced society. The system tries to "solve" this problem by using propaganda to make people WANT the decisions that have been made for them, but even if this "solution" were completely successful in making people feel better, it would be demeaning.

118. Conservatives and some others advocate more "local autonomy." Local communities once did have autonomy, but such autonomy becomes less and less possible as local communities become more enmeshed with and dependent on large-scale systems like public utilities, computer networks, highway systems, the mass communications media and the modern health-care system. Also operating against autonomy is the fact that technology applied in one location often affects people at other locations far away. Thus pesticide or chemical use near a creek may contaminate the water supply hundreds of miles downstream, and the greenhouse effect affects the whole world.

119. The system does not and cannot exist to satisfy human needs. Instead, it is human behavior that has to be modified to fit the needs of the system. This has nothing to do with the political or social ideology that may pretend to guide the technological system. It is not the fault of capitalism and it is not the fault of socialism. It is the fault of technology, because the system is guided not by ideology but by technical necessity. Of course the system does satisfy many human needs, but generally speaking it does this

only to the extent that it is to the advantage of the system to do it. It is the needs of the system that are paramount, not those of the human being. For example, the system provides people with food because the system couldn't function if everyone starved; it attends to people's psychological needs whenever it can CONVENIENTLY do so, because it couldn't function if too many people became depressed or rebellious. But the system, for good, solid, practical reasons, must exert constant pressure on people to mold their behavior to the needs of the system. Too much waste accumulating? The government, the media, the educational system, environmentalists, everyone inundates us with a mass of propaganda about recycling. Need more technical personnel? A chorus of voices exhorts kids to study science. No one stops to ask whether it is inhumane to force adolescents to spend the bulk of their time studying subjects that most of them hate. When skilled workers are put out of a job by technical advances and have to undergo "retraining," no one asks whether it is humiliating for them to be pushed around in this way. It is simply taken for granted that everyone must bow to technical necessity. And for good reason: If human needs were put before technical necessity there would be economic problems, unemployment, shortages or worse. The concept of "mental health" in our society is defined largely by the extent to which an individual behaves in accord with the needs of the system, and does so without showing signs of stress.

120. Efforts to make room for a sense of purpose and for autonomy within the system are no better than a joke. For example, one company, instead of having each of its employees assemble only one section of a catalogue, had each assemble a whole catalogue, and this was supposed to give them a sense of purpose and achievement. Some companies have tried to give their employees more autonomy in their work, but for practical reasons this usually can be done only to a very limited extent, and in any case employees are never given autonomy as to ultimate goals—their "autonomous" efforts can never be directed toward goals that they select personally, but only toward their employer's goals, such as the survival and growth of the company. Any company would soon go out of business if it permitted its employees to act otherwise. Similarly, in any enterprise within a socialist system, workers must direct their efforts toward the goals of the enterprise, otherwise the enterprise will not serve its purpose as part of the system. Once again, for purely technical reasons it is not possible for most individuals or small groups to have much autonomy in industrial society. Even the small business owner commonly has only limited autonomy. Apart from the necessity of government regulation, he is restricted by the fact that he must fit into the economic system and conform to its requirements. For instance, when someone develops a new technology, the small-business person often has to use that technology whether he wants to or not, in order to remain competitive.

THE 'BAD' PARTS OF TECHNOLOGY CANNOT BE SEPARATED FROM THE 'GOOD' PARTS

121. A further reason why industrial society cannot be reformed in favor of freedom is that modern technology is a unified system in which all parts are dependent on one another. You can't get rid of the "bad" parts of technology and retain only the "good" parts. Take

modern medicine, for example. Progress in medical science depends on progress in chemistry, physics, biology, computer science and other fields. Advanced medical treatments require expensive, high-tech equipment that can be made available only by a technologically progressive, economically rich society. Clearly you can't have much progress in medicine without the whole technological system and everything that goes with it.

122. Even if medical progress could be maintained without the rest of the technological system, it would by itself bring certain evils. Suppose for example that a cure for diabetes is discovered. People with a genetic tendency to diabetes will then be able to survive and reproduce as well as anyone else. Natural selection against genes for diabetes will cease and such genes will spread throughout the population. (This may be occurring to some extent already, since diabetes, while not curable, can be controlled through use of insulin.) The same thing will happen with many other diseases susceptibility to which is affected by genetic factors (e.g., childhood cancer), resulting in massive genetic degradation of the population. The only solution will be some sort of eugenics program or extensive genetic engineering of human beings, so that man in the future will no longer be a creation of nature, or of chance, or of God (depending on your religious or philosophical opinions), but a manufactured product.

123. If you think that big government interferes in your life too much NOW, just wait till the government starts regulating the genetic constitution of your children. Such regulation will inevitably follow the introduction of genetic engineering of human beings, because the consequences of unregulated genetic engineering would be disastrous.

124. The usual response to such concerns is to talk about "medical ethics." But a code of ethics would not serve to protect freedom in the face of medical progress; it would only make matters worse. A code of ethics applicable to genetic engineering would be in effect a means of regulating the genetic constitution of human beings. Somebody (probably the upper middle class, mostly) would decide that such and such applications of genetic engineering were "ethical" and others were not, so that in effect they would be imposing their own values on the genetic constitution of the population at large. Even if a code of ethics were chosen on a completely democratic basis, the majority would be imposing their own values on any minorities who might have a different idea of what constituted an "ethical" use of genetic engineering. The only code of ethics that would truly protect freedom would be one that prohibited ANY genetic engineering of human beings, and you can be sure that no such code will ever be applied in a technological society. No code that reduced genetic engineering to a minor role could stand up for long, because the temptation presented by the immense power of biotechnology would be irresistible, especially since to the majority of people many of its applications will seem obviously and unequivocally good (eliminating physical and mental diseases, giving people the abilities they need to get along in today's world). Inevitably, genetic engineering will be used extensively, but only in ways consistent with the needs of the industrial-technological system.

TECHNOLOGY IS A MORE POWERFUL SOCIAL FORCE THAN THE ASPIRATION FOR FREEDOM

125. It is not possible to make a LASTING compromise between technology and freedom, because technology is by far the more powerful social force and continually encroaches on freedom through REPEATED compromises. Imagine the case of two neighbors, each of whom at the outset owns the same amount of land, but one of whom is more powerful than the other. The powerful one demands a piece of the other's land. The weak one refuses. The powerful one says, "OK, let's compromise. Give me half of what I asked." The weak one has little choice but to give in. Some time later the powerful neighbor demands another piece of land, again there is a compromise, and so forth. By forcing a long series of compromises on the weaker man, the powerful one eventually gets all of his land. So it goes in the conflict between technology and freedom.

126. Let us explain why technology is a more powerful social force than the aspiration for freedom.

127. A technological advance that appears not to threaten freedom often turns out to threaten it very seriously later on. For example, consider motorized transport. A walking man formerly could go where he pleased, go at his own pace without observing any traffic regulations, and was independent of technological support-systems. When motor vehicles were introduced they appeared to increase man's freedom. They took no freedom away from the walking man, no one had to have an automobile if he didn't want one, and anyone who did choose to buy an automobile could travel much faster and farther than a walking man. But the introduction of motorized transport soon changed society in such a way as to restrict greatly man's freedom of locomotion. When automobiles became numerous, it became necessary to regulate their use extensively. In a car, especially in densely populated areas, one cannot just go where one likes at one's own pace; one's movement is governed by the flow of traffic and by various traffic laws. One is tied down by various obligations: license requirements, driver test, renewing registration, insurance, maintenance required for safety, monthly payments on purchase price. Moreover, the use of motorized transport is no longer optional. Since the introduction of motorized transport the arrangement of our cities has changed in such a way that the majority of people no longer live within walking distance of their place of employment, shopping areas and recreational opportunities, so that they HAVE TO depend on the automobile for transportation. Or else they must use public transportation, in which case they have even less control over their own movement than when driving a car. Even the walker's freedom is now greatly restricted. In the city he continually has to stop to wait for traffic lights that are designed mainly to serve auto traffic. In the country, motor traffic makes it dangerous and unpleasant to walk along the highway. (Note this important point that we have just illustrated with the case of motorized transport: When a new item of technology is introduced as an option that an individual can accept or not as he chooses, it does not necessarily REMAIN optional. In many cases the new technology changes society in such a way that people eventually find themselves FORCED to use it.)

128. While technological progress AS A WHOLE continually narrows our sphere of freedom, each new technical advance CONSIDERED BY ITSELF appears to be desirable. Electricity, indoor plumbing, rapid long-distance communications...how could one argue against any of these things, or against any other of the innumerable technical advances that have made modern society? It would have been absurd to resist the introduction of the telephone, for example. It offered many advantages and no disadvantages. Yet, as we explained in paragraphs 59-76, all these technical advances taken together have created a world in which the average man's fate is no longer in his own hands or in the hands of his neighbors and friends, but in those of politicians, corporation executives and remote, anonymous technicians and bureaucrats whom he as an individual has no power to influence. The same process will continue in the future. Take genetic engineering, for example. Few people will resist the introduction of a genetic technique that eliminates a hereditary disease. It does no apparent harm and prevents much suffering. Yet a large number of genetic improvements taken together will make the human being into an engineered product rather than a free creation of chance (or of God, or whatever, depending on your religious beliefs).

129. Another reason why technology is such a powerful social force is that, within the context of a given society, technological progress marches in only one direction; it can never be reversed. Once a technical innovation has been introduced, people usually become dependent on it, so that they can never again do without it, unless it is replaced by some still more advanced innovation. Not only do people become dependent as individuals on a new item of technology, but, even more, the system as a whole becomes dependent on it. (Imagine what would happen to the system today if computers, for example, were eliminated.) Thus the system can move in only one direction, toward greater technologization. Technology repeatedly forces freedom to take a step back but technology can never take a step back—short of the overthrow of the whole technological system.

130. Technology advances with great rapidity and threatens freedom at many different points at the same time (crowding, rules and regulations, increasing dependence of individuals on large organizations, propaganda and other psychological techniques, genetic engineering, invasion of privacy through surveillance devices and computers, etc.). To hold back any ONE of the threats to freedom would require a long and difficult social struggle. Those who want to protect freedom are overwhelmed by the sheer number of new attacks and the rapidity with which they develop, hence they become apathetic and no longer resist. To fight each of the threats separately would be futile. Success can be hoped for only by fighting the technological system as a whole; but that is revolution, not reform.

131. Technicians (we use this term in its broad sense to describe all those who perform a specialized task that requires training) tend to be so involved in their work (their surrogate activity) that when a conflict arises between their technical work and freedom, they almost always decide in favor of their technical work. This is obvious in the case of scientists, but it also appears elsewhere: Educators, humanitarian groups, conservation organizations do not hesitate to use propaganda or other psychological techniques to help them achieve their laudable ends. Corporations and government agencies,

when they find it useful, do not hesitate to collect information about individuals without regard to their privacy. Law enforcement agencies are frequently inconvenienced by the constitutional rights of suspects and often of completely innocent persons, and they do whatever they can do legally (or sometimes illegally) to restrict or circumvent those rights. Most of these educators, government officials and law officers believe in freedom, privacy and constitutional rights, but when these conflict with their work, they usually feel that their work is more important.

132. It is well known that people generally work better and more persistently when striving for a reward than when attempting to avoid a punishment or negative outcome. Scientists and other technicians are motivated mainly by the rewards they get through their work. But those who oppose technological invasions of freedom are working to avoid a negative outcome; consequently there are few who work persistently and well at this discouraging task. If reformers ever achieved a signal victory that seemed to set up a solid barrier against further erosion of freedom through technical progress, most would tend to relax and turn their attention to more agreeable pursuits. But the scientists would remain busy in their laboratories, and technology as it progressed would find ways, in spite of any barriers, to exert more and more control over individuals and make them always more dependent on the system.

133. No social arrangements, whether laws, institutions, customs or ethical codes, can provide permanent protection against technology. History shows that all social arrangements are transitory; they all change or break down eventually. But technological advances are permanent within the context of a given civilization. Suppose for example that it were possible to arrive at some social arrangement that would prevent genetic engineering from being applied to human beings, or prevent it from being applied in such a way as to threaten freedom and dignity. Still, the technology would remain, waiting. Sooner or later the social arrangement would break down. Probably sooner, given the pace of change in our society. Then genetic engineering would begin to invade our sphere of freedom, and this invasion would be irreversible (short of a breakdown of technological civilization itself). Any illusions about achieving anything permanent through social arrangements should be dispelled by what is currently happening with environmental legislation. A few years ago it seemed that there were secure legal barriers preventing at least SOME of the worst forms of environmental degradation. A change in the political wind, and those barriers begin to crumble.

134. For all of the foregoing reasons, technology is a more powerful social force than the aspiration for freedom. But this statement requires an important qualification. It appears that during the next several decades the industrial-technological system will be undergoing severe stresses due to economic and environmental problems, and especially due to problems of human behavior (alienation, rebellion, hostility, a variety of social and psychological difficulties). We hope that the stresses through which the system is likely to pass will cause it to break down, or at least will weaken it sufficiently so that a revolution against it becomes possible. If such a revolution occurs and is successful, then at that particular moment the aspiration for freedom will have proved more powerful than technology.

CONTROL OF HUMAN BEHAVIOR

143. Since the beginning of civilization, organized societies have had to put pressures on human beings for the sake of the functioning of the social organism. The kinds of pressures vary greatly from one society to another. Some of the pressures are physical (poor diet, excessive labor, environmental pollution), some are psychological (noise, crowding, forcing human behavior into the mold that society requires). In the past, human nature has been approximately constant, or at any rate has varied only within certain bounds. Consequently, societies have been able to push people only up to certain limits. When the limit of human endurance has been passed, things start going wrong: rebellion, or crime, or corruption, or evasion of work, or depression and other mental problems, or an elevated death rate, or a declining birth rate or something else, so that either the society breaks down, or its functioning becomes too inefficient and it is (quickly or gradually, through conquest, attrition or evolution) replaced by some more efficient form of society.

144. Thus human nature has in the past put certain limits on the development of societies. People could be pushed only so far and no farther. But today this may be changing, because modern technology is developing ways of modifying human beings.

145. Imagine a society that subjects people to conditions that make them terribly unhappy, then gives them drugs to take away their unhappiness. Science fiction? It is already happening to some extent in our own society. It is well known that the rate of clinical depression has been greatly increasing in recent decades. We believe that this is due to disruption of the power process, as explained in paragraphs 59-76. But even if we are wrong, the increasing rate of depression is certainly the result of SOME conditions that exist in today's society. Instead of removing the conditions that make people depressed, modern society gives them antidepressant drugs. In effect, antidepressants are a means of modifying an individual's internal state in such a way as to enable him to tolerate social conditions that he would otherwise find intolerable. (Yes, we know that depression is often of purely genetic origin. We are referring here to those cases in which environment plays the predominant role.)

146. Drugs that affect the mind are only one example of the new methods of controlling human behavior that modern society is developing. Let us look at some of the other methods.

147. To start with, there are the techniques of surveillance. Hidden video cameras are now used in most stores and in many other places, computers are used to collect and process vast amounts of information about individuals. Information so obtained greatly increases the effectiveness of physical coercion (i.e., law enforcement). Then there are the methods of propaganda, for which the mass communications media provide effective vehicles. Efficient techniques have been developed for winning elections, selling products, influencing public opinion. The entertainment industry serves as an important psychological tool of the system, possibly even when it is dishing out large amounts of sex and violence. Entertainment provides modern man with an essential means of escape. While absorbed in television, videos, etc., he can forget stress, anxiety, frus-

tration, dissatisfaction. Many primitive peoples, when they don't have any work to do, are quite content to sit for hours at a time doing nothing at all, because they are at peace with themselves and their world. But most modern people must be constantly occupied or entertained, otherwise they get "bored," i.e., they get fidgety, uneasy, irritable.

148. Other techniques strike deeper than the foregoing. Education is no longer a simple affair of paddling a kid's behind when he doesn't know his lessons and patting him on the head when he does know them. It is becoming a scientific technique for controlling the child's development. Sylvan Learning Centers, for example, have had great success in motivating children to study, and psychological techniques are also used with more or less success in many conventional schools. "Parenting" techniques that are taught to parents are designed to make children accept the fundamental values of the system and behave in ways that the system finds desirable. "Mental health" programs, "intervention" techniques, psychotherapy and so forth are ostensibly designed to benefit individuals, but in practice they usually serve as methods for inducing individuals to think and behave as the system requires. (There is no contradiction here; an individual whose attitudes or behavior bring him into conflict with the system is up against a force that is too powerful for him to conquer or escape from, hence he is likely to suffer from stress, frustration, defeat. His path will be much easier if he thinks and behaves as the system requires. In that sense the system is acting for the benefit of the individual when it brainwashes him into conformity.) Child abuse in its gross and obvious forms is disapproved in most if not all cultures. Tormenting a child for a trivial reason or no reason at all is something that appalls almost everyone. But many psychologists interpret the concept of abuse much more broadly. Is spanking, when used as part of a rational and consistent system of discipline, a form of abuse? The question will ultimately be decided by whether or not spanking tends to produce behavior that makes a person fit in well with the existing system of society. In practice, the word "abuse" tends to be interpreted to include any method of child-rearing that produces behavior inconvenient for the system. Thus, when they go beyond the prevention of obvious, senseless cruelty, programs for preventing "child abuse" are directed toward the control of human behavior on behalf of the system.

149. Presumably, research will continue to increase the effectiveness of psychological techniques for controlling human behavior. But we think it is unlikely that psychological techniques alone will be sufficient to adjust human beings to the kind of society that technology is creating. Biological methods probably will have to be used. We have already mentioned the use of drugs in this connection. Neurology may provide other avenues for modifying the human mind. Genetic engineering of human beings is already beginning to occur in the form of "gene therapy," and there is no reason to assume that such methods will not eventually be used to modify those aspects of the body that affect mental functioning.

150. As we mentioned in paragraph 134, industrial society seems likely to be entering a period of severe stress, due in part to problems of human behavior and in part to economic and environmental problems. And a considerable proportion of the system's economic and environmental problems result from the way human beings behave.

Alienation, low self-esteem, depression, hostility, rebellion; children who won't study, youth gangs, illegal drug use, rape, child abuse, other crimes, unsafe sex, teen pregnancy, population growth, political corruption, race hatred, ethnic rivalry, bitter ideological conflict (e.g., pro-choice vs pro-life), political extremism, terrorism, sabotage, anti-government groups, hate groups. All these threaten the very survival of the system. The system will therefore be FORCED to use every practical means of controlling human behavior.

151. The social disruption that we see today is certainly not the result of mere chance. It can only be a result of the conditions of life that the system imposes on people. (We have argued that the most important of these conditions is disruption of the power process.) If the system succeeds in imposing sufficient control over human behavior to assure its own survival, a new watershed in human history will have been passed. Whereas formerly the limits of human endurance have imposed limits on the development of societies (as we explained in paragraphs 143, 144), industrial-technological society will be able to pass those limits by modifying human beings, whether by psychological methods or biological methods or both. In the future, social systems will not be adjusted to suit the needs of human beings. Instead, human beings will be adjusted to suit the needs of the system.

152. Generally speaking, technological control over human behavior will probably not be introduced with a totalitarian intention or even through a conscious desire to restrict human freedom. Each new step in the assertion of control over the human mind will be taken as a rational response to a problem that faces society, such as curing alcoholism, reducing the crime rate or inducing young people to study science and engineering. In many cases there will be a humanitarian justification. For example, when a psychiatrist prescribes an anti-depressant for a depressed patient, he is clearly doing that individual a favor. It would be inhumane to withhold the drug from someone who needs it. When parents send their children to Sylvan Learning Centers to have them manipulated into becoming enthusiastic about their studies, they do so from concern for their children's welfare. It may be that some of these parents wish that one didn't have to have specialized training to get a job and that their kid didn't have to be brainwashed into becoming a computer nerd. But what can they do? They can't change society, and their child may be unemployable if he doesn't have certain skills. So they send him to Sylvan.

153. Thus control over human behavior will be introduced not by a calculated decision of the authorities but through a process of social evolution (RAPID evolution, however). The process will be impossible to resist, because each advance, considered by itself, will appear to be beneficial, or at least the evil involved in making the advance will seem to be less than that which would result from not making it. (See paragraph 127.) Propaganda for example is used for many good purposes, such as discouraging child abuse or race hatred. Sex education is obviously useful, yet the effect of sex education (to the extent that it is successful) is to take the shaping of sexual attitudes away from the family and put it into the hands of the state as represented by the public school system.

154. Suppose a biological trait is discovered that increases the likelihood that a child will grow up to be a criminal, and suppose some sort of gene therapy can remove this

trait. Of course most parents whose children possess the trait will have them undergo the therapy. It would be inhumane to do otherwise, since the child would probably have a miserable life if he grew up to be a criminal. But many or most primitive societies have a low crime rate in comparison with that of our society, even though they have neither high-tech methods of child-rearing nor harsh systems of punishment. Since there is no reason to suppose that more modern men than primitive men have innate predatory tendencies, the high crime rate of our society must be due the pressures that modern conditions put on people, to which many cannot or will not adjust. Thus a treatment designed to remove potential criminal tendencies is at least in part a way of re-engineering people so that they suit the requirements of the system.

155. Our society tends to regard as a "sickness" any mode of thought or behavior that is inconvenient for the system, and this is plausible because when an individual doesn't fit into the system it causes pain to the individual as well as problems for the system. Thus the manipulation of an individual to adjust him to the system is seen a "cure" for a "sickness" and therefore as good.

156. In paragraph 127 we pointed out that if the use of a new item of technology is INITIALLY optional, it does not necessarily REMAIN optional, because the new technology tends to change society in such a way that it becomes difficult or impossible for an individual to function without using that technology. This applies also to the technology of human behavior. In a world in which most children are put through a program to make them enthusiastic about studying, a parent will almost be forced to put his kid through such a program, because if he does not, then the kid will grow up to be, comparatively speaking, an ignoramus and therefore unemployable. Or suppose a biological treatment is discovered that, without undesirable side-effects, will greatly reduce the psychological stress from which so many people suffer in our society. If large numbers of people choose to undergo the treatment, then the general level of stress in society will be reduced, so that it will be possible for the system to increase the stress-producing pressures. In fact, something like this seems to have happened already with one of our society's most important psychological tools for enabling people to reduce (or at least temporarily escape from) stress, namely, mass entertainment (see paragraph 147). Our use of mass entertainment is "optional": No law requires us to watch television, listen to the radio, read magazines. Yet mass entertainment is a means of escape and stress-reduction on which most of us have become dependent. Everyone complains about the trashiness of television, but almost everyone watches it. A few have kicked the TV habit, but it would be a rare person who could get along today without using ANY form of mass entertainment. (Yet until quite recently in human history most people got along very nicely with no other entertainment than that which each local community created for itself.) Without the entertainment industry the system probably would not have been able to get away with putting as much stress-producing pressure on us as it does.

157. Assuming that industrial society survives, it is likely that technology will eventually acquire something approaching complete control over human behavior. It has been established beyond any rational doubt that human thought and behavior have a largely biological basis. As experimenters have demonstrated, feelings such as hunger,

pleasure, anger and fear can be turned on and off by electrical stimulation of appropriate parts of the brain. Memories can be destroyed by damaging parts of the brain or they can be brought to the surface by electrical stimulation. Hallucinations can be induced or moods changed by drugs. There may or may not be an immaterial human soul, but if there is one it clearly is less powerful that the biological mechanisms of human behavior. For if that were not the case then researchers would not be able so easily to manipulate human feelings and behavior with drugs and electrical currents.

158. It presumably would be impractical for all people to have electrodes inserted in their heads so that they could be controlled by the authorities. But the fact that human thoughts and feelings are so open to biological intervention shows that the problem of controlling human behavior is mainly a technical problem; a problem of neurons, hormones and complex molecules; the kind of problem that is accessible to scientific attack. Given the outstanding record of our society in solving technical problems, it is overwhelmingly probable that great advances will be made in the control of human behavior.

159. Will public resistance prevent the introduction of technological control of human behavior? It certainly would if an attempt were made to introduce such control all at once. But since technological control will be introduced through a long sequence of small advances, there will be no rational and effective public resistance. (See paragraphs 127, 132, 153.)

160. To those who think that all this sounds like science fiction, we point out that yesterday's science fiction is today's fact. The Industrial Revolution has radically altered man's environment and way of life, and it is only to be expected that as technology is increasingly applied to the human body and mind, man himself will be altered as radically as his environment and way of life have been.

HUMAN RACE AT A CROSSROADS

161. But we have gotten ahead of our story. It is one thing to develop in the laboratory a series of psychological or biological techniques for manipulating human behavior and quite another to integrate these techniques into a functioning social system. The latter problem is the more difficult of the two. For example, while the techniques of educational psychology doubtless work quite well in the "lab schools" where they are developed, it is not necessarily easy to apply them effectively throughout our educational system. We all know what many of our schools are like. The teachers are too busy taking knives and guns away from the kids to subject them to the latest techniques for making them into computer nerds. Thus, in spite of all its technical advances relating to human behavior, the system to date has not been impressively successful in controlling human beings. The people whose behavior is fairly well under the control of the system are those of the type that might be called "bourgeois." But there are growing numbers of people who in one way or another are rebels against the system: welfare leeches, youth gangs, cultists, satanists, Nazis, radical environmentalists, militiamen, etc.

162. The system is currently engaged in a desperate struggle to overcome certain problems that threaten its survival, among which the problems of human behavior are the most important. If the system succeeds in acquiring sufficient control over human behavior quickly enough, it will probably survive. Otherwise it will break down. We think the issue will most likely be resolved within the next several decades, say 40 to 100 years.

163. Suppose the system survives the crisis of the next several decades. By that time it will have to have solved, or at least brought under control, the principal problems that confront it, in particular that of "socializing" human beings; that is, making people sufficiently docile so that their behavior no longer threatens the system. That being accomplished, it does not appear that there would be any further obstacle to the development of technology, and it would presumably advance toward its logical conclusion, which is complete control over everything on earth, including human beings and all other important organisms. The system may become a unitary, monolithic organization, or it may be more or less fragmented and consist of a number of organizations coexisting in a relationship that includes elements of both cooperation and competition, just as today the government, the corporations and other large organizations both cooperate and compete with one another. Human freedom mostly will have vanished, because individuals and small groups will be impotent vis-a-vis large organizations armed with supertechnology and an arsenal of advanced psychological and biological tools for manipulating human beings, besides instruments of surveillance and physical coercion. Only a small number of people will have any real power, and even these probably will have only very limited freedom, because their behavior too will be regulated; just as today our politicians and corporation executives can retain their positions of power only as long as their behavior remains within certain fairly narrow limits.

164. Don't imagine that the system will stop developing further techniques for controlling human beings and nature once the crisis of the next few decides is over and increasing control is no longer necessary for the system's survival. On the contrary, once the hard times are over the system will increase its control over people and nature more rapidly, because it will no longer be hampered by difficulties of the kind that it is currently experiencing. Survival is not the principal motive for extending control. As we explained in paragraphs 87-90, technicians and scientists carry on their work largely as a surrogate activity; that is, they satisfy their need for power by solving technical problems. They will continue to do this with unabated enthusiasm, and among the most interesting and challenging problems for them to solve will be those of understanding the human body and mind and intervening in their development. For the "good of humanity," of course.

165. But suppose on the other hand that the stresses of the coming decades prove to be too much for the system. If the system breaks down there may be a period of chaos, a "time of troubles" such as those that history has recorded at various epochs in the past. It is impossible to predict what would emerge from such a time of troubles, but at any rate the human race would be given a new chance. The greatest danger is that industrial society may begin to reconstitute itself within the first few years after the

breakdown. Certainly there will be many people (power-hungry types especially) who will be anxious to get the factories running again.

166. Therefore two tasks confront those who hate the servitude to which the industrial system is reducing the human race. First, we must work to heighten the social stresses within the system so as to increase the likelihood that it will break down or be weakened sufficiently so that a revolution against it becomes possible. Second, it is necessary to develop and propagate an ideology that opposes technology and the industrial society if and when the system becomes sufficiently weakened. And such an ideology will help to assure that, if and when industrial society breaks down, its remnants will be smashed beyond repair, so that the system cannot be reconstituted. The factories should be destroyed, technical books burned, etc.

HUMAN SUFFERING

167. The industrial system will not break down purely as a result of revolutionary action. It will not be vulnerable to revolutionary attack unless its own internal problems of development lead it into very serious difficulties. So if the system breaks down it will do so either spontaneously, or through a process that is in part spontaneous but helped along by revolutionaries. If the breakdown is sudden, many people will die, since the world's population has become so overblown that it cannot even feed itself any longer without advanced technology. Even if the breakdown is gradual enough so that reduction of the population can occur more through lowering of the birth rate than through elevation of the death rate, the process of de-industrialization probably will be very chaotic and involve much suffering. It is naive to think it likely that technology can be phased out in a smoothly managed, orderly way, especially since the technophiles will fight stubbornly at every step. Is it therefore cruel to work for the breakdown of the system? Maybe, but maybe not. In the first place, revolutionaries will not be able to break the system down unless it is already in enough trouble so that there would be a good chance of its eventually breaking down by itself anyway; and the bigger the system grows, the more disastrous the consequences of its breakdown will be; so it may be that revolutionaries, by hastening the onset of the breakdown, will be reducing the extent of the disaster.

168. In the second place, one has to balance struggle and death against the loss of freedom and dignity. To many of us, freedom and dignity are more important than a long life or avoidance of physical pain. Besides, we all have to die some time, and it may be better to die fighting for survival, or for a cause, than to live a long but empty and purposeless life.

169. In the third place, it is not at all certain that survival of the system will lead to less suffering than breakdown of the system would. The system has already caused, and is continuing to cause, immense suffering all over the world. Ancient cultures, that for hundreds of years gave people a satisfactory relationship with each other and with their

environment, have been shattered by contact with industrial society, and the result has been a whole catalogue of economic, environmental, social and psychological problems. One of the effects of the intrusion of industrial society has been that over much of the world traditional controls on population have been thrown out of balance. Hence the population explosion, with all that that implies. Then there is the psychological suffering that is widespread throughout the supposedly fortunate countries of the West (see paragraphs 44, 45). No one knows what will happen as a result of ozone depletion, the greenhouse effect and other environmental problems that cannot yet be foreseen. And, as nuclear proliferation has shown, new technology cannot be kept out of the hands of dictators and irresponsible Third World nations. Would you like to speculate about what Iraq or North Korea will do with genetic engineering?

170. "Oh!" say the technophiles, "Science is going to fix all that! We will conquer famine, eliminate psychological suffering, make everybody healthy and happy!" Yeah, sure. That's what they said 200 years ago. The Industrial Revolution was supposed to eliminate poverty, make everybody happy, etc. The actual result has been quite different. The technophiles are hopelessly naive (or self-deceiving) in their understanding of social problems. They are unaware of (or choose to ignore) the fact that when large changes, even seemingly beneficial ones, are introduced into a society, they lead to a long sequence of other changes, most of which are impossible to predict (paragraph 103). The result is disruption of the society. So it is very probable that in their attempts to end poverty and disease, engineer docile, happy personalities and so forth, the technophiles will create social systems that are terribly troubled, even more so than the present once. For example, the scientists boast that they will end famine by creating new, genetically engineered food plants. But this will allow the human population to keep expanding indefinitely, and it is well known that crowding leads to increased stress and aggression. This is merely one example of the PREDICTABLE problems that will arise. We emphasize that, as past experience has shown, technical progress will lead to other new problems that CANNOT be predicted in advance (paragraph 103). In fact, ever since the Industrial Revolution technology has been creating new problems for society far more rapidly than it has been solving old ones. Thus it will take a long and difficult period of trial and error for the technophiles to work the bugs out of their Brave New World (if they ever do). In the meantime there will be great suffering. So it is not at all clear that the survival of industrial society would involve less suffering than the breakdown of that society would. Technology has gotten the human race into a fix from which there is not likely to be any easy escape.

THE FUTURE

171. But suppose now that industrial society does survive the next several decades and that the bugs do eventually get worked out of the system, so that it functions smoothly. What kind of system will it be? We will consider several possibilities.

172. First let us postulate that the computer scientists succeed in developing intelligent machines that can do all things better than human beings can do them. In that case presumably all work will be done by vast, highly organized systems of machines and no human effort will be necessary. Either of two cases might occur. The machines might be permitted to make all of their own decisions without human oversight, or else human control over the machines might be retained.

173. If the machines are permitted to make all their own decisions we can't make any conjecture as to the results, because it is impossible to guess how such machines might behave. We only point out that the fate of the human race would be at the mercy of the machines. It might be argued that the human race would never be foolish enough to hand over all power to the machines. But we are suggesting neither that the human race would voluntarily turn power over to the machines nor that the machines would willfully seize power. What we do suggest is that the human race might easily permit itself to drift into a position of such dependence on the machines that it would have no practical choice but to accept all of the machines' decisions. As society and the problems that face it become more and more complex and as machines become more and more intelligent, people will let machines make more and more of their decisions for them, simply because machine-made decisions will bring better results than man-made ones. Eventually a stage may be reached at which the decisions necessary to keep the system running will be so complex that human beings will be incapable of making them intelligently. At that stage the machines will be in effective control. People won't be able to just turn the machines off, because they will be so dependent on them that turning them off would amount to suicide.

174. On the other hand it is possible that human control over the machines may be retained. In that case the average man may have control over certain private machines of his own, such as his car or his personal computer, but control over large systems of machines will be in the hands of a tiny elite — just as it is today, but with two differences. Due to improved techniques the elite will have greater control over the masses; and because human work will no longer be necessary the masses will be superfluous, a useless burden on the system. If the elite is ruthless they may simply decide to exterminate the mass of humanity. If they are humane they may use propaganda or other psychological or biological techniques to reduce the birth rate until the mass of humanity becomes extinct, leaving the world to the elite. Or, if the elite consists of soft-hearted liberals, they may decide to play the role of good shepherds to the rest of the human race. They will see to it that everyone's physical needs are satisfied, that all children are raised under psychologically hygienic conditions, that everyone has a wholesome hobby to keep him busy, and that anyone who may become dissatisfied undergoes "treatment" to cure his "problem." Of course, life will be so purposeless that people will have to be biologically or psychologically engineered either to remove their need for the power process or to make them "sublimate" their drive for power into some harmless hobby. These engineered human beings may be happy in such a society, but they most certainly will not be free. They will have been reduced to the status of domestic animals.

175. But suppose now that the computer scientists do not succeed in developing artificial intelligence, so that human work remains necessary. Even so, machines will take care of more and more of the simpler tasks so that there will be an increasing surplus of human workers at the lower levels of ability. (We see this happening already. There are many people who find it difficult or impossible to get work, because for intellectual or psychological reasons they cannot acquire the level of training necessary to make themselves useful in the present system.) On those who are employed, ever-increasing demands will be placed: They will need more and more training, more and more ability, and will have to be ever more reliable, conforming and docile, because they will be more and more like cells of a giant organism. Their tasks will be increasingly specialized, so that their work will be, in a sense, out of touch with the real world, being concentrated on one tiny slice of reality. The system will have to use any means that it can, whether psychological or biological, to engineer people to be docile, to have the abilities that the system requires and to "sublimate" their drive for power into some specialized task. But the statement that the people of such a society will have to be docile may require qualification. The society may find competitiveness useful, provided that ways are found of directing competitiveness into channels that serve the needs of the system. We can imagine a future society in which there is endless competition for positions of prestige and power. But no more than a very few people will ever reach the top, where the only real power is (see end of paragraph 163). Very repellent is a society in which a person can satisfy his need for power only by pushing large numbers of other people out of the way and depriving them of THEIR opportunity for power.

176. One can envision scenarios that incorporate aspects of more than one of the possibilities that we have just discussed. For instance, it may be that machines will take over most of the work that is of real, practical importance, but that human beings will be kept busy by being given relatively unimportant work. It has been suggested, for example, that a great development of the service industries might provide work for human beings. Thus people would spend their time shining each other's shoes, driving each other around in taxicabs, making handicrafts for one another, waiting on each other's tables, etc. This seems to us a thoroughly contemptible way for the human race to end up, and we doubt that many people would find fulfilling lives in such pointless busy-work. They would seek other, dangerous outlets (drugs, crime, "cults," hate groups) unless they were biologically or psychologically engineered to adapt them to such a way of life.

177. Needless to say, the scenarios outlined above do not exhaust all the possibilities. They only indicate the kinds of outcomes that seem to us most likely. But we can envision no plausible scenarios that are any more palatable than the ones we've just described. It is overwhelmingly probable that if the industrial-technological system survives the next 40 to 100 years, it will by that time have developed certain general characteristics: Individuals (at least those of the "bourgeois" type, who are integrated into the system and make it run, and who therefore have all the power) will be more dependent than ever on large organizations; they will be more "socialized" than ever and their physical and mental qualities to a significant extent (possibly to a very great extent) will be those that are engineered into them rather than being the results of chance (or of God's will, or

whatever); and whatever may be left of wild nature will be reduced to remnants preserved for scientific study and kept under the supervision and management of scientists (hence it will no longer be truly wild). In the long run (say a few centuries from now) it is likely that neither the human race nor any other important organisms will exist as we know them today, because once you start modifying organisms through genetic engineering there is no reason to stop at any particular point, so that the modifications will probably continue until man and other organisms have been utterly transformed.

178. Whatever else may be the case, it is certain that technology is creating for human beings a new physical and social environment radically different from the spectrum of environments to which natural selection has adapted the human race physically and psychologically. If man is not adjusted to this new environment by being artificially re-engineered, then he will be adapted to it through a long and painful process of natural selection. The former is far more likely than the latter.

179. It would be better to dump the whole stinking system and take the consequences.

STRATEGY

180. The technophiles are taking us all on an utterly reckless ride into the unknown. Many people understand something of what technological progress is doing to us yet take a passive attitude toward it because they think it is inevitable. But we (FC) don't think it is inevitable. We think it can be stopped, and we will give here some indications of how to go about stopping it.

181. As we stated in paragraph 166, the two main tasks for the present are to promote social stress and instability in industrial society and to develop and propagate an ideology that opposes technology and the industrial system. When the system becomes sufficiently stressed and unstable, a revolution against technology may be possible. The pattern would be similar to that of the French and Russian Revolutions. French society and Russian society, for several decades prior to their respective revolutions, showed increasing signs of stress and weakness. Meanwhile, ideologies were being developed that offered a new world view that was quite different from the old one. In the Russian case revolutionaries were actively working to undermine the old order. Then, when the old system was put under sufficient additional stress (by financial crisis in France, by military defeat in Russia) it was swept away by revolution. What we propose is something along the same lines.

182. It will be objected that the French and Russian Revolutions were failures. But most revolutions have two goals. One is to destroy an old form of society and the other is to set up the new form of society envisioned by the revolutionaries. The French and Russian revolutionaries failed (fortunately!) to create the new kind of society of which they dreamed, but they were quite successful in destroying the old society. We have no illusions about the feasibility of creating a new, ideal form of society. Our goal is only to destroy the existing form of society.

CONFRONTING TECHNOLOGY

183. But an ideology, in order to gain enthusiastic support, must have a positive ideal as well as a negative one; it must be FOR something as well as AGAINST something. The positive ideal that we propose is Nature. That is, WILD nature: those aspects of the functioning of the Earth and its living things that are independent of human management and free of human interference and control. And with wild nature we include human nature, by which we mean those aspects of the functioning of the human individual that are not subject to regulation by organized society but are products of chance, or free will, or God (depending on your religious or philosophical opinions).

184. Nature makes a perfect counter-ideal to technology for several reasons. Nature (that which is outside the power of the system) is the opposite of technology (which seeks to expand indefinitely the power of the system). Most people will agree that nature is beautiful; certainly it has tremendous popular appeal. The radical environmentalists ALREADY hold an ideology that exalts nature and opposes technology. It is not necessary for the sake of nature to set up some chimerical utopia or any new kind of social order. Nature takes care of itself: It was a spontaneous creation that existed long before any human society, and for countless centuries many different kinds of human societies coexisted with nature without doing it an excessive amount of damage. Only with the Industrial Revolution did the effect of human society on nature become really devastating. To relieve the pressure on nature it is not necessary to create a special kind of social system, it is only necessary to get rid of industrial society. Granted, this will not solve all problems. Industrial society has already done tremendous damage to nature and it will take a very long time for the scars to heal. Besides, even pre-industrial societies can do significant damage to nature. Nevertheless, getting rid of industrial society will accomplish a great deal. It will relieve the worst of the pressure on nature so that the scars can begin to heal. It will remove the capacity of organized society to keep increasing its control over nature (including human nature). Whatever kind of society may exist after the demise of the industrial system, it is certain that most people will live close to nature, because in the absence of advanced technology there is no other way that people CAN live. To feed themselves they must be peasants or herdsmen or fishermen or hunters, etc. And, generally speaking, local autonomy should tend to increase, because lack of advanced technology and rapid communications will limit the capacity of governments or other large organizations to control local communities.

185. As for the negative consequences of eliminating industrial society—well, you can't eat your cake and have it too. To gain one thing you have to sacrifice another.

186. Most people hate psychological conflict. For this reason they avoid doing any serious thinking about difficult social issues, and they like to have such issues presented to them in simple, black-and-white terms: THIS is all good and THAT is all bad. The revolutionary ideology should therefore be developed on two levels.

187. On the more sophisticated level the ideology should address itself to people who are intelligent, thoughtful and rational. The object should be to create a core of people who will be opposed to the industrial system on a rational, thought-out basis, with full

appreciation of the problems and ambiguities involved, and of the price that has to be paid for getting rid of the system. It is particularly important to attract people of this type, as they are capable people and will be instrumental in influencing others. These people should be addressed on as rational a level as possible. Facts should never intentionally be distorted and intemperate language should be avoided. This does not mean that no appeal can be made to the emotions, but in making such appeal care should be taken to avoid misrepresenting the truth or doing anything else that would destroy the intellectual respectability of the ideology.

188. On a second level, the ideology should be propagated in a simplified form that will enable the unthinking majority to see the conflict of technology vs nature in unambiguous terms. But even on this second level the ideology should not be expressed in language that is so cheap, intemperate or irrational that it alienates people of the thoughtful and rational type. Cheap, intemperate propaganda sometimes achieves impressive short-term gains, but it will be more advantageous in the long run to keep the loyalty of a small number of intelligently committed people than to arouse the passions of an unthinking, fickle mob who will change their attitude as soon as someone comes along with a better propaganda gimmick. However, propaganda of the rabble-rousing type may be necessary when the system is nearing the point of collapse and there is a final struggle between rival ideologies to determine which will become dominant when the old world-view goes under.

189. Prior to that final struggle, the revolutionaries should not expect to have a majority of people on their side. History is made by active, determined minorities, not by the majority, which seldom has a clear and consistent idea of what it really wants. Until the time comes for the final push toward revolution, the task of revolutionaries will be less to win the shallow support of the majority than to build a small core of deeply committed people. As for the majority, it will be enough to make them aware of the existence of the new ideology and remind them of it frequently; though of course it will be desirable to get majority support to the extent that this can be done without weakening the core of seriously committed people.

190. Any kind of social conflict helps to destabilize the system, but one should be careful about what kind of conflict one encourages. The line of conflict should be drawn between the mass of the people and the power-holding elite of industrial society (politicians, scientists, upper-level business executives, government officials, etc.). It should NOT be drawn between the revolutionaries and the mass of the people. For example, it would be bad strategy for the revolutionaries to condemn Americans for their habits of consumption. Instead, the average American should be portrayed as a victim of the advertising and marketing industry, which has suckered him into buying a lot of junk that he doesn't need and that is very poor compensation for his lost freedom. Either approach is consistent with the facts. It is merely a matter of attitude whether you blame the advertising industry for manipulating the public or blame the public for allowing itself to be manipulated. As a matter of strategy one should generally avoid blaming the public.

191. One should think twice before encouraging any other social conflict than that between the power-holding elite (which wields technology) and the general public (over which technology exerts its power). For one thing, other conflicts tend to distract attention from the important conflicts (between power-elite and ordinary people, between technology and nature); for another thing, other conflicts may actually tend to encourage technologization, because each side in such a conflict wants to use technological power to gain advantages over its adversary. This is clearly seen in rivalries between nations. It also appears in ethnic conflicts within nations. For example, in America many black leaders are anxious to gain power for African Americans by placing black individuals in the technological power-elite. They want there to be many black government officials, scientists, corporation executives and so forth. In this way they are helping to absorb the African American subculture into the technological system. Generally speaking, one should encourage only those social conflicts that can be fitted into the framework of the conflicts of power-elite vs ordinary people, technology vs nature.

192. But the way to discourage ethnic conflict is NOT through militant advocacy of minority rights (see paragraphs 21, 29). Instead, the revolutionaries should emphasize that although minorities do suffer more or less disadvantage, this disadvantage is of peripheral significance. Our real enemy is the industrial-technological system, and in the struggle against the system, ethnic distinctions are of no importance.

193. The kind of revolution we have in mind will not necessarily involve an armed uprising against any government. It may or may not involve physical violence, but it will not be a POLITICAL revolution. Its focus will be on technology and economics, not politics.

194. Probably the revolutionaries should even AVOID assuming political power, whether by legal or illegal means, until the industrial system is stressed to the danger point and has proved itself to be a failure in the eyes of most people. Suppose for example that some "green" party should win control of the United States Congress in an election. In order to avoid betraying or watering down their own ideology they would have to take vigorous measures to turn economic growth into economic shrinkage. To the average man the results would appear disastrous: There would be massive unemployment, shortages of commodities, etc. Even if the grosser ill effects could be avoided through superhumanly skillful management, still people would have to begin giving up the luxuries to which they have become addicted. Dissatisfaction would grow, the "green" party would be voted out of office and the revolutionaries would have suffered a severe setback. For this reason the revolutionaries should not try to acquire political power until the system has gotten itself into such a mess that any hardships will be seen as resulting from the failures of the industrial system itself and not from the policies of the revolutionaries. The revolution against technology will probably have to be a revolution by outsiders, a revolution from below and not from above.

195. The revolution must be international and worldwide. It cannot be carried out on a nation-by-nation basis. Whenever it is suggested that the United States, for example,

should cut back on technological progress or economic growth, people get hysterical and start screaming that if we fall behind in technology the Japanese will get ahead of us. Holy robots! The world will fly off its orbit if the Japanese ever sell more cars than we do! (Nationalism is a great promoter of technology.) More reasonably, it is argued that if the relatively democratic nations of the world fall behind in technology while nasty, dictatorial nations like China, Vietnam and North Korea continue to progress, eventually the dictators may come to dominate the world. That is why the industrial system should be attacked in all nations simultaneously, to the extent that this may be possible. True, there is no assurance that the industrial system can be destroyed at approximately the same time all over the world, and it is even conceivable that the attempt to overthrow the system could lead instead to the domination of the system by dictators. That is a risk that has to be taken. And it is worth taking, since the difference between a "democratic" industrial system and one controlled by dictators is small compared with the difference between an industrial system and a non-industrial one. It might even be argued that an industrial system controlled by dictators would be preferable, because dictator-controlled systems usually have proved inefficient, hence they are presumably more likely to break down. Look at Cuba.

196. Revolutionaries might consider favoring measures that tend to bind the world economy into a unified whole. Free trade agreements like NAFTA and GATT are probably harmful to the environment in the short run, but in the long run they may perhaps be advantageous because they foster economic interdependence between nations. It will be easier to destroy the industrial system on a worldwide basis if the world economy is so unified that its breakdown in any one major nation will lead to its breakdown in all industrialized nations.

197. Some people take the line that modern man has too much power, too much control over nature; they argue for a more passive attitude on the part of the human race. At best these people are expressing themselves unclearly, because they fail to distinguish between power for LARGE ORGANIZATIONS and power for INDIVIDUALS and SMALL GROUPS. It is a mistake to argue for powerlessness and passivity, because people NEED power. Modern man as a collective entity—that is, the industrial system—has immense power over nature, and we (FC) regard this as evil. But modern INDIVIDUALS and SMALL GROUPS OF INDIVIDUALS have far less power than primitive man ever did. Generally speaking, the vast power of "modern man" over nature is exercised not by individuals or small groups but by large organizations. To the extent that the average modern INDIVIDUAL can wield the power of technology, he is permitted to do so only within narrow limits and only under the supervision and control of the system. (You need a license for everything and with the license come rules and regulations.) The individual has only those technological powers with which the system chooses to provide him. His PERSONAL power over nature is slight.

198. Primitive INDIVIDUALS and SMALL GROUPS actually had considerable power over nature; or maybe it would be better to say power WITHIN nature. When primitive man needed food he knew how to find and prepare edible roots, how to track

game and take it with homemade weapons. He knew how to protect himself from heat cold, rain, dangerous animals, etc. But primitive man did relatively little damage to nature because the COLLECTIVE power of primitive society was negligible compared to the COLLECTIVE power of industrial society.

199. Instead of arguing for powerlessness and passivity, one should argue that the power of the INDUSTRIAL SYSTEM should be broken, and that this will greatly INCREASE the power and freedom of INDIVIDUALS and SMALL GROUPS.

200. Until the industrial system has been thoroughly wrecked, the destruction of that system must be the revolutionaries ONLY goal. Other goals would distract attention and energy from the main goal. More importantly, if the revolutionaries permit themselves to have any other goal than the destruction of technology they will be tempted to use technology as a tool for reaching that other goal. If they give in to that temptation, they will fall right back into the technological trap, because modern technology is a unified, tightly organized system, so that in order to retain SOME technology, one finds oneself obliged to retain MOST technology, hence one ends up sacrificing only token amounts of technology.

201. Suppose for example that the revolutionaries took "social justice" as a goal. Human nature being what it is, social justice would not come about spontaneously; it would have to be enforced. In order to enforce it the revolutionaries would have to retain central organization and control. For that they would need rapid long-distance transportation and communication, and therefore all the technology needed to support the transportation and communication systems. To feed and clothe poor people they would have to use agricultural and manufacturing technology. And so forth. So that the attempt to insure social justice would force them to retain most parts of the technological system. Not that we have anything against social justice, but it must not be allowed to interfere with the effort to get rid of the technological system.

202. It would be hopeless for revolutionaries to try to attack the system without using SOME modern technology. If nothing else they must use the communications media to spread their message. But they should use modern technology for only ONE purpose: to attack the technological system.

203. Imagine an alcoholic sitting with a barrel of wine in front of him. Suppose he starts saying to himself, "Wine isn't bad for you if used in moderation. Why, they say small amounts of wine, are even good for you! It won't do me any harm, if I take just one little drink..." Well you know what is going to happen. Never forget that the human race with technology is just like an alcoholic with a barrel of wine.

PART VI: TECHNOTOPIA?

NOTES TOWARD A NEO-LUDDITE MANIFESTO

CHELLIS GLENDINNING

Ms. Glendinning (1947-) is an author, activist, and clinical psychologist. Her books include the 1990 publication *When Technology Wounds*. This essay dates from the same year.

Why Neo-Luddism?

Most students of European history dismiss the Luddites of 19th century England as "reckless machine-smashers" and "vandals" worthy of mention only for their daring tactics. Probing beyond this interpretation, though, we find a complex, thoughtful, and little-understood social movement whose roots lay in a clash between two worldviews.

The worldview that 19th century Luddites challenged was that of laissez-faire capitalism with its increasing amalgamation of power, resources, and wealth, rationalized by its emphasis on "progress."

The worldview they supported was an older, more decentralized one espousing the interconnectedness of work, community, and family through craft guilds, village networks, and townships. They saw the new machines that owners introduced into their workplaces—the gig mills and shearing frames—as threats not only to their jobs, but to the quality of their lives and the structure of the communities they loved. In the end, destroying these machines was a last-ditch effort by a desperate people whose world lay on the verge of destruction.

The current controversy over technology is reminiscent of that of the Luddite period. We too are being barraged by a new generation of technologies—two-way television, fiber optics, biotechnology, superconductivity, fusion energy, space weapons, supercomputers. We too are witnessing protest against the onslaught. A group of Berkeley students recently gathered in Sproul Plaza to kick and smash television sets as an act of "therapy for the victims of technology." A Los Angeles business woman hiked onto Vandenberg Air Force Base and beat a weapons-related computer with a crowbar, bolt cutters, hammer, and cordless drill. Villagers in India resist the bulldozers cutting down their forests by wrapping their bodies around tree trunks. People living near the Narita airport in Japan sit on the tarmac to prevent airplanes from taking off and landing. West Germans climb up the smokestacks of factories to protest emissions that are causing acid rain, which is killing the Black Forest.

Such acts echo the concerns and commitment of the 19th century Luddites. Neo-

Luddites are 20th century citizens—activists, workers, neighbors, social critics, and scholars—who question the predominant modern worldview, which preaches that unbridled technology represents progress. Neo-Luddites have the courage to gaze at the full catastrophe of our century: The technologies created and disseminated by modern Western societies are out of control and desecrating the fragile fabric of life on Earth. Like the early Luddites, we too are a desperate people seeking to protect the livelihoods, communities, and families we love, which lie on the verge of destruction.

What Is Technology?

Just as recent social movements have challenged the idea that current models of gender roles, economic organizations, and family structures are not necessarily "normal" or "natural," so the Neo-Luddite movement has come to acknowledge that technological progress and the kinds of technologies produced in our society are not simply "the way things are."

As philosopher Lewis Mumford pointed out, technology consists of more than machines. It includes the techniques of operation and the social organizations that make a particular machine workable. In essence, a technology reflects a worldview. Which particular forms of technology—machines, techniques, and social organiza-tions—are spawned by a particular worldview depend on its perception of life, death, human potential, and the relationship of humans to one another and to nature.

In contrast to the worldviews of a majority of cultures around the world (especially those of indigenous people), the view that lies at the foundation of modern technological society encourages a mechanistic approach to life: to rational thinking, efficiency, utili-tarianism, scientific detachment, and the belief that the human place in nature is one of ownership and supremacy. The kinds of technologies that result include nuclear power plants, laser beams, and satellites. This worldview has created and promoted the military-industrial-scientific-media complex, multinational corporations, and urban sprawl.

Stopping the destruction brought by such technologies requires not just regulating or eliminating individual items like pesticides or nuclear weapons. It requires new ways of thinking about humanity and new ways of relating to life. It requires the cre-ation of a new worldview.

Principles of Neo-Luddism

1) *Neo-Luddites are not anti-technology.* Technology is intrinsic to human creativity and culture. What we oppose are the kinds of technologies that are, at root, destructive of human lives and communities. We also reject technologies that emanate from a world-view that sees rationality as the key to human potential, material acquisition as the key to human fulfillment, and technological development as the key to social progress.

2) *All technologies are political.* As social critic Jerry Mander writes in *Four Arguments for the Elimination of Television*, technologies are not neutral tools that can

be used for good or evil depending on who uses them. They are entities that have been consciously structured to reflect and serve specific powerful interests in specific historical situations. The technologies created by mass technological society are those that serve the perpetuation of mass technological society. They tend to be structured for short-term efficiency, ease of production, distribution, marketing, and profit potential—or for war-making. As a result, they tend to create rigid social systems and institutions that people do not understand and cannot change or control.

As Mander points out, television does not just bring entertainment and information to households across the globe. It offers corporations a surefire method of expanding their markets and controlling social and political thought. (It also breaks down family communications and narrows people's experience of life by mediating reality and lowering their span of attention.)

Similarly, the Dalkon Shield intrauterine device did not just make birth control easier for women. It created tremendous profits for corporate entrepreneurs at a time when the largest generation ever born in the United States was coming of age and oral contraceptives were in disfavor. (It also damaged hundreds of thousands of women by causing septic abortions, pelvic inflammatory disease, torn uteruses, sterility, and death.)

3) *The personal view of technology is dangerously limited.* The often-heard message "but I couldn't live without my word processor" denies the wider consequences of widespread use of computers (toxic contamination of workers in electronic plants and the solidifying of corporate power through exclusive access to new information in data bases).

As Mander points out, producers and disseminators of technologies tend to introduce their creations in upbeat, utopian terms. Pesticides will increase yields to feed a hungry planet! Nuclear energy will be "too cheap to meter." The pill will liberate women! Learning to critique technology demands fully examining its sociological context, economic ramifications, and political meanings. It involves asking not just what is gained—but what is lost, and by whom. It involves looking at the introduction of technologies from the perspective not only of human use, but of their impact on other living beings, natural systems, and the environment.

Program for the Future

1) As a move toward dealing with the consequences of modern technologies and preventing further destruction of life, *we favor the dismantling of the following destructive technologies*:

- Nuclear technologies—which cause disease and death at every stage of the fuel cycle;
- Chemical technologies—which re-pattern natural processes through the creation of synthetic, often poisonous chemicals and leave behind toxic and undisposable wastes;
- Genetic engineering technologies—which create dangerous mutagens that when released into the biosphere threaten us with unprecedented risks;

- Television—which functions as a centralized mind-controlling force, disrupts community life, and poisons the environment;
- Electromagnetic technologies—whose radiation alters the natural electrical dynamic of living beings, causing stress and disease; and
- Computer technologies—which cause disease and death in their manufacture and use, enhance centralized political power, and remove people from direct experience of life.

2) *We favor a search for new technological forms.* As political scientist Langdon Winner advocates in *Autonomous Technology*, we favor the adoption of technologies by the people directly involved in their use—not by scientists, engineers, and entrepreneurs who gain financially from mass production and distribution of their inventions and who know little about the context in which their technologies are used.

We favor the creation of technologies that are of a scale and structure that make them understandable to the people who use them and are affected by them. We favor the creation of technologies built with a high degree of flexibility so that they do not impose a rigid and irreversible imprint on their users, and we favor the creation of technologies that foster independence from technological addiction and promise political freedom, economic justice, and ecological balance.

3) *We favor the creation of technologies in which politics, morality, ecology, and technics are merged for the benefit of life on Earth:*

- Community-based energy sources utilizing solar, wind, and water technologies—which are renewable and enhance both community relations and respect for nature;
- Organic, biological technologies in agriculture, engineering, architecture, art, medicine, transportation, and defense—which derive directly from natural models and systems;
- Conflict resolution technologies — which emphasize cooperation, understanding, and continuity of relationship; and
- Decentralized social technologies — which encourage participation, responsibility, and empowerment.

4) *We favor the development of a life-enhancing worldview in Western technological societies.* We hope to instill a perception of life, death, and human potential into technological societies that will integrate the human need for creative expression, spiritual experience, and community with the capacity for rational thought and functionality. We perceive the human role not as the dominator of other species and planetary biology, but as integrated into the natural world with appreciation for the sacredness of all life.

We foresee a sustainable future for humanity if and when Western technological societies restructure their mechanistic projections and foster the creation of machines, techniques, and social organizations that respect both human dignity and nature's wholeness. In progressing towards such a transition, we are aware: We have nothing to lose except a way of living that leads to the destruction of all life. We have a world to gain.

INTELLIGENT ROBOTS OR CYBORGS

KEVIN WARWICK

Warwick (1954-) is a professor of cybernetics and robotics at the University of Reading, UK. The following article was published in 2004.

I've got a chip on my shoulder. Well that's what some people say. In fact, it's not far from the truth. On the morning of March 14, 2002, I found myself lying on the operating table in Theatre 1 at the Radcliffe Infirmary, Oxford. I was ready to have a small array of one hundred electrodes fired into the main nervous system in my left arm. I knew that once the array was in place, we would head back to my lab at Reading University and try to link my nervous system directly to a computer. It was an operation I didn't need for medical reasons but purely for research—in order to find out.

The operation had been going for about an hour and a half before the neurosurgeons were ready to fire into my nervous system. Peter Teddy, the consultant neurosurgeon, called for a large microscope, which was wheeled forward and rotated into position over my arm. Out of the corner of my eye, one of the TV monitors flicked through channels, displaying the view through the microscope. If I had turned my head slightly, I could have seen the monitor clearly, but I wasn't sure that I wanted to. Mark Gasson, one of my researchers, said that you could clearly see both the array and the exposed nerve fibers and they looked just as we had expected. Carefully and painstakingly, Peter maneuvered the array head into position over the nerves.

Peter doused my nerves with more local anesthetic. I could feel it swishing around as it touched parts of my arm that were not fully anesthetized until then. Next, the impactor unit (a fancy name for a pneumatic hammer) was switched on and brought into position. The unit was charged with compressed air, the head of the impactor sucked up and then forced down, just like a hammer. The head had to hit the array in order to force the electrodes into my nerves, like hammering a nail into a piece of wood.

"Kevin, can you wiggle your fingers?" asked Peter. I tried and found that I could. "Right, then. Are you ready to go for it?" As Peter asked the second question, I knew this could easily be the last time I would be able to move or feel some, or all, of my fingers. But this was what I was here for. "Let's go for it, I replied. Peter nodded. There was a sense of hush and expectancy in the operating theater. All eyes, except my own, were on either the impactor or the TV monitor.

"Okay. Here we go" said Peter. I heard a loud click as the impactor fired and felt a ping of electricity in my thumb, but nothing too much. Peter announced that the array had not gone in. "Let's try again; he said. Peter checked that I was ready and fired the impactor for a second time. Once more, it failed. Peter asked me to move my fingers, which I did. Everything was still okay. He tried once more, but to no avail.

The equipment was checked for leaky pipes, but everything seemed to be in order. Peter tried again, but it still didn't work, and once more, but no luck. All in all, Peter tried seven or eight times to force in the array without success. There was nothing I could do but feel disappointed, tremendously sad. We had come so far. It had taken so much time and effort and people to get everything together, and here we were, falling at the final hurdle. All that was left was to pull out the array, stitch me up, and we might as well go home.

But then it was spotted that the two connecting tubes on the impactor unit had been connected wrong, so that the unit had been sucking when it should have been blowing. It was merely dropping onto the array head by means of gravity. The effect was about the same as trying to crack open a coconut with a teaspoon, so the connecting tubes were switched over.

"Okay, Kevin, are you ready?" asked Peter once more. "Go for it,'" I replied. Peter then fired the impactor. This time the click was much louder, and I felt a bolt of electricity down the inside of my thumb—zap! Peter checked once more that I could move my fingers and had not lost any feeling and announced that the array had gone in successfully.

Relief poured out from everyone there. To make sure that we really had a connection, Peter used the impactor to hit the array twice more. Each time I felt a large zap on the inside of my thumb. At last, the array was sitting comfortably, and the incision could be closed up.

All of this might seem to be rather a lot to go through for a scientific experiment. Taking on such dangers is not the usual fare for a university professor. So what was it that made me do it? What was I hoping to achieve? For the answer to that, you have to step back and look at my history with robots, in particular investigating the intelligence of robots—how it compares with that of humans, what this experiment means as far as human existence is concerned, why it's important.

Humans are certainly good at doing one thing, and that is being human. In our present *Homo sapiens* form, we have been around for 100,000 years, not a significantly long time in comparison to planet earth or indeed many of the other living creatures. One thing appears to be clear, and that is that evolution keeps on going, and things change. Over a period of time, creatures adapt. Either they become more successful and survive, or they die out. History is littered with species of one type or another that were once extremely powerful but are with us no longer. As far as humans go the big question is: What of our future? Will we adapt, survive, and retain our position, or is it likely that something more powerful than humans will emerge to steal the spotlight?

If we compare human abilities, in particular our intelligence, with other creatures on earth, on the whole we come out ahead. As a result, humans have, without doubt, enjoyed a wonderful period of relative dominance over other species. We have used our position to treat many other creatures with utter contempt. In some cases, we merely destroy the habitat of other living creatures; others we keep captive. The most "fortunate" we farm, killing them for food, for political reasons, or, unfortunately on many occasions, just for fun. Indeed, ending the lives of creatures less powerful than ourselves is for some the social norm.

As far as we know, humans are the only creatures who have explored and can exist in virtually all regions of the world. In the past fifty years, we have even continued our explorations beyond our own planet. So what is it about humans that make this domination possible? What is it that we have that makes us "better" than other creatures? Many creatures exist that are, without assistance, faster or stronger than humans. So how have we gained our position? The answer has to be our intelligence, in all its intricacies.

In many cases, physically superior people are controlled by others, possibly because they are intellectually more capable. Machines, such as robots, have been created to do things that humans either do not want to do or because they can outperform humans due to their strength, speed, accuracy, and reliability. This works well when the machine is programmed and merely carries out the tasks—in particular, when it has little or nothing that could be construed as intelligence.

Humans in many parts of the world have in fact become dependent on machines of one type or another. Most of us could not imagine life without a phone, washing machine, or automobile.

For many in the Western world, the same is now true of credit cards and the Internet. Today we live in a machine-based technological world. We rely on machines for our way of life; we trust machines; we ask them questions and rely on their answers. We have grown to expect a certain standard of life that can be attained only with their help. Human progress is seen in terms of what new machines can do for us now or what more they will enable us to do in the future. Not only do we physically exist alongside the machines, but our bodies have also been tuned to rely on the standards that are provided by technology. The food we eat, the water we drink, and, in many cases, even the air we breathe depends on machines for the quality provided. Our bodies are delicately adapted to deal with the bacteria, germs, and microbes present in the technological world of today.

It is a basic rule of evolution that species must adapt as the environment changes if they are going to survive. But through technology, humans have, to a certain extent, modified the environment as needed. Each machine usually needs someone to decide when and how it should operate. A human is needed to switch the machine on. Once it is switched on, we rely on the machine to go through its programmed sequence of events in order to achieve the desired goal. Once the steering wheel of an automobile is turned, it is expected that the automobile itself will turn in that direction. In each case, we expect the machine to carry out a specific task once we have told it to do so, with us calling the shots.

In recent years, however, robots and computers are being used to make the decisions as to whether another machine should carry out certain actions. The decision might be quite a simple one, for example, switching the heating on in a house, or as complex as buying stocks. Lately machines have been employed not only because of their physical capabilities but also so they can handle our decision making. As time passes, machines are being allowed to take on more roles and this necessarily leads us to ask, How far will it go?

Nowadays, machines such as robots with computers for brains raise serious questions as to whether humans will continue to control the future. If a robot can outthink a human—and is allowed to act on these thoughts—then a rather different picture emerges. Importantly, because human intelligence is key to our status on earth, we must ask, Is it possible for robots to be more intelligent than humans? If the answer is yes, then potentially we have a serious problem on our hands. Would intelligent robots be content with us, less intelligent beings, telling them what to do, particularly when they would know better? While it is fine to have machines that are more physically capable than humans, it is quite another thing altogether for them to be more intelligent. Humans have evolved in a natural world and have now created—in the West, at least—a technological world in which we operate and on which our existence depends. We have a cozy relationship with the machines of today. They are extremely helpful, take on many of the burdens of everyday life, and allow us to do things that we would otherwise find impossible, like flying. But the performance measures and the intelligence required in today's technological world are very different from those in the natural world that existed several thousand years ago.

Over the next few years, due to the relatively slow process of biological evolution, human brains are likely to stay roughly the same size, with roughly the same number of brain cells that they possess now. Perhaps on average, there will be a slight increase, but this will be relatively insignificant. Even the most conservative estimates indicate that it will not be long before stand-alone computers outpower the human brain. Rodney Brooks of MIT said recently that "the amount of computational power in a personal computer will surpass that in a human brain sometime in the next twenty years." Meanwhile, Ian Pearson of British Telecom Labs predicts that it is more likely to be within ten years. It must also be remembered that computers do not usually tend to operate on their own; they are almost invariably networked. What, then, is the brain power of a network? There can be little doubt that it will not be too long before, in terms of share brain power, computers will outperform humans.

It is therefore important to consider which aspects of intelligence should really be considered useful. If we were to choose those aspects, which were important for life and human domination over other species, if we knew what they were, then what would happen if a machine could surpass humans in those areas? Would a machine not at least question why it was doing certain things for a human, a less intelligent being?

It could be argued that no matter how intelligent machines get, as long as there is a human who can switch it on, or off, humans will stay in control. But think again. Even now, would it be possible to switch off the Internet completely? We might be able to disconnect a few computers, but so many people depend on it for their daily existence that as a practical reality, it is simply not possible to turn it off.

In truth, it is difficult to conceive realistically of any important aspect of human intelligence in which robots will not be able to exceed in the near future. At this point you maybe saying, "Yes, but robots will never be..."—filling in the gap with words like *conscious*, *self-aware*, *creative*, or perhaps *emotional*. What you actually mean is that they may not exhibit such characteristics in the same way that a human does. It is incorrect to conclude that because robots are unlikely to be approximately equivalent to humans, they will always be subservient to us. The fact is that we already have robots that in their own way exhibit these characteristics themselves—not human self-awareness but robot self-awareness, for example.

It is because robots are different, because they have distinct advantages, many of which we know about already, that they can be better than we are. In this way, robots can dominate humans physically through their superior intelligence. What matters clearly is performance. For some time, we have been able to witness computers performing feats that we consider important aspects of intelligence, such as mathematical equations or fact retrieval, easily outperforming humans in doing so. We might then say, *Well, if a computer can beat a human, it can't be such an intelligent act after all.* Do we keep making that excuse until we have run out of intelligent acts?

So is that it? Will it be the case that in the next twenty or thirty years, robots will take over and probably treat us just as we treat less intelligent creatures now? Will we face a future in which robots keep humans on farms, keep humans as pets, or simply kill us for fun on a nice sporting robot weekend? Or is there an alternative somewhere ahead?

Humans have, until now, been fairly successful in evolutionary terms. We have to be honest, though: we are very restricted in what we can do. Obviously, we have distinct physical limitations; that is why technology has been employed to help us. But we are also restricted in how we sense the world, with only five basic senses at our disposal. Technology has been used to help us out thus far, by converting information in the ultraviolet, infrared and X-ray spectra into visual forms that our brains can understand.

But there are two factors that are perhaps far more important than the way in which we sense the world. First, human brains have evolved to perceive and understand the world in terms of three dimensions—four if you include time as a dimension. This suffocates human thinking and beliefs about what is possible and what is not. Meanwhile, machines have the ability to process multidimensional information; they have the potential to perceive the world in hundreds of dimensions.

In fact, we now have enormous problems in attempting to deal with the plethora of information that technology throws up. Even greater than this (and we must be completely honest on this one), in comparison with technology, the way in which humans communicate is so poor that it is embarrassing. Human speech is serial, error prone, and an incredibly slow way of communicating because of our use of mechanical, low-speed sound waves. Our coding procedures, called languages, severely restrict our intellectual abilities as all other thoughts and ideas must be transformed into speech signals that do not accurately represent the original concept.

So let us look at a possible alternative future scenario to one in which intelligent robots rule the planet. Is it possible for us to conceive of a creature that starts out as a human but takes on technological capabilities in order to upgrade their potential? In science-fiction terms, such creatures are referred to as cyborgs—cybernetic organisms, part human and part machine. Would it be possible for such a creature to understand the world in many more dimensions, to have the memory and mathematical capabilities of a computer, to sense the world in all sorts of ways, to have physical abilities much greater than any human and, most important of all, to communicate not by means of speech signals but by thought signals alone? This truly would be a more natural form of evolution. We can look at a future in which it is not intelligent robots that dominate the planet or humans but rather cyborgs acting as upgraded humans.

How is this to be achieved? Quite simply, with silicon chip technology implanted into the human body—merely one or two very small devices that connect directly with the human nervous system and brain. In this way, electronic signals in silicon become electrochemical signals in the human body, and vice versa. Ultimately, no doubt, the device will be implanted by means of a swift and simple injection. At first, microscale devices will be employable, but if the technology develops as expected, this will probably be surpassed by nanoscale connections. What we are looking at here is not extra memory being connected directly to the brain but implants used as interfaces linking the human nervous system to a remote computer network by radio.

Humans will need to be educated in the ways in which it is best to use this technology. In the fullness of time, our children's children will look back with wonder at how their ancestors could have been so primitive as to communicate by means of silly little noises called speech.

With brains linked to technology, there will be no need to learn much: Why should we, when the computer can do so much better? There will be no need to remember anything because computers have better storage facilities. When it is required to recall something, then it will be possible simply to download the necessary piece of information. It will even be possible to relive memories that you didn't yourself experience in the first place.

Perhaps you are thinking, *Well this is all well and good, but it is not really going to happen; it is merely from the pages of science fiction, a flight of fancy.* Well, think again. It was the first step on this path that I took in my operation of March 14. As a result, by means of a radio transmitter/receiver unit for a period of over three months, my nervous system was linked directly with a computer on a regular basis. We were able to transmit signals from my nervous system. During that time, we carried out a range of experiments.

Signals from the nervous system in my left arm were used to control the movement of a remote robot hand. As I opened and closed my own hand, the neural signals from my brain that carried out the action were also used to control the robot hand. My body was effectively extended with the computer to include a robot hand.

Next, my nervous system signals were used to drive around a small-wheeled robot: left, right, forward, backward.

The same signals were then used to decide on the mood of a group of small robots called the diddybots. With my hand open, the diddybots acted in a friendly fashion, flocking together. With my hand closed, the diddybots tried to get away from each other as fast as they could.

Using my own neural signals; I could control the movement of a virtual me inside a virtual house. I could move the virtual me from room to room. Inside each room, I could select objects on a menu and choose to operate specific pieces of technology. I could switch on lights or a coffee maker. But when I did, actual lights or a real coffee maker was activated. I could control my local environment merely by signals from my nervous system.

Messages from my nervous system were used to control the appearance of jewelry worn by my wife. As I squeezed her hand, the jewelry turned from bright red to a luminous blue. Ultimately it will be moods and feelings that direct the jewelry.

With my hair shaved off and electrodes positioned around my head, we were able to link and associate signals picked up in my brain with those being witnessed in my arm. I wore a blindfold, and the output from ultrasonic sensors fitted to my cap was fed down directly onto my nervous system. As I moved toward an object, I felt more and more pulses of current on my nervous system. As I moved away from the object, the pulses died down. In this way, I could move around and avoid objects without any visual input. I had a sixth sense, a bat-like sense, with signals traveling directly from the ultrasonic sensors, along my nervous system to my brain.

On May 20, 2002, I traveled to Columbia University in New York, and from there, an Internet link was created between my nervous system in New York and my laboratory in Reading University in the U.K. I was able to move the robot hand around again using neural signals via the Internet. This time, I was in New York but the robot hand was in the U.K. As a cyborg, my nervous system extended across the Internet. We were also able to send pulses of current from Reading University via the Internet to New York. When three pulses were sent, my fingers were stimulated three times.

Back in the U.K. a few weeks later, touch sensors on the fingers of the robot hand were employed to provide signals directly onto my nervous system, allowing me to detect just how much force the hand was applying. With a blindfold on, I could get the robot hand to grip an object with just sufficient force. By employing the same method I had used earlier to move around a small wheeled robot, using signals from my nervous system I was able to drive myself around on a wheelchair: forward, backward, left, right. I wasn't a good driver. In the final experiment, my wife, Irena, had electrodes pushed into her median nerve as well. Signals were picked up from her nervous system, transmitted across the Internet, and played down onto my nervous system. The same thing happened in reverse. When she opened and closed her hand four times, I felt four pulses on my nervous system. We had successfully achieved, in a very basic way, the first direct nervous-system-to-nervous-system communication. Obviously, with implants positioned directly in the brain, rather than merely the nervous system, this will result in a primary form of thought communication.

Far from being science fiction, all these things are science fact; they have been achieved in practice. Quite clearly, the next few years will see many more such trials. Brain implants linking memory and motor functions of a human brain to a computer are immediate steps. Now that a cyborg path has been successfully trodden, the route to a cyborg society is definitely on the horizon. In this way, humans will be able to evolve by harnessing the superintelligence and extra abilities offered by the machines of the future. It will not be a case of robots acting against humans but rather one in which we join with them. It will not be necessary for everyone to become a cyborg. There may be those who would prefer to remain a human.

If you are happy with your lot as a human, then so be it; it's your choice. But remember this: just as us humans split from our chimpanzee cousins many years ago, so cyborgs will split from humans. Those who elect to remain human will become a mere subspecies.

With extrasensory abilities, a high-performance means of communication and the best of human and machine brains it will be cyborgs that take control. For me, there is no argument. My goal is clear: I want to be a cyborg.

WHY THE FUTURE DOESN'T NEED US

BILL JOY

Bill Joy (1954-) is a computer scientist and co-founder of Sun Microsystems. This article appeared in *Wired* magazine in 2000.

From the moment I became involved in the creation of new technologies, their ethical dimensions have concerned me, but it was only in the autumn of 1998 that I became anxiously aware of how great are the dangers facing us in the 21st century. I can date the onset of my unease to the day I met Ray Kurzweil, the deservedly famous inventor of the first reading machine for the blind and many other amazing things.

Ray and I were both speakers at George Gilder's "Telecosm" conference, and I encountered him by chance in the bar of the hotel after both our sessions were over. I was sitting with John Searle, a Berkeley philosopher who studies consciousness. While we were talking, Ray approached and a conversation began, the subject of which haunts me to this day.

I had missed Ray's talk and the subsequent panel that Ray and John had been on, and they now picked right up where they'd left off, with Ray saying that the rate of improvement of technology was going to accelerate and that we were going to become robots or fuse with robots or something like that, and John countering that this couldn't happen, because the robots couldn't be conscious.

While I had heard such talk before, I had always felt sentient robots were in the realm of science fiction. But now, from someone I respected, I was hearing a strong argument that they were a near-term possibility. I was taken aback, especially given Ray's proven ability to imagine and create the future. I already knew that new technologies like genetic engineering and nanotechnology were giving us the power to remake the world, but a realistic and imminent scenario for intelligent robots surprised me.

It's easy to get jaded about such breakthroughs. We hear in the news almost every day of some kind of technological or scientific advance. Yet this was no ordinary prediction. In the hotel bar, Ray gave me a partial preprint of his then-forthcoming book *The Age of Spiritual Machines,* which outlined a utopia he foresaw—one in which humans gained near immortality by becoming one with robotic technology. On reading it, my sense of unease only intensified; I felt sure he had to be understating the dangers, understating the probability of a bad outcome along this path.

I found myself most troubled by a passage detailing a *dys*topian scenario:

THE NEW LUDDITE CHALLENGE

First let us postulate that the computer scientists succeed in developing intelligent machines that can do all things better than human beings can do them. In that case presumably all work will be done by vast, highly organized systems of machines and no human effort will be necessary. Either of two cases might occur. The machines might be permitted to make all of their own decisions without human oversight, or else human control over the machines might be retained.

If the machines are permitted to make all their own decisions, we can't make any conjectures as to the results, because it is impossible to guess how such machines might behave. We only point out that the fate of the human race would be at the mercy of the machines. It might be argued that the human race would never be foolish enough to hand over all the power to the machines. But we are suggesting neither that the human race would voluntarily turn power over to the machines nor that the machines would willfully seize power. What we do suggest is that the human race might easily permit itself to drift into a position of such dependence on the machines that it would have no practical choice but to accept all of the machines' decisions. As society and the problems that face it become more and more complex and machines become more and more intelligent, people will let machines make more of their decisions for them, simply because machine-made decisions will bring better results than man-made ones. Eventually a stage may be reached at which the decisions necessary to keep the system running will be so complex that human beings will be incapable of making them intelligently. At that stage the machines will be in effective control. People won't be able to just turn the machines off, because they will be so dependent on them that turning them off would amount to suicide.

On the other hand it is possible that human control over the machines may be retained. In that case the average man may have control over certain private machines of his own, such as his car or his personal computer, but control over large systems of machines will be in the hands of a tiny elite — just as it is today, but with two differences. Due to improved techniques the elite will have greater control over the masses; and because human work will no longer be necessary the masses will be superfluous, a useless burden on the system. If the elite is ruthless they may simply decide to exterminate the mass of humanity. If they are humane they may use propaganda or other psychological or biological techniques to reduce the birth rate until the mass of humanity becomes extinct, leaving the world to the elite. Or, if the elite consists of soft-hearted liberals, they may decide to play the role of good shepherds to the rest of the human race. They will see to it that everyone's physical needs are satisfied, that all children are raised under psychologically hygienic conditions, that everyone has a wholesome hobby to keep him busy, and that anyone who may become dissatisfied undergoes "treatment" to cure his "problem." Of course, life will be so purposeless that people will have to be biologically or psychologically engineered either to remove their need for the power process or make them "sublimate" their drive for power into some harmless hobby. These engineered human beings may be happy in such a society, but they will most certainly not be free. They will have been reduced to the status of domestic animals.

In the book, you don't discover until you turn the page that the author of this passage is Theodore Kaczynski—the Unabomber [ISAIF paragraphs 172-174]. I am no apologist for Kaczynski. His bombs killed three people during a 17-year terror campaign and wounded many others. One of his bombs gravely injured my friend David Gelernter, one of the most brilliant and visionary computer scientists of our time. Like many of my colleagues, I felt that I could easily have been the Unabomber's next target.

Kaczynski's actions were murderous and, in my view, criminally insane. He is clearly a Luddite, but simply saying this does not dismiss his argument; as difficult as it is for me to acknowledge, I saw some merit in the reasoning in this single passage. I felt compelled to confront it.

Kaczynski's dystopian vision describes unintended consequences, a well-known problem with the design and use of technology, and one that is clearly related to Murphy's law— "Anything that can go wrong, will." (Actually, this is Finagle's law, which in itself shows that Finagle was right.) Our overuse of antibiotics has led to what may be the biggest such problem so far: the emergence of antibiotic-resistant and much more dangerous bacteria. Similar things happened when attempts to eliminate malarial mosquitoes using DDT caused them to acquire DDT resistance; malarial parasites likewise acquired multi-drug-resistant genes.

The cause of many such surprises seems clear: The systems involved are complex, involving interaction among and feedback between many parts. Any changes to such a system will cascade in ways that are difficult to predict; this is especially true when human actions are involved.

I started showing friends the Kaczynski quote from *The Age of Spiritual Machines*; I would hand them Kurzweil's book, let them read the quote, and then watch their reaction as they discovered who had written it. At around the same time, I found Hans Moravec's book *Robot: Mere Machine to Transcendent Mind*. Moravec is one of the leaders in robotics research, and was a founder of the world's largest robotics research program, at Carnegie Mellon University. *Robot* gave me more material to try out on my friends—material surprisingly supportive of Kaczynski's argument. For example:

The Short Run (Early 2000s)

Biological species almost never survive encounters with superior competitors. Ten million years ago, South and North America were separated by a sunken Panama isthmus. South America, like Australia today, was populated by marsupial mammals, including pouched equivalents of rats, deer, and tigers. When the isthmus connecting North and South America rose, it took only a few thousand years for the northern placental species, with slightly more effective metabolisms and reproductive and nervous systems, to displace and eliminate almost all the southern marsupials.

In a completely free marketplace, superior robots would surely affect humans as North American placentals affected South American marsupials (and as humans have affected countless species). Robotic industries would compete vigorously among themselves for matter, energy, and space, incidentally driving their price beyond human reach. Unable to afford the necessities of life, biological humans would be squeezed out of existence.

There is probably some breathing room, because we do not live in a completely free marketplace. Government coerces nonmarket behavior, especially by collecting taxes. Judiciously applied, governmental coercion could support human populations in high style on the fruits of robot labor, perhaps for a long while.

A textbook dystopia—and Moravec is just getting wound up. He goes on to discuss how our main job in the 21st century will be "ensuring continued cooperation from the robot industries" by passing laws decreeing that they be "nice," and to describe how seriously dangerous a human can be "once transformed into an unbounded super-intelligent robot." Moravec's view is that the robots will eventually succeed us—that humans clearly face extinction.

I decided it was time to talk to my friend Danny Hillis. Danny became famous as the cofounder of Thinking Machines Corporation, which built a very powerful parallel supercomputer. Despite my current job title of Chief Scientist at Sun Microsystems, I am more a computer architect than a scientist, and I respect Danny's knowledge of the information and physical sciences more than that of any other single person I know. Danny is also a highly regarded futurist who thinks long-term—four years ago he started the Long Now Foundation, which is building a clock designed to last 10,000 years, in an attempt to draw attention to the pitifully short attention span of our society.

So I flew to Los Angeles for the express purpose of having dinner with Danny and his wife, Pati. I went through my now-familiar routine, trotting out the ideas and passages that I found so disturbing. Danny's answer—directed specifically at Kurzweil's scenario of humans merging with robots—came swiftly, and quite surprised me. He said, simply, that the changes would come gradually, and that we would get used to them.

But I guess I wasn't totally surprised. I had seen a quote from Danny in Kurzweil's book in which he said, "I'm as fond of my body as anyone, but if I can be 200 with a body of silicon, I'll take it." It seemed that he was at peace with this process and its attendant risks, while I was not.

While talking and thinking about Kurzweil, Kaczynski, and Moravec, I suddenly remembered a novel I had read almost 20 years ago—*The White Plague,* by Frank

Herbert—in which a molecular biologist is driven insane by the senseless murder of his family. To seek revenge he constructs and disseminates a new and highly contagious plague that kills widely but selectively. (We're lucky Kaczynski was a mathematician, not a molecular biologist.) I was also reminded of the Borg of *Star Trek,* a hive of partly biological, partly robotic creatures with a strong destructive streak. Borg-like disasters are a staple of science fiction, so why hadn't I been more concerned about such robotic dystopias earlier? Why weren't other people more concerned about these nightmarish scenarios?

Part of the answer certainly lies in our attitude toward the new—in our bias toward instant familiarity and unquestioning acceptance. Accustomed to living with almost routine scientific breakthroughs, we have yet to come to terms with the fact that the most compelling 21st-century technologies—robotics, genetic engineering, and nanotechnology—pose a different threat than the technologies that have come before. Specifically, robots, engineered organisms, and nanobots share a dangerous amplifying factor: They can self-replicate. A bomb is blown up only once—but one bot can become many, and quickly get out of control.

Much of my work over the past 25 years has been on computer networking, where the sending and receiving of messages creates the opportunity for out-of-control replication. But while replication in a computer or a computer network can be a nuisance, at worst it disables a machine or takes down a network or network service. Uncontrolled self-replication in these newer technologies runs a much greater risk: a risk of substantial damage in the physical world.

Each of these technologies also offers untold promise: The vision of near immortality that Kurzweil sees in his robot dreams drives us forward; genetic engineering may soon provide treatments, if not outright cures, for most diseases; and nanotechnology and nanomedicine can address yet more ills. Together they could significantly extend our average life span and improve the quality of our lives. Yet, with each of these technologies, a sequence of small, individually sensible advances leads to an accumulation of great power and, concomitantly, great danger.

What was different in the 20th century? Certainly, the technologies underlying the weapons of mass destruction (WMD)—nuclear, biological, and chemical (NBC)—were powerful, and the weapons an enormous threat. But building nuclear weapons required, at least for a time, access to both rare—indeed, effectively unavailable—raw materials and highly protected information; biological and chemical weapons programs also tended to require large-scale activities.

The 21st-century technologies—genetics, nanotechnology, and robotics (GNR)—are so powerful that they can spawn whole new classes of accidents and abuses. Most dangerously, for the first time, these accidents and abuses are widely within the reach of individuals or small groups. They will not require large facilities or rare raw materials. Knowledge alone will enable the use of them.

Thus we have the possibility not just of weapons of mass destruction but of knowledge-enabled mass destruction (KMD), this destructiveness hugely amplified by the power of self-replication.

I think it is no exaggeration to say we are on the cusp of the further perfection of extreme evil, an evil whose possibility spreads well beyond that which weapons of mass destruction bequeathed to the nation-states, on to a surprising and terrible empowerment of extreme individuals.

Nothing about the way I got involved with computers suggested to me that I was going to be facing these kinds of issues.

My life has been driven by a deep need to ask questions and find answers. When I was 3, I was already reading, so my father took me to the elementary school, where I sat on the principal's lap and read him a story. I started school early, later skipped a grade, and escaped into books—I was incredibly motivated to learn. I asked lots of questions, often driving adults to distraction.

As a teenager I was very interested in science and technology. I wanted to be a ham radio operator but didn't have the money to buy the equipment. Ham radio was the Internet of its time: very addictive, and quite solitary. Money issues aside, my mother put her foot down—I was not to be a ham; I was antisocial enough already.

I may not have had many close friends, but I was awash in ideas. By high school, I had discovered the great science fiction writers. I remember especially Heinlein's *Have Spacesuit Will Travel* and Asimov's *I, Robot,* with its Three Laws of Robotics. I was enchanted by the descriptions of space travel, and wanted to have a telescope to look at the stars; since I had no money to buy or make one, I checked books on telescope-making out of the library and read about making them instead. I soared in my imagination.

Thursday nights my parents went bowling, and we kids stayed home alone. It was the night of Gene Roddenberry's original *Star Trek,* and the program made a big impression on me. I came to accept its notion that humans had a future in space, Western-style, with big heroes and adventures. Roddenberry's vision of the centuries to come was one with strong moral values, embodied in codes like the Prime Directive: to not interfere in the development of less technologically advanced civilizations. This had an incredible appeal to me; ethical humans, not robots, dominated this future, and I took Roddenberry's dream as part of my own.

I excelled in mathematics in high school, and when I went to the University of Michigan as an undergraduate engineering student I took the advanced curriculum of the mathematics majors. Solving math problems was an exciting challenge, but when I discovered computers I found something much more interesting: a machine into which you could put a program that attempted to solve a problem, after which the machine quickly checked the solution. The computer had a clear notion of correct and incorrect, true and false. Were my ideas correct? The machine could tell me. This was very seductive.

I was lucky enough to get a job programming early supercomputers and discovered the amazing power of large machines to numerically simulate advanced designs. When I went to graduate school at UC Berkeley in the mid-1970s, I started staying up late, often all night, inventing new worlds inside the machines. Solving problems. Writing the code that argued so strongly to be written.

In *The Agony and the Ecstasy,* Irving Stone's biographical novel of Michelangelo, Stone described vividly how Michelangelo released the statues from the stone, "breaking the marble spell," carving from the images in his mind. In my most ecstatic moments, the software in the computer emerged in the same way. Once I had imagined it in my mind I felt that it was already there in the machine, waiting to be released. Staying up all night seemed a small price to pay to free it—to give the ideas concrete form.

After a few years at Berkeley I started to send out some of the software I had written—an instructional Pascal system, Unix utilities, and a text editor called vi (which is still, to my surprise, widely used more than 20 years later)—to others who had similar small PDP-11 and VAX minicomputers. These adventures in software eventually turned into the Berkeley version of the Unix operating system, which became a personal "success disaster"—so many people wanted it that I never finished my PhD. Instead I got a job working for Darpa putting Berkeley Unix on the Internet and fixing it to be reliable and to run large research applications well. This was all great fun and very rewarding. And, frankly, I saw no robots here, or anywhere near.

Still, by the early 1980s, I was drowning. The Unix releases were very successful, and my little project of one soon had money and some staff, but the problem at Berkeley was always office space rather than money—there wasn't room for the help the project needed, so when the other founders of Sun Microsystems showed up I jumped at the chance to join them. At Sun, the long hours continued into the early days of workstations and personal computers, and I have enjoyed participating in the creation of advanced microprocessor technologies and Internet technologies such as Java and Jini.

From all this, I trust it is clear that I am not a Luddite. I have always, rather, had a strong belief in the value of the scientific search for truth and in the ability of great engineering to bring material progress. The Industrial Revolution has immeasurably improved everyone's life over the last couple hundred years, and I always expected my career to involve the building of worthwhile solutions to real problems, one problem at a time.

I have not been disappointed. My work has had more impact than I had ever hoped for and has been more widely used than I could have reasonably expected. I have spent the last 20 years still trying to figure out how to make computers as reliable as I want them to be (they are not nearly there yet) and how to make them simple to use (a goal that has met with even less relative success). Despite some progress, the problems that remain seem even more daunting.

But while I was aware of the moral dilemmas surrounding technology's consequences in fields like weapons research, I did not expect that I would confront such issues in my own field, or at least not so soon.

Perhaps it is always hard to see the bigger impact while you are in the vortex of a change. Failing to understand the consequences of our inventions while we are in the rapture of discovery and innovation seems to be a common fault of scientists and technologists; we have long been driven by the overarching desire to know that is the nature of science's quest, not stopping to notice that the progress to newer and more powerful technologies can take on a life of its own.

I have long realized that the big advances in information technology come not from the work of computer scientists, computer architects, or electrical engineers, but from that of physical scientists. The physicists Stephen Wolfram and Brosl Hasslacher introduced me, in the early 1980s, to chaos theory and nonlinear systems. In the 1990s, I learned about complex systems from conversations with Danny Hillis, the biologist

Stuart Kauffman, the Nobel-laureate physicist Murray Gell-Mann, and others. Most recently, Hasslacher and the electrical engineer and device physicist Mark Reed have been giving me insight into the incredible possibilities of molecular electronics.

In my own work, as co-designer of three microprocessor architectures—SPARC, picoJava, and MAJC—and as the designer of several implementations thereof, I've been afforded a deep and firsthand acquaintance with Moore's law. For decades, Moore's law has correctly predicted the exponential rate of improvement of semiconductor technology. Until last year I believed that the rate of advances predicted by Moore's law might continue only until roughly 2010, when some physical limits would begin to be reached. It was not obvious to me that a new technology would arrive in time to keep performance advancing smoothly.

But because of the recent rapid and radical progress in molecular electronics— where individual atoms and molecules replace lithographically drawn transistors— and related nanoscale technologies, we should be able to meet or exceed the Moore's law rate of progress for another 30 years. By 2030, we are likely to be able to build machines, in quantity, a million times as powerful as the personal computers of today—sufficient to implement the dreams of Kurzweil and Moravec.

As this enormous computing power is combined with the manipulative advances of the physical sciences and the new, deep understandings in genetics, enormous transformative power is being unleashed. These combinations open up the opportunity to completely redesign the world, for better or worse: The replicating and evolving processes that have been confined to the natural world are about to become realms of human endeavor.

In designing software and microprocessors, I have never had the feeling that I was designing an intelligent machine. The software and hardware is so fragile and the capabilities of the machine to "think" so clearly absent that, even as a possibility, this has always seemed very far in the future.

But now, with the prospect of human-level computing power in about 30 years, a new idea suggests itself: that I may be working to create tools which will enable the construction of the technology that may replace our species. How do I feel about this? Very uncomfortable. Having struggled my entire career to build reliable software systems, it seems to me more than likely that this future will not work out as well as some people may imagine. My personal experience suggests we tend to overestimate our design abilities.

Given the incredible power of these new technologies, shouldn't we be asking how we can best coexist with them? And if our own extinction is a likely, or even possible, outcome of our technological development, shouldn't we proceed with great caution?

The dream of robotics is, first, that intelligent machines can do our work for us, allowing us lives of leisure, restoring us to Eden. Yet in his history of such ideas, *Darwin Among the Machines,* George Dyson warns: "In the game of life and evolution there are three players at the table: human beings, nature, and machines. I am firmly on the side of nature. But nature, I suspect, is on the side of the machines." As we have seen, Moravec agrees, believing we may well not survive the encounter with the superior robot species.

How soon could such an intelligent robot be built? The coming advances in computing power seem to make it possible by 2030. And once an intelligent robot exists, it is only a small step to a robot species—to an intelligent robot that can make evolved copies of itself.

A second dream of robotics is that we will gradually replace ourselves with our robotic technology, achieving near immortality by downloading our consciousnesses; it is this process that Danny Hillis thinks we will gradually get used to and that Ray Kurzweil elegantly details in *The Age of Spiritual Machines*.

But if we are downloaded into our technology, what are the chances that we will thereafter be ourselves or even human? It seems to me far more likely that a robotic existence would not be like a human one in any sense that we understand, that the robots would in no sense be our children, that on this path our humanity may well be lost.

Genetic engineering promises to revolutionize agriculture by increasing crop yields while reducing the use of pesticides; to create tens of thousands of novel species of bacteria, plants, viruses, and animals; to replace reproduction, or supplement it, with cloning; to create cures for many diseases, increasing our life span and our quality of life; and much, much more. We now know with certainty that these profound changes in the biological sciences are imminent and will challenge all our notions of what life is.

Technologies such as human cloning have in particular raised our awareness of the profound ethical and moral issues we face. If, for example, we were to reengineer ourselves into several separate and unequal species using the power of genetic engineering, then we would threaten the notion of equality that is the very cornerstone of our democracy.

Given the incredible power of genetic engineering, it's no surprise that there are significant safety issues in its use. My friend Amory Lovins recently cowrote, along with Hunter Lovins, an editorial that provides an ecological view of some of these dangers. Among their concerns: that "the new botany aligns the development of plants with their economic, not evolutionary, success." Amory's long career has been focused on energy and resource efficiency by taking a whole-system view of human-made systems; such a whole-system view often finds simple, smart solutions to otherwise seemingly difficult problems, and is usefully applied here as well.

After reading the Lovins' editorial, I saw an op-ed by Gregg Easterbrook in *The New York Times* (November 19, 1999) about genetically engineered crops, under the headline: "Food for the Future: Someday, rice will have built-in vitamin A. Unless the Luddites win."

Are Amory and Hunter Lovins Luddites? Certainly not. I believe we all would agree that golden rice, with its built-in vitamin A, is probably a good thing, if developed with proper care and respect for the likely dangers in moving genes across species boundaries.

Awareness of the dangers inherent in genetic engineering is beginning to grow, as reflected in the Lovins' editorial. The general public is aware of, and uneasy about, genetically modified foods, and seems to be rejecting the notion that such foods should be permitted to be unlabeled.

But genetic engineering technology is already very far along. As the Lovins note, the USDA has already approved about 50 genetically engineered crops for unlimited release; more than half of the world's soybeans and a third of its corn now contain genes spliced in from other forms of life.

While there are many important issues here, my own major concern with genetic engineering is narrower: that it gives the power—whether militarily, accidentally, or in a deliberate terrorist act—to create a White Plague.

The many wonders of nanotechnology were first imagined by the Nobel-laureate physicist Richard Feynman in a speech he gave in 1959, subsequently published under the title "There's Plenty of Room at the Bottom." The book that made a big impression on me, in the mid-'80s, was Eric Drexler's *Engines of Creation,* in which he described beautifully how manipulation of matter at the atomic level could create a utopian future of abundance, where just about everything could be made cheaply, and almost any imaginable disease or physical problem could be solved using nanotechnology and artificial intelligences.

A subsequent book, *Unbounding the Future: The Nanotechnology Revolution,* which Drexler cowrote, imagines some of the changes that might take place in a world where we had molecular-level "assemblers." Assemblers could make possible incredibly low-cost solar power, cures for cancer and the common cold by augmentation of the human immune system, essentially complete cleanup of the environment, incredibly inexpensive pocket supercomputers—in fact, any product would be manufacturable by assemblers at a cost no greater than that of wood—spaceflight more accessible than transoceanic travel today, and restoration of extinct species.

I remember feeling good about nanotechnology after reading *Engines of Creation.* As a technologist, it gave me a sense of calm—that is, nanotechnology showed us that incredible progress was possible, and indeed perhaps inevitable. If nanotechnology was our future, then I didn't feel pressed to solve so many problems in the present. I would get to Drexler's utopian future in due time; I might as well enjoy life more in the here and now. It didn't make sense, given his vision, to stay up all night, all the time.

Drexler's vision also led to a lot of good fun. I would occasionally get to describe the wonders of nanotechnology to others who had not heard of it. After teasing them with all the things Drexler described I would give a homework assignment of my own: "Use nanotechnology to create a vampire; for extra credit create an antidote."

———————————

With these wonders came clear dangers, of which I was acutely aware. As I said at a nanotechnology conference in 1989, "We can't simply do our science and not worry about these ethical issues." But my subsequent conversations with physicists convinced me that nanotechnology might not even work—or, at least, it wouldn't work anytime soon. Shortly thereafter I moved to Colorado, to a skunk works I had set up, and the focus of my work shifted to software for the Internet, specifically on ideas that became Java and Jini.

Then, last summer, Brosl Hasslacher told me that nanoscale molecular electronics was now practical. This was *new* news, at least to me, and I think to many people—and it radically changed my opinion about nanotechnology. It sent me back to *Engines of Creation.* Rereading Drexler's work after more than 10 years, I was dismayed to realize how little I had remembered of its lengthy section called "Dangers and Hopes," including a discussion of how nanotechnologies can become "engines of destruction." Indeed, in my rereading of this cautionary material today, I am struck by how naive some of Drexler's safeguard proposals seem, and how much greater I judge the dangers to be now than even he seemed to then.

The enabling breakthrough to assemblers seems quite likely within the next 20 years. Molecular electronics—the new subfield of nanotechnology where individual molecules are circuit elements—should mature quickly and become enormously lucrative within this decade, causing a large incremental investment in all nanotechnologies.

Unfortunately, as with nuclear technology, it is far easier to create destructive uses for nanotechnology than constructive ones. Nanotechnology has clear military and terrorist uses, and you need not be suicidal to release a massively destructive nanotechnological device—such devices can be built to be selectively destructive, affecting, for example, only a certain geographical area or a group of people who are genetically distinct.

An immediate consequence of the Faustian bargain in obtaining the great power of nanotechnology is that we run a grave risk—the risk that we might destroy the biosphere on which all life depends.

As Drexler explained:

> "Plants" with "leaves" no more efficient than today's solar cells could out-compete real plants, crowding the biosphere with an inedible foliage. Tough omnivorous "bacteria" could out-compete real bacteria: They could spread like blowing pollen, replicate swiftly, and reduce the biosphere to dust in a matter of days. Dangerous replicators could easily be too tough, small, and rapidly spreading to stop—at least if we make no preparation. We have trouble enough controlling viruses and fruit flies.

> Among the cognoscenti of nanotechnology, this threat has become known as the "gray goo problem." Though masses of uncontrolled replicators need not be gray or gooey, the term "gray goo" emphasizes that replicators able to obliterate life might be less inspiring than a single species of crabgrass. They might be superior in an evolutionary sense, but this need not make them valuable.

> The gray goo threat makes one thing perfectly clear: We cannot afford certain kinds of accidents with replicating assemblers.

> Gray goo would surely be a depressing ending to our human adventure on Earth, far worse than mere fire or ice, and one that could stem from a simple laboratory accident.

Oops.

It is most of all the power of destructive self-replication in genetics, nanotechnology, and robotics (GNR) that should give us pause. Self-replication is the modus operandi of genetic engineering, which uses the machinery of the cell to replicate its designs, and the prime danger underlying gray goo in nanotechnology. Stories of run-amok robots like the Borg, replicating or mutating to escape from the ethical constraints imposed on them by their creators, are well established in our science fiction books and movies. It is even possible that self-replication may be more fundamental than we thought, and hence harder—or even impossible—to control. A recent article by Stuart Kauffman in *Nature* titled "Self-Replication: Even Peptides Do It" discusses the discovery that a 32-amino-acid peptide can "autocatalyse its own synthesis." We don't know how widespread this ability is, but Kauffman notes that it may hint at "a route to self-reproducing molecular systems on a basis far wider than Watson-Crick base-pairing."

In truth, we have had in hand for years clear warnings of the dangers inherent in widespread knowledge of GNR technologies—of the possibility of knowledge alone enabling mass destruction. But these warnings haven't been widely publicized; the public discussions have been clearly inadequate. There is no profit in publicizing the dangers.

The nuclear, biological, and chemical (NBC) technologies used in 20th-century weapons of mass destruction were and are largely military, developed in government laboratories. In sharp contrast, the 21st-century GNR technologies have clear commercial uses and are being developed almost exclusively by corporate enterprises. In this age of triumphant commercialism, technology—with science as its handmaiden—is delivering a series of almost magical inventions that are the most phenomenally lucrative ever seen. We are aggressively pursuing the promises of these new technologies within the now-unchallenged system of global capitalism and its manifold financial incentives and competitive pressures.

This is the first moment in the history of our planet when any species, by its own voluntary actions, has become a danger to itself—as well as to vast numbers of others.

> "It might be a familiar progression, transpiring on many worlds—a planet, newly formed, placidly revolves around its star; life slowly forms; a kaleidoscopic procession of creatures evolves; intelligence emerges which, at least up to a point, confers enormous survival value; and then technology is invented. It dawns on them that there are such things as laws of Nature, that these laws can be revealed by experiment, and that knowledge of these laws can be made both to save and to take lives, both on unprecedented scales. Science, they recognize, grants immense powers. In a flash, they create world-altering contrivances. Some planetary civilizations see their way through, place limits on what may and what must not be done, and safely pass through the time of perils. Others, not so lucky or so prudent, perish."

That is Carl Sagan, writing in 1994, in *Pale Blue Dot,* a book describing his vision of the human future in space. I am only now realizing how deep his insight was, and how sorely I miss, and will miss, his voice. For all its eloquence, Sagan's contribution was not least that of simple common sense—an attribute that, along with humility, many of the leading advocates of the 21st-century technologies seem to lack.

I remember from my childhood that my grandmother was strongly against the overuse of antibiotics. She had worked since before the first World War as a nurse and had a commonsense attitude that taking antibiotics, unless they were absolutely necessary, was bad for you.

It is not that she was an enemy of progress. She saw much progress in an almost 70-year nursing career; my grandfather, a diabetic, benefited greatly from the improved treatments that became available in his lifetime. But she, like many level-headed people, would probably think it greatly arrogant for us, now, to be designing a robotic "replacement species," when we obviously have so much trouble making relatively simple things work, and so much trouble managing—or even under-standing—ourselves.

I realize now that she had an awareness of the nature of the order of life, and of the necessity of living with and respecting that order. With this respect comes a necessary humility that we, with our early-21st-century chutzpah, lack at our peril. The common-sense view, grounded in this respect, is often right, in advance of the scientific evidence. The clear fragility and inefficiencies of the human-made systems we have built should give us all pause; the fragility of the systems I have worked on certainly humbles me.

We should have learned a lesson from the making of the first atomic bomb and the resulting arms race. We didn't do well then, and the parallels to our current situation are troubling.

The effort to build the first atomic bomb was led by the brilliant physicist J. Robert Oppenheimer. Oppenheimer was not naturally interested in politics but became painfully aware of what he perceived as the grave threat to Western civilization from the Third Reich, a threat surely grave because of the possibility that Hitler might obtain nuclear weapons. Energized by this concern, he brought his strong intellect, passion for physics, and charismatic leadership skills to Los Alamos and led a rapid and successful effort by an incredible collection of great minds to quickly invent the bomb.

What is striking is how this effort continued so naturally after the initial impetus was removed. In a meeting shortly after V-E Day with some physicists who felt that perhaps the effort should stop, Oppenheimer argued to continue. His stated reason seems a bit strange: not because of the fear of large casualties from an invasion of Japan, but because the United Nations, which was soon to be formed, should have foreknowledge of atomic weapons. A more likely reason the project continued is the momentum that had built up—the first atomic test, Trinity, was nearly at hand.

We know that in preparing this first atomic test the physicists proceeded despite a large number of possible dangers. They were initially worried, based on a calculation by Edward Teller, that an atomic explosion might set fire to the atmosphere. A revised calculation reduced the danger of destroying the world to a three-in-a-million chance. (Teller says he was later able to dismiss the prospect of atmospheric ignition entirely.) Oppenheimer, though, was sufficiently concerned about the result of Trinity that he arranged for a possible evacuation of the southwest part of the state of New Mexico. And, of course, there was the clear danger of starting a nuclear arms race.

Within a month of that first, successful test, two atomic bombs destroyed Hiroshima and Nagasaki. Some scientists had suggested that the bomb simply be demonstrated, rather than dropped on Japanese cities—saying that this would greatly improve the chances for arms control after the war—but to no avail. With the tragedy of Pearl Harbor still fresh in Americans' minds, it would have been very difficult for President Truman to order a demonstration of the weapons rather than use them as he did—the desire to quickly end the war and save the lives that would have been lost in any invasion of Japan was very strong. Yet the overriding truth was probably very simple: As the physicist Freeman Dyson later said, "The reason that it was dropped was just that nobody had the courage or the foresight to say no."

It's important to realize how shocked the physicists were in the aftermath of the bombing of Hiroshima, on August 6, 1945. They describe a series of waves of emotion: first, a sense of fulfillment that the bomb worked, then horror at all the people that had been killed, and then a convincing feeling that on no account should another bomb be dropped. Yet of course another bomb was dropped, on Nagasaki, only three days after the bombing of Hiroshima.

In November 1945, three months after the atomic bombings, Oppenheimer stood firmly behind the scientific attitude, saying, "It is not possible to be a scientist unless you believe that the knowledge of the world, and the power which this gives, is a thing which is of intrinsic value to humanity, and that you are using it to help in the spread of knowledge and are willing to take the consequences."

Oppenheimer went on to work, with others, on the Acheson-Lilienthal report, which, as Richard Rhodes says in his recent book *Visions of Technology,* "found a way to prevent a clandestine nuclear arms race without resorting to armed world government"; their suggestion was a form of relinquishment of nuclear weapons work by nation-states to an international agency.

This proposal led to the Baruch Plan, which was submitted to the United Nations in June 1946 but never adopted (perhaps because, as Rhodes suggests, Bernard Baruch had "insisted on burdening the plan with conventional sanctions," thereby inevitably dooming it, even though it would "almost certainly have been rejected by Stalinist Russia anyway"). Other efforts to promote sensible steps toward internationalizing nuclear power to prevent an arms race ran afoul either of US politics and internal distrust, or distrust by the Soviets. The opportunity to avoid the arms race was lost, and very quickly.

Two years later, in 1948, Oppenheimer seemed to have reached another stage in his thinking, saying, "In some sort of crude sense which no vulgarity, no humor, no overstatement can quite extinguish, the physicists have known sin; and this is a knowledge they cannot lose."

In 1949, the Soviets exploded an atom bomb. By 1955, both the US and the Soviet Union had tested hydrogen bombs suitable for delivery by aircraft. And so the nuclear arms race began.

Nearly 20 years ago, in the documentary *The Day After Trinity,* Freeman Dyson summarized the scientific attitudes that brought us to the nuclear precipice:

I have felt it myself. The glitter of nuclear weapons. It is irresistible if you come to them as a scientist. To feel it's there in your hands, to release this energy that fuels the stars, to let it do your bidding. To perform these miracles, to lift a million tons of rock into the sky. It is something that gives people an illusion of illimitable power, and it is, in some ways, responsible for all our troubles—this, what you might call technical arrogance, that overcomes people when they see what they can do with their minds.

Now, as then, we are creators of new technologies and stars of the imagined future, driven—this time by great financial rewards and global competition—despite the clear dangers, hardly evaluating what it may be like to try to live in a world that is the realistic outcome of what we are creating and imagining.

In 1947, *The Bulletin of the Atomic Scientists* began putting a Doomsday Clock on its cover. For more than 50 years, it has shown an estimate of the relative nuclear danger we have faced, reflecting the changing international conditions. The hands on the clock have moved 15 times and today, standing at nine minutes to midnight, reflect continuing and real danger from nuclear weapons. The recent addition of India and Pakistan to the list of nuclear powers has increased the threat of failure of the nonproliferation goal, and this danger was reflected by moving the hands closer to midnight in 1998.

In our time, how much danger do we face, not just from nuclear weapons, but from all of these technologies? How high are the extinction risks?

The philosopher John Leslie has studied this question and concluded that the risk of human extinction is at least 30 percent, while Ray Kurzweil believes we have "a better than even chance of making it through," with the caveat that he has "always been accused of being an optimist." Not only are these estimates not encouraging, but they do not include the probability of many horrid outcomes that lie short of extinction.

Faced with such assessments, some serious people are already suggesting that we simply move beyond Earth as quickly as possible. We would colonize the galaxy using von Neumann probes, which hop from star system to star system, replicating as they go. This step will almost certainly be necessary 5 billion years from now (or sooner if our solar system is disastrously impacted by the impending collision of our galaxy with the Andromeda galaxy within the next 3 billion years), but if we take Kurzweil and Moravec at their word it might be necessary by the middle of this century.

What are the moral implications here? If we must move beyond Earth this quickly in order for the species to survive, who accepts the responsibility for the fate of those (most of us, after all) who are left behind? And even if we scatter to the stars, isn't it likely that we may take our problems with us or find, later, that they have followed us? The fate of our species on Earth and our fate in the galaxy seem inextricably linked.

Another idea is to erect a series of shields to defend against each of the dangerous technologies. The Strategic Defense Initiative, proposed by the Reagan administration, was an attempt to design such a shield against the threat of a nuclear attack from the Soviet Union. But as Arthur C. Clarke, who was privy to discussions about the project, observed: "Though it might be possible, at vast expense, to construct local defense systems that would 'only' let through a few percent of ballistic missiles, the much touted idea of a national umbrella was nonsense. Luis Alvarez, perhaps the greatest experimental physicist of this century, remarked to me that the advocates of such schemes were 'very bright guys with no common sense.'"

Clarke continued: "Looking into my often cloudy crystal ball, I suspect that a total defense might indeed be possible in a century or so. But the technology involved would produce, as a by-product, weapons so terrible that no one would bother with anything as primitive as ballistic missiles."

In *Engines of Creation,* Eric Drexler proposed that we build an active nanotechnological shield—a form of immune system for the biosphere—to defend against dangerous replicators of all kinds that might escape from laboratories or otherwise be maliciously created. But the shield he proposed would itself be extremely dangerous—nothing could prevent it from developing autoimmune problems and attacking the biosphere itself.

Similar difficulties apply to the construction of shields against robotics and genetic engineering. These technologies are too powerful to be shielded against in the time frame of interest; even if it were possible to implement defensive shields, the side effects of their development would be at least as dangerous as the technologies we are trying to protect against.

These possibilities are all thus either undesirable or unachievable or both. The only realistic alternative I see is *relinquishment*: to limit development of the technologies that are too dangerous, by limiting our pursuit of certain kinds of knowledge.

Yes, I know, knowledge is good, as is the search for new truths. We have been seeking knowledge since ancient times. Aristotle opened his *Metaphysics* with the simple statement: "All men by nature desire to know." We have, as a bedrock value in our society, long agreed on the value of open access to information, and recognize the problems that arise with attempts to restrict access to and development of knowledge. In recent times, we have come to revere scientific knowledge.

But despite the strong historical precedents, if open access to and unlimited development of knowledge henceforth puts us all in clear danger of extinction, then common sense demands that we reexamine even these basic, long-held beliefs.

It was Nietzsche who warned us, at the end of the 19th century, not only that God is dead but that

"faith in science, which after all exists undeniably, cannot owe its origin to a calculus of utility; it must have originated *in spite of* the fact that the disutility and dangerousness of the 'will to truth,' of 'truth at any price' is proved to it constantly."

(Gay Science, 344)

It is this further danger that we now fully face—the consequences of our truth-seeking. The truth that science seeks can certainly be considered a dangerous substitute for God if it is likely to lead to our extinction.

If we could agree, as a species, what we wanted, where we were headed, and why, then we would make our future much less dangerous—then we might understand what we can and should relinquish. Otherwise, we can easily imagine an arms race developing over GNR technologies, as it did with the NBC technologies in the 20th century. This is perhaps the greatest risk, for once such a race begins, it's very hard to end it. This time—unlike during the Manhattan Project—we aren't in a war, facing an implacable enemy that is threatening our civilization; we are driven, instead, by our habits, our desires, our economic system, and our competitive need to know.

I believe that we all wish our course could be determined by our collective values, ethics, and morals. If we had gained more collective wisdom over the past few thousand years, then a dialogue to this end would be more practical, and the incredible powers we are about to unleash would not be nearly so troubling.

One would think we might be driven to such a dialogue by our instinct for self-preservation. Individuals clearly have this desire, yet as a species our behavior seems to be not in our favor. In dealing with the nuclear threat, we often spoke dishonestly to ourselves and to each other, thereby greatly increasing the risks. Whether this was politically motivated, or because we chose not to think ahead, or because when faced with such grave threats we acted irrationally out of fear, I do not know, but it does not bode well.

The new Pandora's boxes of genetics, nanotechnology, and robotics are almost open, yet we seem hardly to have noticed. Ideas can't be put back in a box; unlike uranium or plutonium, they don't need to be mined and refined, and they can be freely copied. Once they are out, they are out. Churchill remarked, in a famous left-handed compliment, that the American people and their leaders "invariably do the right thing, after they have examined every other alternative." In this case, however, we must act more presciently, as to do the right thing only at last may be to lose the chance to do it at all.

As Thoreau said, "We do not ride on the railroad; it rides upon us"; and this is what we must fight, in our time. The question is, indeed, Which is to be master? Will we survive our technologies?

We are being propelled into this new century with no plan, no control, no brakes. Have we already gone too far down the path to alter course? I don't believe so, but we aren't trying yet, and the last chance to assert control—the fail-safe point—is rapidly approaching. We have our first pet robots, as well as commercially available genetic engineering techniques, and our nanoscale techniques are advancing rapidly. While the development of these technologies proceeds through a number of steps, it isn't necessarily the case—as happened in the Manhattan Project and the Trinity test—that the last step in proving a technology is large and hard. The breakthrough to wild self-replication in robotics, genetic engineering, or nanotechnology could come suddenly, reprising the surprise we felt when we learned of the cloning of a mammal.

And yet I believe we do have a strong and solid basis for hope. Our attempts to deal with weapons of mass destruction in the last century provide a shining example of relinquishment for us to consider: the unilateral US abandonment, without preconditions, of the development of biological weapons. This relinquishment stemmed from the realization that while it would take an enormous effort to create these terrible weapons, they could from then on easily be duplicated and fall into the hands of rogue nations or terrorist groups.

The clear conclusion was that we would create additional threats to ourselves by pursuing these weapons, and that we would be more secure if we did not pursue them. We have embodied our relinquishment of biological and chemical weapons in the 1972 Biological Weapons Convention (BWC) and the 1993 Chemical Weapons Convention (CWC).

As for the continuing sizable threat from nuclear weapons, which we have lived with now for more than 50 years, the US Senate's recent rejection of the Comprehensive Test Ban Treaty makes it clear relinquishing nuclear weapons will not be politically easy. But we have a unique opportunity, with the end of the Cold War, to avert a multipolar arms race. Building on the BWC and CWC relinquishments, successful abolition of nuclear weapons could help us build toward a habit of relinquishing dangerous technologies. (Actually, by getting rid of all but 100 nuclear weapons worldwide—roughly the total destructive power of World War II and a considerably easier task—we could eliminate this extinction threat.)

Verifying relinquishment will be a difficult problem, but not an unsolvable one. We are fortunate to have already done a lot of relevant work in the context of the BWC and other treaties. Our major task will be to apply this to technologies that are naturally much more commercial than military. The substantial need here is for transparency, as difficulty of verification is directly proportional to the difficulty of distinguishing relinquished from legitimate activities.

I frankly believe that the situation in 1945 was simpler than the one we now face: The nuclear technologies were reasonably separable into commercial and military uses, and monitoring was aided by the nature of atomic tests and the ease with which radioactivity could be measured. Research on military applications could be performed at national laboratories such as Los Alamos, with the results kept secret as long as possible.

The GNR technologies do not divide clearly into commercial and military uses; given their potential in the market, it's hard to imagine pursuing them only in national laboratories. With their widespread commercial pursuit, enforcing relinquishment will require a verification regime similar to that for biological weapons, but on an unprecedented scale. This, inevitably, will raise tensions between our individual privacy and desire for proprietary information, and the need for verification to protect us all. We will undoubtedly encounter strong resistance to this loss of privacy and freedom of action.

Verifying the relinquishment of certain GNR technologies will have to occur in cyberspace as well as at physical facilities. The critical issue will be to make the necessary transparency acceptable in a world of proprietary information, presumably by providing new forms of protection for intellectual property.

Verifying compliance will also require that scientists and engineers adopt a strong code of ethical conduct, resembling the Hippocratic oath, and that they have the courage to whistleblow as necessary, even at high personal cost. This would answer the call—50 years after Hiroshima—by the Nobel laureate Hans Bethe, one of the most senior of the surviving members of the Manhattan Project, that all scientists "cease and desist from work creating, developing, improving, and manufacturing nuclear weapons and other weapons of potential mass destruction." In the 21st century, this requires vigilance and personal responsibility by those who would work on both NBC and GNR technologies to avoid implementing weapons of mass destruction and knowledge-enabled mass destruction.

Thoreau also said that we will be "rich in proportion to the number of things which we can afford to let alone." We each seek to be happy, but it would seem worthwhile to question whether we need to take such a high risk of total destruction to gain yet more knowledge and yet more things; common sense says that there is a limit to our material needs—and that certain knowledge is too dangerous and is best forgone.

Neither should we pursue near immortality without considering the costs, without considering the commensurate increase in the risk of extinction. Immortality, while perhaps the original, is certainly not the only possible utopian dream.

I recently had the good fortune to meet the distinguished author and scholar Jacques Attali, whose book *Lignes d'horizons* (*Millennium,* in the English translation) helped inspire the Java and Jini approach to the coming age of pervasive computing, as previously described in this magazine. In his new book *Fraternités,* Attali describes how our dreams of utopia have changed over time:

> At the dawn of societies, men saw their passage on Earth as nothing more than a labyrinth of pain, at the end of which stood a door leading, via their death, to the company of gods and to *Eternity.* With the Hebrews and then the Greeks, some men dared free themselves from theological demands and dream of an ideal City where *Liberty* would flourish. Others, noting the evolution of the market society, understood that the liberty of some would entail the alienation of others, and they sought *Equality.*

Jacques helped me understand how these three different utopian goals exist in tension in our society today. He goes on to describe a fourth utopia, *Fraternity,* whose foundation is altruism. Fraternity alone associates individual happiness with the happiness of others, affording the promise of self-sustainment.

This crystallized for me my problem with Kurzweil's dream. A technological approach to Eternity—near immortality through robotics—may not be the most desirable utopia, and its pursuit brings clear dangers. Maybe we should rethink our utopian choices.

Where can we look for a new ethical basis to set our course? I have found the ideas in the book *Ethics for the New Millennium,* by the Dalai Lama, to be very helpful. As is perhaps well known but little heeded, the Dalai Lama argues that the most important thing is for us to conduct our lives with love and compassion for others, and that our societies need to develop a stronger notion of universal responsibility and of our interdependency; he proposes a standard of positive ethical conduct for individuals and societies that seems consonant with Attali's Fraternity utopia.

The Dalai Lama further argues that we must understand what it is that makes people happy, and acknowledge the strong evidence that neither material progress nor the pursuit of the power of knowledge is the key—that there are limits to what science and the scientific pursuit alone can do.

Our Western notion of happiness seems to come from the Greeks, who defined it as "the exercise of vital powers along lines of excellence in a life affording them scope."

Clearly, we need to find meaningful challenges and sufficient scope in our lives if we are to be happy in whatever is to come. But I believe we must find alternative outlets for our creative forces, beyond the culture of perpetual economic growth; this growth has largely been a blessing for several hundred years, but it has not brought us unalloyed happiness, and we must now choose between the pursuit of unrestricted and undirected growth through science and technology and the clear accompanying dangers.

It is now more than a year since my first encounter with Ray Kurzweil and John Searle. I see around me cause for hope in the voices for caution and relinquishment and in those people I have discovered who are as concerned as I am about our current predicament. I feel, too, a deepened sense of personal responsibility—not for the work I have already done, but for the work that I might yet do, at the confluence of the sciences.

But many other people who know about the dangers still seem strangely silent. When pressed, they trot out the "this is nothing new" riposte—as if awareness of what could happen is response enough. They tell me, There are universities filled with bioethicists who study this stuff all day long. They say, All this has been written about before, and by experts. They complain, Your worries and your arguments are already old hat.

I don't know where these people hide their fear. As an architect of complex systems I enter this arena as a generalist. But should this diminish my concerns? I am aware of how much has been written about, talked about, and lectured about so authoritatively. But does this mean it has reached people? Does this mean we can discount the dangers before us?

Knowing is not a rationale for not acting. Can we doubt that knowledge has become a weapon we wield against ourselves?

The experiences of the atomic scientists clearly show the need to take personal responsibility, the danger that things will move too fast, and the way in which a process can take on a life of its own. We can, as they did, create insurmountable problems in almost no time flat. We must do more thinking up front if we are not to be similarly surprised and shocked by the consequences of our inventions.

My continuing professional work is on improving the reliability of software. Software is a tool, and as a toolbuilder I must struggle with the uses to which the tools I make are put. I have always believed that making software more reliable, given its many uses, will make the world a safer and better place; if I were to come to believe the opposite, then I would be morally obligated to stop this work. I can now imagine such a day may come.

This all leaves me not angry but at least a bit melancholic. Henceforth, for me, progress will be somewhat bittersweet.

Each of us has our precious things, and as we care for them we locate the essence of our humanity. In the end, it is because of our great capacity for caring that I remain optimistic we will confront the dangerous issues now before us.

My immediate hope is to participate in a much larger discussion of the issues raised here, with people from many different backgrounds, in settings not predisposed to fear or favor technology for its own sake.

As a start, I have twice raised many of these issues at events sponsored by the Aspen Institute and have separately proposed that the American Academy of Arts and Sciences take them up as an extension of its work with the Pugwash Conferences. (These have been held since 1957 to discuss arms control, especially of nuclear weapons, and to formulate workable policies.)

It's unfortunate that the Pugwash meetings started only well after the nuclear genie was out of the bottle—roughly 15 years too late. We are also getting a belated start on seriously addressing the issues around 21st-century technologies—the prevention of knowledge-enabled mass destruction—and further delay seems unacceptable.

So I'm still searching; there are many more things to learn. Whether we are to succeed or fail, to survive or fall victim to these technologies, is not yet decided. I'm up late again—it's almost 6 am. I'm trying to imagine some better answers, to break the spell and free them from the stone.

PROMISE AND PERIL

RAY KURZWEIL

Kurzweil (1948-) is an inventor, writer, and futurist. He is the author of *The Age of Intelligent Machines* (1990), *The Age of Spiritual Machines* (1998), and *The Singularity is Near* (2005). The following article was published in 2003.

Consider these articles we'd rather not see on the Web:

- Impress Your Enemies: How to Build Your Own Atomic Bomb from Readily Available Materials.
- How to Modify the Influenza Virus in Your College Laboratory to Release Snake Venom.
- Ten Easy Modifications to the E. coli Virus.
- How to Modify Smallpox to Counteract the Smallpox Vaccine.
- How to Build a Self-guiding, Low-Flying Airplane Using an Inexpensive Aircraft, GPS, and a Notebook Computer.

Or how about the following:

- The Genome of Ten Leading Pathogens.
- The Floor Plans of Leading Skyscrapers.
- The Layout of U.S. Nuclear Reactors.
- Personal Health Information on 100 Million Americans.
- The Customer Lists of Top Pornography Sites.

Anyone posting the first item above is almost certain to get a quick visit from the FBI, as did Nate Ciccolo, a fifteen-year-old high school student, in March 2000. For a school science project, he built a papier-mâché model of an atomic bomb that turned out to be disturbingly accurate. In the ensuing media storm, Nate told ABC News, "Someone just sort of mentioned, you know, you can go on the Internet now and get information. And I, sort of, wasn't exactly up to date on things. Try it. I went on there and a couple of clicks and I was right there."

Of course, Nate didn't possess the key ingredient, plutonium, nor did he have any intention of acquiring it, but the report created shock waves in the media, not to mention among the authorities who worry about nuclear proliferation. Nate had reported finding 563 web pages on atomic bomb designs. The publicity resulted in an urgent effort to remove this information. Unfortunately, trying to get rid of information on the Internet is akin to trying to sweep back the ocean with a broom. The information continues to be easily available today. I won't provide any URLs in this essay, but they are not hard to find.

Although the actual article titles above are fictitious, one can find extensive information on the Internet about all of these topics. The web is an extraordinary research tool. In my own experience as a technologist and author, I've found that research which used to require half a day at the library can now typically be accomplished in a couple of minutes. This has enormous and obvious benefits for advancing beneficial technologies, but is also empowering those whose values are inimical to the mainstream of society.

My urgent concern with this issue goes back at least a couple of decades. When I wrote my first book, *The Age of Intelligent Machines*, in the mid-1980s, I was deeply concerned with the ability of genetic engineering, then an emerging technology, to allow those skilled in the art and with access to fairly widely available equipment to modify bacterial and viral pathogens to create new diseases. In malevolent or merely careless hands, these engineered pathogens could potentially combine a high degree of communicability and destructiveness.

In the 1980s this was not easy to do, but was nonetheless feasible. We now know that bioweapons programs in the Soviet Union and elsewhere were doing exactly this. At the time, I made a conscious decision to not talk about this specter in my book, feeling that I did not want to give the wrong people a destructive idea. I had disturbing visions of a future disaster, with the perpetrators saying that they got the idea from Ray Kurzweil.

Partly as a result of this decision, there was some reasonable criticism that the book emphasized the benefits of future technology while ignoring the downside. Thus, when I wrote *The Age of Spiritual Machines* in the late 1997-98 timeframe, I attempted to cover both promise and peril. There had been sufficient public attention to the perils by that time (for example, the 1995 movie *Outbreak*, which portrays the terror and panic that follow the release of a new viral pathogen) that I felt comfortable in beginning to address the issue publicly.

It was at that time, in September 1998, with a just-finished manuscript, that I ran into Bill Joy, an esteemed and longtime colleague in the high-technology world, in a bar in Lake Tahoe. I had long admired Bill for his pioneering work in creating the leading software language for interactive web systems (Java) and having cofounded Sun Microsystems. But my focus at this brief get-together was not on Bill but on the third person sitting in our small booth: John Searle, an eminent philosopher from the University of California, Berkeley. John had built a career out of defending the deep mysteries of human consciousness from apparent attack by materialists like me (though I deny the characterization). John and I had just finished debating the issue of whether a machine could be conscious; we'd been part of the closing panel of George Gilder's "Telecosm" conference, which was devoted to a discussion of the philosophical implications of *The Age of Spiritual Machines*.

I gave Bill a preliminary manuscript of the book and tried to bring him up to speed on the debate about consciousness that John and I were having. As it turned out,

Bill focused on a completely different issue, specifically the impending dangers to human civilization from three emerging technologies I had presented in the book: genetics, nanotechnology, and robotics, or GNR for short. My discussion of the downsides of future technology alarmed Bill, as he relays in his now-famous cover story for *Wired* magazine, "Why the Future Doesn't Need Us." In the article, Bill describes how he asked his friends in the scientific and technology community whether the projections I was making were credible and was dismayed to discover how close these capabilities were to realization.

Needless to say, Bill's article focused entirely on the downside scenarios, and created a firestorm. Here was one of the technology world's leading figures addressing new and dire emerging dangers from future technology. It was reminiscent of the attention that George Soros, the currency arbitrager and arch-capitalist, received when he made vaguely critical comments about the excesses of unrestrained capitalism, although the "Bill Joy" controversy became far more intense. The *New York Times* cited about 10,000 articles commenting on and discussing Bill's article, more than any other article in the history of commentary on technology issues.

My attempt to relax in a Lake Tahoe lounge ended up fostering two long-term debates. My dialogue with John Searle has continued to this day, and my debate with Bill has taken on a life of its own. Perhaps this is one reason I now avoid hanging out in bars.

Despite Bill's concerns, my reputation as a technology optimist has remained intact, and Bill and I have been invited to a variety of forums to debate the peril and promise, respectively, of future technologies. Although I am expected to take up the "promise" side of the debate, I often end up spending most of my time at these forums defending the feasibility of the dangers. I recall one event at Harvard during which a Nobel Prize–winning biologist dismissed the "N" (nanotechnology) danger by stating that he did not expect to see self-replicating nanoengineered entities for a hundred years.

I replied that indeed, a hundred years matched my own estimate of the amount of progress required—at *today's* rate of progress. However, since my models show that we are doubling the paradigm shift rate (the rate of technological progress) every ten years, we can expect to make a hundred years of progress—at today's rate—in about twenty-five calendar years, which matches the consensus view within the nanotechnology community. Thus are both promise and peril much closer at hand.

My view is that technology has always been a double-edged sword, bringing us longer and healthier life spans, freedom from physical and mental drudgery, and many new creative possibilities on the one hand, while introducing new and salient dangers on the other. Technology empowers both our creative and our destructive natures. Stalin's tanks and Hitler's trains used technology. We benefit from nuclear power, but live today with sufficient nuclear weapons (not all of which appear to be well accounted for) to end all mammalian life on the planet.

Bioengineering holds the promise of making enormous strides in reversing disease and aging processes. However, the means and knowledge it has created, which began to exist in the 1980s, will soon enable an ordinary college bioengineering lab to create unfriendly pathogens more dangerous than nuclear weapons.

As technology accelerates toward the full realization of G (genetic engineering, also known as biotechnology), followed by N (nanotechnology) and ultimately R (robotics, also referred to as "strong" AI—artificial intelligence at human levels and beyond), we will see the same intertwined potentials: a feast of creativity resulting from human intelligence expanded many-fold, combined with many grave new dangers.

Consider the manner in which extremely small robots, or nanobots, are likely to develop. Nanobot technology requires billions or trillions of such intelligent devices to be useful. The most cost-effective way to scale up to such levels is through self-replication, essentially the same approach used in the biological world. And in the same way that biological self-replication gone awry (i.e., cancer) results in biological destruction, a defect in the mechanism curtailing nanobot self-replication would endanger all physical entities, biological or otherwise. Later in this chapter I suggest steps we can take to address this grave risk, but we cannot have complete assurance in any strategy that we devise today.

Other primary concerns include who controls the nanobots and who the nanobots are talking to. Organizations (e.g., governments or extremist groups) or just a clever individual could put trillions of undetectable nanobots in the water or food supply of an individual or of an entire population. These "spy" nanobots could then monitor, influence, and even control our thoughts and actions. In addition to physical spy nanobots, existing nanobots could be influenced through software viruses and other software "hacking" techniques. When there is software running in our brains, issues of privacy and security will take on a new urgency.

My own expectation is that the creative and constructive applications of this technology will dominate, as I believe they do today. I believe we need to vastly increase our investment in developing specific defensive technologies, however. We are at the critical stage today for biotechnology, and we will reach the stage where we need to directly implement defensive technologies for nanotechnology during the late teen years of this century.

The Inevitability of a Transformed Future

The diverse GNR technologies are progressing on many fronts and comprise hundreds of small steps forward, each benign in itself. An examination of the underlying trends, which I have studied for the past quarter-century, shows that full-blown GNR is inevitable.

The motivation for this study came from my interest in inventing. As an inventor in the 1970s, I came to realize that my inventions needed to make sense in terms of the enabling technologies and market forces that would exist when the invention was introduced, which would represent a very different world than when it was conceived. I began to develop models of how distinct technologies—electronics, communications, computer processors, memory, magnetic storage, and the key feature sizes in a range of technologies—developed and how these changes rippled through markets and ultimately our social institutions. I realized that most inventions fail not because they never work, but because their timing is wrong. Inventing is a lot like surfing; you have to anticipate and catch the wave at just the right moment.

In the 1980s my interest in technology trends and implications took on a life of its own, and I began to use my models of these trends to project and anticipate the technologies of the future. This enabled me to invent with the capabilities of the future in mind. I wrote *The Age of Intelligent Machines*, which ended with the specter of machine intelligence becoming indistinguishable from its human progenitors. The book included hundreds of predictions about the 1990s and early 2000s, and my track record of prediction has held up well.

During the 1990s, I gathered empirical data on the apparent acceleration of all information-related technologies and sought to refine the mathematical models underlying these observations. In *The Age of Spiritual Machines*, I introduced improved models of technology and a theory I called "the law of accelerating returns," which explained why technology evolves in an exponential fashion.

The Intuitive Linear View versus the Historical Exponential View

The most important trend this study has revealed concerns the overall pace of technological progress itself. The future is widely misunderstood. Our forebears expected the future to be pretty much like their present, which had been pretty much like their past. Although exponential trends did exist a thousand years ago, they were at that very early stage where an exponential trend is so flat and so slow that it looks like no trend at all. So their expectation of stasis was largely fulfilled. Today, in accordance with the common wisdom, everyone expects continuous technological progress and the social repercussions that follow. But the future will nonetheless be far more surprising than most observers realize because few have truly internalized the implications of the fact that the rate of change itself is accelerating.

Most long-range forecasts of technical feasibility in future time periods dramatically underestimate the power of future developments because they are based on what I call the "intuitive linear view" of history rather than the "historical exponential view." To express this another way, it is not the case that we will experience 100 years of progress in the twenty-first century; rather, we will witness on the order of 20,000 years of progress (again, at *today's* rate of progress).

When people think of a future period, they intuitively assume that the current rate of progress will continue into that future. Even for those who have been around long enough to experience how the pace, increases over time, an unexamined intuition nonetheless provides the impression that progress happens at the rate we have experienced recently. From the mathematician's perspective, a primary reason for this is that an exponential curve approximates a straight line when viewed for a brief duration. It is typical, therefore, for even sophisticated commentators, when considering the future, to extrapolate the current pace of change over the next ten years or hundred years to determine their expectations. This is why I call this way of looking at the future the "intuitive linear view."

But a serious assessment of the history of technology shows that technological change is exponential. In exponential growth, a key measurement such as computational power is multiplied by a constant factor for each unit of time (e.g., doubling every year) rather than just increased incrementally. Exponential growth is a feature of any evolutionary process, of which technology is a prime example. One can examine the data in different ways, on different time scales, and for a wide variety of technologies ranging from electronic to biological, as well as for social implications ranging from the size of the economy to human life span, and the acceleration of progress and growth applies to all.

Indeed, we find not just simple exponential growth, but "double" exponential growth, meaning that the rate of exponential growth is itself growing exponentially. These observations do not rely merely on an assumption of the continuation of Moore's Law (i.e., the exponential shrinking of transistor sizes on an integrated circuit), which I discuss a bit later, but are based on a rich model of diverse technological processes. What the model clearly shows is that technology, particularly the pace of technological change, advances (at least) exponentially, not linearly, and has been doing so since the advent of technology, indeed since the advent of evolution on Earth.

Many scientists and engineers have what my colleague Lucas Hendrich calls "engineer's pessimism." Often engineers or scientists who are so immersed in the difficulties and intricate details of a contemporary challenge fail to appreciate the ultimate long-term implications of their own work and, in particular, the larger field of work that they operate in. Consider the biochemists in 1985 who were skeptical of the announced goal of transcribing the entire genome in a mere fifteen years. These scientists had just spent an entire year transcribing a mere one ten-thousandth of the genome, so even with reasonable anticipated advances, it seemed to them as though it would be hundreds of years, if not longer, before the entire genome could be sequenced.

Or consider the skepticism expressed in the mid-1980s that the Internet would ever be a significant phenomenon, given that it included only tens of thousands of nodes. The fact that the number of nodes was doubling every year, so tens of millions of nodes were likely to exist ten years later, was not appreciated by those struggling with a limited "state of the art" technology in 1985 that permitted adding only a few thousand nodes throughout the world in a year.

I emphasize this point because it is the most important failure that would-be prognosticators make in considering future trends. The vast majority of technology forecasts and forecasters ignore altogether this "historical exponential view" of technological progress. Indeed, almost everyone I meet has a linear view of the future. That is why people tend to overestimate what can be achieved in the short term (because we tend to leave out necessary details) but underestimate what can be achieved in the long term (because the exponential growth is ignored).

The Law of Accelerating Returns

The ongoing acceleration of technology is the implication and the inevitable result of what I call the "law of accelerating returns," which describes the acceleration of the pace and the exponential growth of the products of an evolutionary process. This process includes information-bearing technologies such as computation as well as the accelerating trend toward miniaturization—all the prerequisites for the full realization of GNR.

A wide range of technologies are subject to the law of accelerating returns. The exponential trend that has gained the greatest public recognition has become known as Moore's Law. Gordon Moore, one of the inventors of integrated circuits and then chairman of Intel, noted in the mid-1970s that we could squeeze twice as many transistors on an integrated circuit every 24 months. Given that the electrons have a shorter distance to travel, the transistors change states more quickly. This, along with other techniques, allows the circuits to run faster, providing an overall quadrupling of computational power.

The exponential growth of computing is much broader than Moore's Law, however. If we plot the speed (in instructions per second) per $1,000 (in constant dollars) of 49 famous calculators and computers spanning the entire twentieth century, we note that there were four completely different paradigms providing exponential growth in the price-performance of computing before integrated circuits were invented. Therefore, Moore's Law was not the first but the fifth paradigm of exponential growth of computational power. And it won't be the last. When Moore's Law approaches its limit, now expected before 2020, the exponential growth will continue with three-dimensional molecular computing, a prime example of the application of nanotechnology, which will constitute the sixth paradigm.

When in 1999 I suggested that three-dimensional molecular computing, particularly an approach based on using carbon nanotubes, would become the dominant computing hardware technology in the teen years of this century, that was considered a radical notion. But so much progress has been accomplished in the past four years, with literally dozens of major milestones having been achieved, that this expectation is now a mainstream view"'

The exponential growth of computing is a marvelous quantitative example of the exponentially growing returns from an evolutionary process. We can express the exponential growth of computing in terms of an accelerating pace: it took ninety years to achieve the first million instructions per second (MIPS) per $1,000; now we add one MIPS per $1,000 every day.

The human brain uses a very inefficient electrochemical digital-controlled analog computational process. The bulk of the calculations are done in the interneuronal connections at a speed of only about two hundred calculations per second (in each connection), which is about ten million times slower than contemporary electronic circuits. But the brain gains its prodigious powers from its extremely parallel organization *in three dimensions*. There are many technologies in the wings that build circuitry in three dimensions.

Nanotubes (tubes formed from graphite sheets, consisting of hexagonal arrays of carbon atoms) are good conductors and can be used to build compact circuits; these are already working in laboratories. One cubic inch of nanotube circuitry would theoretically be a million times more powerful than the human brain. There are more than enough new computing technologies now being researched, including three-dimensional silicon chips, spin computing, crystalline computing, DNA computing, and quantum computing, to keep the law of accelerating returns as applied to computation going for a long time.

It is important to distinguish between the S curve (an S stretched to the right, comprising very slow, virtually unnoticeable growth—followed by very rapid growth—followed by a flattening out as the process approaches an asymptote, or limit) that is characteristic of any specific technological paradigm, and the continuing exponential growth that is characteristic of the ongoing evolutionary process of technology. Specific paradigms, such as Moore's Law, do ultimately reach levels at which exponential growth is no longer feasible. That is why Moore's Law is an S curve.

But the growth of computation is an ongoing exponential pattern (at least until we "saturate" the universe with the intelligence of our human-machine civilization, but that is not likely to happen in this century). In accordance with the law of accelerating returns, paradigm shift (also called innovation) turns the S curve of any specific paradigm into a continuing exponential pattern. A new paradigm (e.g., three-dimensional circuits) takes over when the old paradigm approaches its natural limit, which has already happened at least four times in the history of computation. This difference also distinguishes the tool making of nonhuman species, in which the mastery of a tool-making (or tool-using) skill by each animal is characterized by an abruptly ending S-shaped learning curve, from human-created technology, which has followed an exponential pattern of growth and acceleration since its inception.

This "law of accelerating returns" applies to all of technology, indeed to any true evolutionary process, and can be measured with remarkable precision in information-based technologies. There are a great many examples of the exponential growth implied by the law of accelerating returns in technologies, as varied as electronics of all kinds, DNA sequencing, communication speeds, brain scanning, reverse engineering of the brain, the size and scope of human knowledge, and the rapidly shrinking size of technology, which is directly relevant to emergence of nanotechnology.

The future GNR age results not from the exponential explosion of computation alone, but rather from the interplay and myriad synergies that will result from intertwined technological revolutions. Keep in mind that every point on the exponential growth curves underlying this panoply of technologies represents an intense human drama of innovation and competition. It is remarkable, therefore, that these chaotic processes result in such smooth and predictable exponential trends.

Economic Imperative

It is the economic imperative of a competitive marketplace that is fueling the law of accelerating returns and driving technology forward toward the full realization of GNR. In turn, the law of accelerating returns is transforming economic relationships.

We are moving toward nanoscale machines, as well as more intelligent machines, as the result of a myriad of small advances, each with their own particular economic justification. There is a vital economic imperative to create smaller and more intelligent technology. Machines that can more precisely carry out their missions have enormous value, which is why they are being built. There are tens of thousands of projects advancing the various aspects of the law of accelerating returns in diverse incremental ways.

Regardless of near-term business cycles, the support for "high tech" in the business community, and in particular for software advancement, has grown enormously. When I started my optical character recognition (OCR) and speech synthesis company (Kurzweil Computer Products) in 1974, high-tech venture deals totaled approximately $10 million. Even during today's high-tech recession, the figure is a hundred times greater. We would have to repeal capitalism and every visage of economic competition to stop this progression.

The economy (viewed either in total or per capita) has been growing exponentially throughout this century. This underlying exponential growth is a far more powerful force than periodic recessions. Even the Great Depression represents only a minor blip on this pattern of growth. Most importantly, recessions and depressions represent only temporary deviations from the underlying curve. In each case, the economy ends up exactly where it would have been had the recession or depression never occurred.

In addition to GDP, other improvements include productivity (economic output per worker), quality and features of products and services (for example, $1,000 of computation today is more powerful, by a factor of more than a thousand, than $1,000 of computation ten years ago), new products and product categories, and the value of existing goods, which has been increasing at 1.5 percent per year for the past twenty years because of qualitative improvements.

Intertwined Benefits...

Significant portions of our species have already experienced substantial alleviation of the poverty, disease, hard labor, and misfortune that have characterized much of human history. Many of us have the opportunity to gain satisfaction and meaning from our work rather than merely toiling to survive. We have ever more powerful tools to express ourselves. We have worldwide sharing of culture, art, and humankind's exponentially expanding knowledge base.

Ubiquitous N and R are two to three decades away. A prime example of their application will be the deployment of billions of nanobots the size of human blood cells to travel inside the human bloodstream. This technology will be feasible within twenty-five years, based on miniaturization and cost-reduction trends. In addition to scanning the brain to facilitate reverse engineering (that is, analyzing how the brain works in order to copy its design), nanobots will be able to perform a broad variety of diagnostic and therapeutic functions, inside the bloodstream and human body. Robert A. Freitas, for example, has designed robotic replacements for human blood cells that perform hundreds or thousands of times more effectively than their biological counterparts. With Freitas's "respirocytes" (robotic red blood cells), you could do an Olympic sprint for fifteen minutes without taking a breath. His robotic macrophages would be far more effective than our white blood cells at combating pathogens. His DNA repair robot would be able to repair DNA transcription errors and even implement needed DNA changes.

Although realization of Freitas's conceptual designs are two or three decades away, substantial progress has already been achieved on bloodstream-based devices.

For example, one scientist has cured type I diabetes in rats with a nanoengineered device that incorporates pancreatic islet cells. The device has seven-nanometer pores that let insulin out but block the antibodies that destroy these cells. Many innovative projects of this type are already under way.

Clearly, nanobot technology has profound military applications, and any expectation that such uses will be "relinquished" are highly unrealistic. Already, the Department of Defense is developing "smart dust"—tiny robots the size of insects or even smaller. Although not quite nanotechnology millions of these devices can be dropped into enemy territory to provide highly detailed surveillance. The potential application for even smaller nanotechnology-based devices is even greater. Want to find Saddam Hussein or Osama bin Laden? Need to locate hidden weapons of mass destruction? Billions of essentially invisible spies could monitor every square inch of enemy territory, identify every person and every weapon, and even carry out missions to destroy enemy targets. The only way for an enemy to counteract such a force is, of course, with their own nanotechnology. The point is that nanotechnology-based weapons will one day make larger weapons obsolete.

Nanobots will also expand our experiences and our capabilities by providing fully immersive, totally convincing virtual reality. They will take up positions in dose proximity to every interneuronal connection of our sensory organs (e.g., eyes, ears, and skin). We already have the technology that enables electronic devices to communicate with neurons in both directions, and it requires no direct physical contact with the neurons. For example, scientists at the Max Planck Institute have developed "neuron transistors" that can detect the firing of a nearby neuron or, alternatively, can cause a nearby neuron to fire, or suppress it from firing. This amounts to two-way communication between neurons and the electronically based neuron transistors. The institute scientists demonstrated their invention by controlling the movement of a living leech from their computer.

When we want to experience real reality, the nanobots just stay in position (in the capillaries) and do nothing. If we want to enter virtual reality, they suppress all of the inputs coming from the real senses and replace them with the signals that would be appropriate for the virtual environment. You (i.e., your brain) could decide to cause your muscles and limbs to move as you normally would, but the nanobots would intercept these interneuronal signals, suppress your real limbs from moving, and instead cause your virtual limbs to move and provide the appropriate reorientation in the virtual environment.

The primary limitation of nanobot-based virtual reality at this time is only that it's not yet feasible in size and cost. One day, however, the web will provide a panoply of virtual environments to explore. Some will be re-creations of real places; others will be fanciful environments that have no "real" counterpart. Some will be impossible in the physical world, perhaps because they violate the laws of physics. We will be able to "go" to these virtual environments by ourselves, or we will meet other people there, both real people and simulated people. Of course, ultimately there won't be a clear distinction between the two.

By 2030, going to a website will mean entering a full-immersion virtual-reality environment. In addition to encompassing all of the senses, some of these shared environments will include emotional overlays, as the nanobots will be capable of triggering the neurological correlates of emotions, sexual pleasure, and other derivatives of our sensory experience and mental reactions.

In the same way that people today beam their lives from webcams in their bedrooms, "experience beamers" circa 2030 will beam their entire flow of sensory experiences and, if so desired, their emotions and other secondary reactions. We'll be able to plug in (by going to the appropriate website) and experience other people's lives, as in the movie *Being John Malkovich*. We will be able to archive particularly interesting experiences and relive them at any time.

We won't need to wait until 2030 to experience shared virtual-reality environments, though, at least for the visual and auditory senses. Full-immersion visual-auditory environments will be available by the end of this decade, with images written directly onto our retinas by our eyeglasses and contact lenses. All of the electronics for computation, image reconstruction, and a very high-bandwidth wireless connection to the Internet will be embedded in our glasses and woven into our clothing, so computers as distinct objects will disappear.

It's not just the virtual world that will benefit from ubiquitous application of nanobots and fully realized nanotechnology. Portable manufacturing systems will be able to produce virtually any physical product from information for pennies a pound, thereby providing for our physical needs at almost no cost. Nanobots will be able to reverse the environmental destruction left by the first industrial revolution. Nanoengineered fuel cells and solar cells will provide clean energy. Nanobots in our physical bodies will destroy pathogens and cancer cells, repair DNA, and reverse the ravages of aging.

These technologies will become so integral to our health and well-being that we will eventually become indistinguishable from our machine support systems. In fact, in my view, the most significant implication of the development of nanotechnology and related advanced technologies of the twenty-first century will be the merger of biological and nonbiological intelligence. First, it is important to point out that well before the end of the twenty-first century, thinking on nonbiological substrates will dominate. Biological thinking is stuck at 10^{26} calculations per second (for all biological human brains), and that figure will not appreciably change, even with bioengineering changes to our genome.

Nonbiological intelligence, on the other hand, is growing at a double-exponential rate and will vastly exceed biological intelligence well before the middle of this century. In my view, however, this nonbiological intelligence should still be considered human, since it is fully derivative of the human-machine civilization. The merger of these two worlds of intelligence is not merely a merger of biological and nonbiological thinking media but, more importantly, one of method and organization of thinking.

Nanobot technology will be able to expand our minds in virtually any imaginable way. Our brains today are relatively fixed in design. Although we do add patterns of interneuronal connections and neurotransmitter concentrations as a normal part of the learning process, the current overall capacity of the human brain is highly constrained, restricted to a mere hundred trillion connections. Brain implants based on massively distributed intelligent nanobots will ultimately expand our memories a trillionfold, and otherwise vastly improve all of our sensory, pattern recognition, and cognitive abilities. Since the nanobots will be communicating with each other over a wireless local area network, they will be able to create any set of new neural connections, break existing connections (by suppressing neural firing), and create new hybrid biological-nonbiological networks, as well as add vast new nonbiological networks.

Using nanobots as brain extenders will be a significant improvement over surgically installed neural implants, which are beginning to be used today (e.g., ventral posterior nucleus, subthalmic nucleus, and ventral lateral thalamus neural implants to counteract Parkinson's disease and tremors from other neurological disorders; cochlear implants; and the like). Nanobots will be introduced without surgery, essentially just by injecting or even swallowing them. They can all be directed to leave, so the process is easily reversible. They are programmable, in that they can provide virtual reality one minute and a variety of brain extensions the next. They can change their configuration and alter their software. Perhaps most importantly, they are massively distributed and. therefore can take up billions or trillions of positions throughout the brain, whereas a surgically introduced neural implant can be placed in only one or at most a few locations.

...and Dangers

Needless to say, we have already experienced technology's downside. One hundred million people were killed in two world wars during the last century—a scale of mortality made possible by technology. The crude technologies of the first industrial revolution have crowded out many of the species on our planet that existed a century ago. Our centralized technologies (e.g., buildings, cities, airplanes, and power plants) are demonstrably insecure.

The NBC (nuclear, biological, and chemical) technologies of warfare were all used, or threatened to be used, in our recent past. The far more powerful GNR technologies pose what philosopher of science Nick Bostrom calls "existential risks," referring to potential threats to the viability of human civilization itself.

If we manage to get past the concerns about genetically altered designer pathogens, followed by self-replicating entities created through nanotechnology, we will next encounter robots whose intelligence will rival and ultimately exceed our own. Such robots may make great assistants, but who's to say that we can count on them to remain reliably friendly to mere humans?

In my view, "strong AI" (artificial intelligence at human levels and beyond) promises to continue the exponential gains of human civilization. But the dangers are also more profound precisely because of this amplification of intelligence. Intelligence is inherently impossible to control, so the various strategies that have been devised to control nanotechnology won't work for strong AI. There have been discussions and proposals to guide AI development toward "friendly AI." These are useful for discussion, but it is impossible to devise strategies today that will absolutely ensure that future AI embodies human ethics and values.

Relinquishment

In his *Wired* essay and subsequent presentations, Bill Joy eloquently describes the plagues of centuries past and how new self-replicating technologies, such as mutant bioengineered pathogens and "nanobots" run amok, may bring back long-forgotten pestilence. Of course, as Joy graciously acknowledges, it has also been technological advances, such as antibiotics and improved sanitation, that have freed us from the prevalence of such plagues. Suffering in the world continues and demands our steadfast attention. Should we tell the millions of people afflicted with cancer and other devastating conditions that we are canceling the development of all bioengineered treatments because there is a risk that these same technologies may someday be used for malevolent purposes?

Relinquishment is Bill's most controversial recommendation and personal commitment. I do feel that relinquishment at the right level is part of a responsible and constructive response to these genuine perils. The issue, however, is exactly this: at what level are we to relinquish technology?

Ted Kaczynski would have us renounce all of it. This, in my view, is neither desirable nor feasible, and the futility of such a position is only underscored by the senselessness of Kaczynski's deplorable tactics. There are other voices, less reckless than Kaczynski, who are nonetheless arguing for broad-based relinquishment of technology. The environmentalist Bill McKibben takes the position that "environmentalists, must now grapple squarely with the idea of a world that has enough wealth and enough technological capability, and should not pursue more." In my view, this position ignores the extensive suffering that remains in the human world, which we will be in a position to alleviate through continued technological progress.

Another level of relinquishment, one recommended by Joy, would be to forgo certain fields—nanotechnology, for example—that might be regarded as too dangerous. But such sweeping strokes of relinquishment are equally untenable. As I pointed out above, nanotechnology is simply the inevitable end result of the persistent trend toward miniaturization that pervades all of technology. It is far from a single centralized effort, but is being pursued by a myriad of projects with many diverse goals.

One observer wrote:

> A further reason why industrial society cannot be reformed...is that modern technology is a unified system in which all parts are dependent on one another. You can't get rid of the "bad" parts of technology and retain only the "good" parts. Take modern medicine, for example. Progress in medical science depends on progress in chemistry, physics, biology, computer science and other fields. Advanced medical treatments require expensive, high-tech equipment that can be made available only by a technologically progressive, economically rich society. Clearly you can't have much progress in medicine without the whole technological system and everything that goes with it.

The observer, again, is Ted Kaczynski [ISAIF paragraph 121]. Although one will properly resist Kaczynski as an authority, I believe he is correct on the deeply entangled nature of the benefits and risks. He and I clearly part company on our overall assessment of the relative balance between the two, however. Bill Joy and I have debated this issue both publicly and privately, and we both believe that technology will and should progress and that we need to be actively concerned with the dark side. If he and I disagree, it's on the granularity of relinquishment that is both feasible and desirable.

Abandonment of broad areas of technology would only push them underground, where development would continue unimpeded by ethics and regulation. In such a situation, it would be the less stable, less responsible practitioners (e.g., terrorists) who would have all the expertise.

I do think that relinquishment at the right level needs to be part of our ethical response to the dangers of twenty-first-century technologies. One constructive example is the ethical guideline proposed by the Foresight Institute, founded by nanotechnology pioneer Eric Drexler, along with Christine Peterson, that nanotechnologists agree to relinquish the development of physical entities that can self-replicate in a natural environment. Another is a ban on self-replicating physical entities that contain their own codes for self-replication. In what nanotechnologist Ralph Merkle calls the "broadcast architecture," such entities would have to obtain these codes from a centralized secure server, which would guard against undesirable replication. I discuss these guidelines further below.

The broadcast architecture is impossible in the biological world, so it represents at least one way in which nanotechnology can be made safer than biotechnology. In other ways, nanotechnology is potentially more dangerous because nanobots can be physically stronger, and more intelligent, than protein-based entities. It will eventually be possible to combine the two by having nanotechnology provide the codes within biological entities (replacing DNA), in which case biological entities will be able to use the much safer broadcast architecture.

As responsible technologists, our ethics should include such "fine-grained" relinquishment among our professional ethical guidelines. Protections must also include oversight by regulatory bodies, the development of technology-specific "immune" responses, and computer-assisted surveillance by law enforcement organizations. Many people are not aware that our intelligence agencies already use advanced technologies such as automated word spotting to monitor a substantial flow of telephone conversations. As we go forward, balancing our cherished rights of privacy with our need to be protected from the malicious use of powerful twenty-first-century technologies will be one of many profound challenges. This is one reason such issues as an encryption "trap door" (in which law enforcement authorities would have access to otherwise secure information) and the FBI's 'Carnivore' email-snooping system have been controversial.

We can take a small measure of comfort from how our society has dealt with one recent technological challenge. There exists today a new form of fully nonbiological self-replicating entity that did not exist just a few decades ago: the computer virus. When this form of destructive intruder first appeared, strong concerns were voiced that as these software pathogens became more sophisticated, they would have the potential to destroy the computer network medium they live in. Yet the "immune system" that has evolved in response to this challenge has been largely effective. Although destructive self-replicating software entities do cause damage from time to time, the injury is but a small fraction of the benefit we receive from the computers and communication links that harbor them. No one would suggest we do away with computers, local area networks, and the Internet because of software viruses.

One might counter that computer viruses do not have the lethal potential of biological viruses or of destructive nanotechnology. This is not always the case; we rely on software to monitor patients in critical care units, to fly and land airplanes, to guide intelligent weapons in wartime, and to perform other "mission-critical" tasks. To the extent that this assertion is true, however, it only strengthens my argument. The fact that computer viruses are not usually deadly to humans only means that more people are willing to create and release them. It also means that our response to the danger is that much less intense. Conversely, when it comes to self-replicating entities that are potentially lethal on a large scale, our response on all levels will be vastly more serious, as we have seen since September 11.

The Development of Defensive Technologies and the Impact of Regulation

Bill Joy's *Wired* treatise is effective because he paints a picture of future dangers as if they were released on today's unprepared world. The reality is that the sophistication and power of our defensive technologies and knowledge will grow along with the dangers. When we have "gray goo" (unrestrained nanobot replication), we will also have "blue goo" ("police" nanobots that combat the "bad" nanobots). The story of the twenty-first century has not yet been written, so we cannot say with assurance that we will successfully avoid all misuse. But the surest way to prevent the development of defensive technologies would be to relinquish the pursuit of knowledge in broad areas. We have been able to largely control harmful software virus replication because the requisite knowledge is widely available to responsible practitioners. Attempts to restrict this knowledge would have created a far less stable situation. Responses to new challenges would have been far slower, and the balance would have likely shifted toward the more destructive applications (e.g., software viruses).

As we compare the success we have had in controlling engineered software viruses to the coming challenge of controlling engineered biological viruses, we are struck with one salient difference: the software industry is almost completely unregulated. The same is obviously not true for biotechnology. We require scientists developing defensive technologies to follow the existing regulations, which slow down the innovation process at every step. A bioterrorist, however, does not need to put his "innovations" through the FDA. Moreover, under existing regulations and ethical standards, it is impossible to test defenses against bioterrorist agents. Extensive discussion is already under way regarding modifying these regulations to allow for animal models and simulations, since human trials are infeasible.

For reasons I have articulated above, stopping these technologies is not feasible, and pursuit of such broad forms of relinquishment will only distract us from the vital task in front of us. It is quite clearly a race. There is simply no alternative. We cannot relinquish our way out of this challenge. In the software field, defensive technologies have remained a step ahead of offensive ones. With the extensive regulation in the medical field slowing down innovation at each stage, we cannot have the same confidence with regard to the abuse of biotechnology.

In the current environment, when one person dies in gene therapy trials, research can be severely restricted. There is a legitimate need to make biomedical research as safe as possible, but our balancing of risks is completely off. The millions of people who desperately need the advances that will result from gene therapy and other breakthrough biotechnologies appear to carry little political weight against a handful of well-publicized casualties from the inevitable risks of progress.

This equation will become even more stark when we consider the emerging dangers of bioengineered pathogens. What is needed is a change in public attitude in terms of tolerance for needed risk.

Hastening defensive technologies is absolutely vital to our security. We need to streamline regulatory procedures to achieve this, and we also need to greatly increase our investment in defensive technologies explicitly. In the biotechnology field, this means the rapid development of antiviral medications. We will not have time to develop specific countermeasures for each new challenge that comes along. We are close to developing more generalized antiviral technologies, and these need to be accelerated.

I have addressed here the issue of biotechnology because that is the threshold and challenge that we now face. As the threshold for nanotechnology comes closer, we will then need to invest specifically in the development of defensive technologies in that area, including the creation of a nanotechnology-based immune system. Bill Joy and other observers have pointed out that such an immune system would itself be a danger because of the potential of "autoimmune" reactions (i.e., the immune system using its powers to attack the world it is supposed to be defending).

This observation is not a compelling reason to avoid the creation of an immune system, however. No one would argue that humans would be better off without an immune system because of the possibility of autoimmune diseases. Although the immune system can itself be a danger, humans would not last more than a few weeks (barring extraordinary efforts at isolation) without one. The development of a technological immune system for nanotechnology will happen even without explicit efforts to create one. We have effectively done this with regard to software viruses. We created a software virus immune system not through a formal grand design, but rather through our incremental responses to each new challenge and through the development of heuristic algorithms for early detection. We can expect the same thing will happen as challenges from nanotechnology-based dangers emerge. The point for public policy will be to specifically invest in these defensive technologies.

It is premature to develop specific defensive nanotechnologies today, since we can have only a general idea of what we are trying to defend against. It would be similar to the engineering world creating defenses against software viruses before the first virus had been created. There is already fruitful dialogue and discussion on anticipating these issues, however, and significantly expanded investment in these efforts is to be encouraged.

As one example, as I mentioned above, the Foresight Institute has devised a set of ethical standards and strategies for ensuring the development of safe nanotechnology. These guidelines include:

- Artificial replicators must not be capable of replication in a natural, uncontrolled environment.
- Evolution within the context of a self-replicating manufacturing system is discouraged.
- MNT (molecular nanotechnology) designs should specifically limit proliferation and provide traceability of any replicating systems.
- Distribution of molecular manufacturing development capability should be restricted, whenever possible, to responsible actors that have agreed to the guidelines. No such restriction need apply to end products of the development process.

Other strategies proposed by the Foresight Institute include:

- Replication should require materials not found in the natural environment.
- Manufacturing (replication) should be separated from the functionality of end products. Manufacturing devices can create end products but should not be able to replicate themselves, and end products should have no replication capabilities.
- Replication should require codes that are encrypted and time-limited. (The broadcast architecture mentioned earlier is an example.)

These guidelines and strategies are likely to be effective for preventing accidental release of dangerous self-replicating nanotechnology entities. But the intentional design and release of such entities is a more complex and challenging problem. A sufficiently determined and destructive opponent could possibly defeat each of these layers of protections. Take, for example, the broadcast architecture. When properly designed, each entity is unable to replicate without first obtaining replication codes. These codes are not passed on from one replication generation to the next. A modification to such a design could bypass the destruction of the codes, however, and thereby pass them on to the next generation. To counteract that possibility, it has been recommended that the memory for the replication codes be limited to only a subset of the full code so that insufficient memory exists to pass the full set of codes along. But this guideline could be defeated by expanding the size of the replication code memory to incorporate the entire code. Another suggestion is to encrypt the codes and build protections such as time limits into the decryption systems. We can see how easy it has been to defeat protections against unauthorized replications of intellectual property such as music files, however. Once replication codes and protective layers are stripped away, the information can be replicated without these restrictions.

My point is not that protection is impossible. Rather, we need to realize that any level of protection will only work to a certain level of sophistication. The meta-lesson here is that we will need to place society's highest priority during the twenty-first century on continuing to advance defensive technologies and on keeping them one or more steps ahead of destructive technologies. We have seen analogies to this in many areas, including technologies for national defense as well as our largely successful efforts to combat software viruses.

The broadcast architecture won't protect us against abuses of strong AI. The barriers of the broadcast architecture rely on the nanoengineered entities lacking the intelligence to overcome the built-in restrictions. By definition, intelligent entities have the cleverness to easily overcome such barriers. Inherently, there will be no absolute protection other than dominance by friendly AI. Although the argument is subtle, I believe that maintaining an open system for incremental scientific and technological progress, in which each step is subject to market acceptance, will provide the most constructive environment for technology to embody widespread human values. Attempts to control these technologies in dark government programs, along with inevitable underground development, would create an unstable environment in which the dangerous applications would likely become dominant.

One profound trend already well under way that will provide greater stability is the movement from centralized technologies to distributed ones, and from the real world to the virtual world discussed above. Centralized technologies involve an aggregation of resources such as people (e.g., cities and buildings), energy (e.g., nuclear power plants, liquid natural gas and oil tankers, and energy pipelines), transportation (e.g., airplanes and trains), and other resources. Centralized technologies are subject to disruption and disaster. They also tend to be inefficient, wasteful, and harmful to the environment.

Distributed technologies, on the other hand, tend to be flexible, efficient, and relatively benign in their environmental effects. The quintessential distributed technology is the Internet. Despite concerns about viruses, these information-based pathogens are mere nuisances. The Internet is essentially indestructible. If any hub or channel goes down, the information simply routes around it. The Internet is remarkably resilient, a quality that continues to increase with its continued exponential growth.

In energy, we need to move rapidly toward the opposite end of the spectrum of contemporary energy sources, away from the extremely concentrated energy installations we now depend on. In one example of a trend in the right direction, Integrated Fuel Cell Technologies is pioneering microscopic fuel cells that use microelectromechanical systems (MEMS) technology. The fuel cells are manufactured like electronic chips but are actually batteries with an energy-to-size ratio vastly exceeding conventional technology. Ultimately, forms of energy along these lines could power everything from our cell phones to our cars and homes, and would not be subject to disaster or disruption.

As these technologies develop, our need to aggregate people in large buildings and cities will diminish and people will spread out, living where they want and gathering together in virtual reality.

But we don't need to look past today to see the intertwined promise and peril of technological advancement. If we imagine describing the dangers that exist today to people who lived a couple of hundred years ago, they would think it mad to take such risks. On the other hand, how many people in the year 2000 would really want to go back to the short, brutish, disease-filled, poverty-stricken, disaster-prone lives that 99 percent of the human race struggled through a couple of centuries ago? We may romanticize the past, but up until fairly recently, most of humanity lived extremely fragile lives where one all-too-common misfortune could spell disaster. Two hundred years ago, life expectancy for females in the record-holding country (Sweden) was roughly thirty-five years, compared to the longest life expectancy today—almost eighty-five years—enjoyed by Japanese women. Life expectancy for males was roughly thirty-three years to the current seventy-nine years in the record-holding countries. It took half the day to prepare the evening meal, and hard labor character-ized most human activity. There were no social safety nets. Substantial portions of our species still live in this precarious way, which is at least one reason to continue tech-nological progress and the economic enhancement that accompanies it.

People often go through three stages in considering future technologies: awe and wonderment at their potential to overcome age-old problems; a sense of dread at the new set of grave dangers that accompany the new technologies; and, finally (and hopefully), the realization that the only viable and responsible path is to set a careful course that can reap the benefits while managing the dangers.

Technology will remain a double-edged sword. It represents vast power to be used for all humankind's purposes. We have no choice but to work hard to apply these quickening technologies to advance our human values, despite what often appears to be a lack of consensus on what those values should be.

POST SCRIPT – News story (Feb 15, 2005):

INVENTOR PRESERVES SELF TO WITNESS IMMORTALITY

WELLESLEY, Massachusetts (AP) — Ray Kurzweil doesn't tailgate. A man who plans to live forever doesn't take chances with his health on the highway, or any-where else.

As part of his daily routine, Kurzweil ingests 250 supplements, eight to 10 glasses of alkaline water and 10 cups of green tea. He also periodically tracks 40 to 50 fitness indicators, down to his "tactile sensitivity." Adjustments are made as needed. "I do actually fine-tune my programming," he said.

The inventor and computer scientist is serious about his health because if it fails him he might not live long enough to see humanity achieve immortality, a seismic development he predicts in his new book is no more than 20 years away. It's a blink of an eye in history, but long enough for the 56-year-old Kurzweil to pay close heed to his fitness. He urges others to do the same in his book *Fantastic Voyage: Live Long Enough to Live Forever*.

The book is partly a health guide so people can live to benefit from a coming explosion in technology he predicts will make infinite life spans possible.

Kurzweil writes of millions of blood cell-sized robots, which he calls "nanobots," that will keep us forever young by swarming through the body, repairing bones, muscles, arteries and brain cells. Improvements to our genetic coding will be downloaded via the Internet. We won't even need a heart.

The claims are fantastic, but Kurzweil is no crank. He's a recipient of the $500,000 Lemelson-MIT prize, which is billed as a sort of Academy Award for inventors, and he won the 1999 National Medal of Technology Award. He has written on the emergence of intelligent machines in publications ranging from Wired to Time magazine. The Christian Science Monitor has called him a "modern Edison." He was inducted into the Inventors Hall of Fame in 2002. Perhaps the MIT graduate's most famous invention is the first reading machine for the blind that could read any typeface.

During a recent interview in his company offices, Kurzweil sipped green tea and spoke of humanity's coming immortality as if it's as good as done. He sees human intelligence not only conquering its biological limits, including death, but completely mastering the natural world. "In my view, we are not another animal, subject to nature's whim," he said.

Critics say Kurzweil's predictions of immortality are wild fantasies based on unjustifiable leaps from current technology. "I'm not calling Ray a quack, but I am calling his message about immortality in line with the claims of other quacks that are out there," said Thomas Perls, a Boston University aging specialist who studies the genetics of centenarians. Sherwin Nuland, a bioethics professor at Yale University's School of Medicine, calls Kurzweil a "genius" but also says he's a product of a narcissistic age when brilliant people are becoming obsessed with their longevity. "They've forgotten they're acting on the basic biological fear of death and extinction, and it distorts their rational approach to the human condition," Nuland said.

Kurzweil says his critics often fail to appreciate the exponential nature of technological advance, with knowledge doubling year by year so that amazing progress eventually occurs in short periods. His predictions, Kurzweil said, are based on carefully constructed scientific models that have proven accurate. For instance, in his 1990 book, *The Age of Intelligent Machines*, Kurzweil predicted the development of a worldwide computer network and of a computer that could beat a chess champion. "It's not just guesses," he said. "There's a methodology to this."

Kurzweil has been thinking big ever since he was little. At age 8, he developed a miniature theater in which a robotic device moved the scenery. By 16, the New York City native built his own computer and programmed it to compose original melodies.

His interest in health developed out of concern about his own future. Kurzweil's grandfather and father suffered from heart disease, his father dying when he was 22. Kurzweil was diagnosed with Type 2 diabetes in his mid-30s. After insulin treatments were ineffective, Kurzweil devised his own solution, including a drastic cut in fat consumption, allowing him to control his diabetes without insulin. His rigorous health regimen is not excessive, just effective, he says, adding that his worst sickness in the last several years has been mild nasal congestion.

Health science and technology merge

In the past decade, Kurzweil's interests in technology and health sciences have merged as scientists have discovered similarities. "All the genes we have, the 20,000 to 30,000 genes, are little software programs," Kurzweil said.

In his latest book, Kurzweil defines what he calls his three bridges to immortality. The "First Bridge" is the health regimen he describes with co-author Dr. Terry Grossman to keep people fit enough to cross the "Second Bridge," a biotechnological revolution. Kurzweil writes that humanity is on the verge of controlling how genes express themselves and ultimately changing the genes. With such technology, humanity could block disease-causing genes and introduce new ones that would slow or stop the aging process.

The "Third Bridge" is the nanotechnology and artificial intelligence revolution, which Kurzweil predicts will deliver the nanobots that work like repaving crews in our bloodstreams and brains. These intelligent machines will destroy disease, rebuild organs and obliterate known limits on human intelligence, he believes.

Immortality would leave little standing in current society, in which the inevitability of death is foundational to everything from religion to retirement planning. The planet's natural resources would be greatly stressed, and the social order shaken. Kurzweil says he believes new technology will emerge to meet increasing human needs. And he said society will be able to control the advances he predicts as long as it makes decisions openly and democratically, without excessive government interference.

No guarantees

But there are no guarantees, he adds.

Meanwhile, Kurzweil refuses to concede the inevitably of his own death, even if science doesn't advance as quickly as he predicts. "Death is a tragedy," a process of suffering that rids the world of its most tested, experienced members—people whose contributions to science and the arts could only multiply with agelessness, he said.

Kurzweil said he's no "cheerleader" for unlimited scientific progress and added he knows science can't answer questions about why eternal lives are worth living. That's left for philosophers and theologians, he said. But to him there's no question of huge advances in things that make life worth living, such as art, cultural, music and science. "Biological evolution passed the baton of progress to human cultural and technological development," he said.

Lee Silver, a Princeton biologist, said he'd love to believe in the future as Kurzweil sees it, but the problem is, humans are involved. The instinct to preserve individuality, and to gain advantage for yourself and children, would survive any breakthrough into biological immortality—which Silver doesn't think is possible. The gap between the haves and have-nots would widen and Kurzweil's vision of a united humanity would become ever more elusive, he said. "I think it would require a change in human nature," Silver said, "and I don't think people want to do that."

APPENDIX A

HOW TO READ A PHILOSOPHICAL ESSAY

Students new to philosophy may be unprepared for the demands that such reading places on the reader. The very subject matter makes it unlike any other form of reading. Essays are often densely-worded, use technical terms, and address complex topics. Careful and fruitful reading of philosophy is an acquired skill; it takes time to develop! Here are some suggestions on how to tackle philosophical pieces:

Approach each new essay with both *open-minded sympathy*, and, a *bit of skepticism*. The author probably has some important and insightful things to say, but we should be prepared to mentally question him or her.

Read the essay once over, very quickly, just to get the general idea of the author.

Figure out the *rough structure* of the essay – is it organized into 1, 2, or more sections? Are there one or two main ideas? And so on.

Read it a second time, *very slowly and carefully*, underlining or **highlighting** what seem to be the key words, phrases, or ideas. Important: Read....it....slowly!

If you don't understand a particular word, grab a dictionary and look it up! (Even professional philosophers do this from time to time.)

Make brief notes or comments in the margins – good points ('*'), bad points ('?'), questions.

Write out a rough *outline-form* **synopsis** of the essay. Organize it the same way that the author has done it; by section, or by main ideas. Hand-written, bullet-points are fine; 'complete sentences' are *not* necessary. Do *not* make a whole new paper out of it. A friend should be able to read your synopsis and get the basic ideas of the essay. At the end of the synopsis, write a short *personal assessment* of the piece. SEE BELOW for an example.

If, later on, you are preparing for an exam, or to write a paper of your own, re-read the essay a 3rd time, concentrating on the sections you previously highlighted. Use your synopsis as a study guide.

Read each essay *as if you were going to lecture on it tomorrow!* What are the main points? What would you say?

EXAMPLE SYNOPSIS (partial)

Apology
By Plato

The *Apology* is organized into *3 parts*: (I) defense, (II) penalty, (III) farewell.

(I). Defense:
- 2 charges: *corrupting the youth,* and *impiety.*
- 2 accusers: (1) past reputation, (2) Meletus.

    (1) Past reputation:
- "investigates everything" (undermines tradition).
- "turns weaker argument into the stronger"
- "fails to acknowledge the gods" (impiety)
- Never charged a fee.
- Story of the Delphi oracle – 'none are wiser'.
- Oracle was right!
- Made enemies, lives in poverty.

    (2) Meletus:
- "corrupting the youth":
  - o   xxx
  - o   yyy
  - o   zzz
- "impiety: not acknowledging the gods"
  - o   xxx
  - o   yyy
  - o   zzz

    (3) Defense of philosopher's life:
- Only consider whether actions are just, not the consequences.
- xxx

    (4) Conclusion: He will not grovel:
- xxx

(II.)   Penalty: ...

(III.)   Farewell: ...

Assessment: At the end of every synopsis, write a short *personal commentary* on this piece—like a short movie review: Did you like it, or not like it, and why. For example: "I really liked this reading by Plato, because it was clearly written and really shows the conflicts that Socrates faced in ancient Athens…" Or, "I didn't like this piece very much. It was hard to understand, and I couldn't figure out why Socrates would prefer to face death than to plead for forgiveness from the jury…". DON'T write a recap of the piece: "In the *Apology* Socrates was on trial for his beliefs. These beliefs were…
"

Assessment should be *2-3 paragraphs in length*. NOTE: When turning synopses in for grading, any loose sheets should be **stapled together**.

APPENDIX B

COPYRIGHT ACKNOWLEGMENTS

Hillis, D. "Close to the singularity". From *The Third Culture* (1995). Simon and Schuster.

Jensen and Draffan. "Better humans". From *Welcome to the Machine* (2004). Chelsea Green.

Diamond, J. "The worst mistake". *Discover* (May), 1987.

Mumford, L. "Technics and the nature of man". *Technology and Culture*, 7(3), 1966.

White, L. "Historical roots of our ecologic crisis". *Science*, 155(3767), 1967.

Merchant, C. "Mining the earth's womb." From *The Death of Nature* (1980). Harper.

Rousseau, J-J. "Discourse on the Arts and Sciences." (G.D.H. Cole, trans; 1973). Knopf.

Thoreau, H. *Walden* (1992). Knopf.

Marx, K. *Das Kapital* (vol. 1). B. Fowkes, trans (1976). Penguin.

Whitehead, A. *Science and the Modern World* (1953). Free Press.

Orwell, G. *Road to Wigan Pier* (1958). Harcourt, Brace.

Heidegger, M. *The Question concerning Technology* (1977). Harper and Row.

Marcuse, H. *One Dimensional Man* (1964). Beacon Press.

Illich, I. *Tools for Conviviality* (1973). Harper and Row.

Naess, A. *Ecosophy, Community, and Lifestyle* (1989). Cambridge University Press.

Skolimowski, H. "From religious consciousness to technological consciousness." *Teilhard Review*, 1989.

Kaczynski, T. Industrial Society and its Future. No copyright.

Warwick, K. "Intelligent Robots or Cyborgs." In *Writing the Future* (Rothenburg, ed.), 2004. MIT Press.

Joy, B. "Why the future doesn't need us." *Wired* (April), 2000.

Kurzweil, R. "Promise and peril." From *Living with the Genie* (Lightman, ed.). 2003. Island Press.